# About Island Press

Since 1984, the nonprofit Island Press has been stimulating, shaping, and communicating the ideas that are essential for solving environmental problems worldwide. With more than 800 titles in print and some 40 new releases each year, we are the nation's leading publisher on environmental issues. We identify innovative thinkers and emerging trends in the environmental field. We work with world-renowned experts and authors to develop cross-disciplinary solutions to environmental challenges.

Island Press designs and implements coordinated book publication campaigns in order to communicate our critical messages in print, in person, and online using the latest technologies, programs, and the media. Our goal: to reach targeted audiences—scientists, policymakers, environmental advocates, the media, and concerned citizens—who can and will take action to protect the plants and animals that enrich our world, the ecosystems we need to survive, the water we drink, and the air we breathe.

Island Press gratefully acknowledges the support of its work by the Agua Fund, Inc., The Margaret A. Cargill Foundation, Betsy and Jesse Fink Foundation, The William and Flora Hewlett Foundation, The Kresge Foundation, The Forrest and Frances Lattner Foundation, The Andrew W. Mellon Foundation, The Curtis and Edith Munson Foundation, The Overbrook Foundation, The David and Lucile Packard Foundation, The Summit Foundation, Trust for Architectural Easements, The Winslow Foundation, and other generous donors.

# *Saving a Million Species*

## EXTINCTION RISK FROM CLIMATE CHANGE

Edited by Lee Hannah

*Washington | Covelo | London*

Library of Congress Cataloging-in-Publication Data

Saving a million species : extinction risk from climate change / edited by Lee Hannah.
p. cm.
ISBN-13: 978-1-59726-569-0 (cloth)
ISBN-10: 1-59726-569-1 (cloth)
ISBN-13: 978-1-59726-570-6 (paper)
ISBN-10: 1-59726-570-5 (paper)
1. Climatic changes. 2. Global warming. 3. Extinction (Biology)—Environmental aspects. I. Hannah, Lee Jay.
QC902.9.S28 2011
551.6—dc23
2011040327

Printed on recycled, acid-free paper

Manufactured in the United States of America
10 9 8 7 6 5 4 3 2 1

*Keywords*: Island Press, climate change, extinction, extinction risk, biodiversity, freshwater, marine, biology, coral bleaching, species area relationship, million species, conservation

## CONTENTS

# FOREWORD

It is decades since the "eureka moment," as Stephen Schneider dubbed the time I asked him, "What does what you do (climate change) have to do with what I do (biodiversity)?" Since then, relevant science has gone from projections based on paleontological data, through Chris Thomas and colleagues' 2004 forward projection of the extinction consequences of doubling preindustrial carbon dioxide, and beyond. Today there seems to be a rush of new observations of climate-induced biological change almost daily.

It is within this trajectory of growing understanding that the editors and authors have assembled this important volume. It makes a critical contribution. Even though the initial increments of change are mostly (but not entirely) small, they at least partly unveil what can lie beyond. At early stages it is often very difficult to differentiate between linear and exponential change, and it can be easy to ignore the profoundly important. In addition, the inherently (and necessarily) conservative approach of the Intergovernmental Panel on Climate Change (IPCC) has reinforced that to the point of consistently underestimating the extent of the biodiversity–climate change problem.

Impacts on the living planet are already pulling away from that phase and demonstrating the change to be exponential. There is already abrupt change at the ecosystem level. Coral reefs are a clear example as they experience "bleaching" events in which diversity, productivity, and human benefit crash. Unknown prior to 1984, they now are commonplace and chronic. We are seeing similar threshold change in the coniferous forests of western North America: milder winters and longer summers have tipped the balance in favor of native bark beetles such that in many places 70 percent of the trees are dead. These two changes are tantamount to ecosystem failure with major and as yet only superficially understood consequences for constituent biodiversity.

There is an imperative need to move beyond the initial phase of climate change science in which physical science has dominated. Biology must become central to climate change science and policy formulation. The planet does not work just as a physical system, but rather as a linked biological and physical system; that reality needs to become fundamental to the way we pursue the science and derive policy recommendations.

It is important to remember that the effects we are currently seeing are caused by global warming of a bit more than 0.75 degree Celsius. The inconclusive policy debates in Cancún and previously in Copenhagen were primarily around reining in greenhouse gas emissions to stop climate change before or at 2.0 degrees Celsius. It doesn't take a lot of thinking to realize that 2 degrees Celsius is too much for ecosystems and that we should choose a less disruptive goal of, say, 1.5 degrees Celsius.

The awful reality is that if we decide to stop at 2 degrees Celsius, global emissions have to peak in 2016. Many would say there is no hope of even stopping at 2 degrees Celsius. Also, many of the models project much greater increases in greenhouse gases (double and triple preindustrial levels) and consequent global temperature. The folly of allowing this from a biological point of view is well laid out in this volume.

Just a single example is "Amazon dieback," in which the forest would die back and be replaced by savannah vegetation in the south and east-southeast of Amazonia. First projected to occur at 2.5 degrees Celsius by the Hadley Center's general climate model, subsequent refinement lowered the threshold to 2 degrees Celsius (the very target discussed by the negotiators). This would represent a staggering loss of biological diversity, serious impacts on the indigenous and other peoples living in that part of the Amazon, and a huge release of more carbon to the atmosphere. Ominously, there was probably a preview in 2005, when the Amazon had the greatest drought in recorded history. This was repeated with an even more intense drought in 2010.

Further, because the high wall of the Andes stops the westward movement of moisture, it both generates the rainfall that largely creates the Amazon river system and sends airborne moisture to the north and south of the Amazon. Termed "rivers in the sky," the moisture provides essential rainfall to agro-industry south of the Amazon and further—at least as far as northern Argentina. What is clear is that the effects approach a continental scale.

The picture is further complicated by other impacts on the hydrological system of the Amazon, namely those of deforestation and fire. The World Bank invested a million dollars in a modeling study looking at the combined effects of deforestation, fire, and climate change. The disturbing result was the suggestion of a tipping point to dieback at around 20 percent deforestation, with the current extent being about 18 percent. In this case there is an obvious policy response, namely to aggressively reforest in the south and east to build back a margin of safety.

On top of these issues is the acidification of the oceans as a consequence of the higher atmospheric concentrations of carbon dioxide. Our understanding of the implications for marine species and ecosystems is rudimentary at best, as laid out in the oceans chapter. When it comes to "solutions," those that address temperature must be viewed mostly skeptically not only because they address a symptom and not the cause, but also because they ignore the effects of acidification of the marine two-thirds of the planet.

This volume essentially makes the case that the planet is a biophysical system and that the biological elements make it very important to limit climate change and further buildup of greenhouse gases. It also takes us beyond that to thinking about ways to use the planetary biophysical system to pull carbon dioxide out of the atmosphere. That has happened twice in the history of life on Earth—the first with life and green plants emerging onto land, and the second with the advent of modern flowering plants. It shows that biology has very great power. The problem from a practical (and very anthropocentric) point of view is that we don't have tens of millions of years to achieve this a third time.

It is rarely expressed this way, but roughly half of the excess carbon dioxide in the atmosphere actually comes from the destruction and degradation of ecosystems over the past three centuries. That means we can choose to manage our planet as a biophysical system and pull significant amounts of carbon dioxide out of the atmosphere by ecosystem restoration on a planetary scale. In one sense, it is tantamount to joining forces with, rather than working counter to, Gaia (in a nonmystical sense). This is not the place for details of a debate over how much carbon dioxide could be sequestered in such fashion and what that might mean in terms of land use decisions. Suffice it to say it seems that roughly 50 parts per million of carbon dioxide could be pulled out of the atmosphere in a half-century period without impinging on land necessary for agricultural production to support 9 billion people.

This is not the place for further elaboration—however essential it might be—of additional nonbiological solutions for lowering atmospheric carbon dioxide. What the editors and authors who have created this volume have achieved is to make an impregnable case for the sensitivity of the biology of our planet to the changes we are creating in the atmosphere. And by implication, they lay the basis for engaging the biology of the planet in a grand solution. Were we to recognize and seize on this, it would be a magnificent new chapter in organic evolution, one in which consciousness (itself a consequence of evolution) rises above its immediate and petty concerns to create a better future for humans and all life on Earth.

This, of necessity, would require a very different worldview—namely, one in which the limits of the global biophysical system are embraced, but simultaneously, the system's power is engaged to achieve a sustainable outcome for humanity and life on Earth. Interestingly, studies of the 2 billion poor of the planet indicate they depend very directly and to great degree (38–89 percent) on goods and services from biodiversity and ecosystems. A global commitment to re-green the "emerald planet" would recognize that a similar dependence and respect is important for all humanity.

Thomas E. Lovejoy

# PART I

# Introduction

In this first section of the book, we look at the 2004 publication that sparked worldwide interest in extinction risk from climate change. The overall purpose and scope of the book are described in chapter 1, with an overview of the chapter structures and the reason for each part of the book. Chris D. Thomas, lead author on the 2004 study, then explains the research, the limitations of the methods used at the time, and the importance of the findings. The final chapter of this part reviews the policy implications of extinction risk from climate change, from the UK House of Commons and the US Senate to international climate treaties and beyond.

# Chapter 1

## *Are a Million Species at Risk?*

Lee Hannah

The research paper "Extinction Risk from Climate Change" created front-page headlines around the world when it appeared as the cover story of *Nature* in January 2004 (Thomas et al., 2004). The notion that climate change could drive more than a million species to extinction captured popular imagination and the attention of policy makers. The story was covered by CNN, ABC News, NBC News, NPR, and major newspapers and magazines in Europe and the United States and was the subject of debate in the House of Commons and in the US Senate.

An unprecedented round of scientific critique quickly followed the huge popular interest in the story. *Nature* itself published three articles challenging fundamental points of the paper (Harte et al., 2004; Buckley and Roughgarden, 2004; Thuiller et al., 2004), while publications refining or debating the underlying science continue to appear in top research journals. This welter of publications makes for a diverse literature not easily synthesized or accessed, despite the critical policy implications of the research. Most important, the variety of critiques leaves unresolved the major question: What is the extinction risk associated with climate change, and how many species may perish?

*Saving a Million Species* addresses this important question by synthesizing the literature, by having leaders in the field refine the original estimates of extinction risk, and by drawing on these authors to elaborate the science, conservation, and policy implications of this research in an accessible format.

This book will speak to conservationists, researchers, teachers, undergraduate to graduate students, and policy makers interested in a clear explanation of the science behind the headline-grabbing estimates. Unpacking the research behind the headlines reveals complex chains of causation, with many taxa facing unique challenges. These stories reveal connections among climate change and many other alterations to the natural world, from rain forest destruction to overfishing. The chapters of the book are organized into six parts, each bringing to bear the insights of a relevant discipline. This multiplicity of perspectives breaks down the monolithic large numbers and reveals the complexity of the problem. It allows solutions to begin to take shape.

The six sections explore evidence from the past and present, estimates of future risk from modeling and taxonomic perspectives, and finally the conservation and policy implications. The chapters in part I introduce the original research and its critiques. Part II examines the research published since the 2004 article to refine estimation techniques. Part III explores extinctions documented in the contemporary record—the first of the extinctions due to human-induced climate change. Part IV examines extinctions from past natural climate change. New risk estimates from modeling of future climate change are presented in part V. The sixth and final section addresses the conservation and policy implications of the estimates: What does extinction risk imply for the future of biodiversity and global cooperation on action to curtail climate change?

## Science behind the Hype

The saga of extinction risk from climate change began in London in 2002. The International Union for Conservation of Nature (IUCN) called a meeting of international climate change experts there to discuss the threat posed by climate change. The IUCN is a membership organization representing conservation groups from government agencies to nonprofit organizations. They assess and determine the global list of species threatened with extinction. IUCN officers were concerned about the threat climate change posed but didn't know what to do about it. They called in international researchers for advice.

The researchers at the IUCN meeting found a remarkable thing. They all were engaged in modeling of changes in species ranges in different parts of the world—locations as different as Australia and the Amazon. But all were finding high numbers of species losing suitable range, even in relatively mild midcentury (2050) scenarios. And all were finding significant numbers of species losing all of their suitable habitat. Climate change might be a much more serious threat to species' survival than anyone had previously imagined.

But they needed some way to compare their diverse results. The key question was how to estimate extinctions from models of shrinking range size. Chris Thomas, then of the University of Leeds (now at the University of York), had an idea. What if you used the well established species-area relationship (SAR) to estimate extinctions from range-size models? The SAR had previously been used to estimate extinctions based on decreasing forest size. Why not disaggregate that relationship and apply it to range loss in individual species? There were several ways to do that, and it turned out they all showed high levels of extinction risk—almost always in double digits, some estimates as high as 30 or 40 percent.

The London group agreed that these were important results and coalesced behind the leadership of Thomas to produce a research paper describing the results. Thomas had a strong track record with *Nature,* the most well respected journal in the field, so the manuscript was submitted there. After some hard review questions and revisions, *Nature* accepted the paper and scheduled it for publication in early 2004. Thomas developed a press release describing the research, in which he extrapolated the results of the paper to all species on Earth, to give the media a sense of the scale of the problem. The research results, and that extrapolation, created the widespread media and public interest in extinction risk from climate change.

Thomas's estimate was built on straightforward math. The extinction risk estimates in the research had a wide range of values, but midrange values showed 18–34 percent of species becoming extinct. There are a wide range of estimates of the number of species on the planet, but 10 million is a midrange value, with about half of those in the oceans. Because the areas modeled in the research were all terrestrial, Thomas excluded the marine species. The math was then simple—18–34 percent of 5 million terrestrial species is 900,000 to 1.7 million species extinctions. Thus, 1 million species is a lower-end estimate of the risk of extinction due to climate change, and it is likely that the total is more than 1 million. That's a large number in anyone's definition, and attracted worldwide attention as a result. Interestingly, despite catalyzing the 2002 meeting, the IUCN never fully bought into the results. The 2004 paper didn't follow the specific rules IUCN uses to create the Red List of species threatened with extinction. The IUCN rules were developed to flag species under immediate threat. The short-term time horizons in those rules are poorly suited to assessing threats such as climate change, which happen now but have effects years or decades in the future. IUCN Red List specialists have published a paper (Akçakaya et al., 2006) making it clear that the methods of Thomas et al. didn't meet the existing international criteria. The IUCN continues to struggle to incorporate climate change into their threat assessments.

## Why Should We Care?

The public cared about the research results because extinction is a threat that ordinary people care about and relate to. Researchers often refer to biodiversity, which is a more abstract concept, less widely grasped by nonspecialists. The extinction risk paper translated results into terms to which everyone could relate.

Anyone concerned about conservation sees extinction as a critical yardstick, because it indicates irreversible loss. Biologists care because loss of a species is the loss of an entire evolutionary history and unique set of biological attributes—information that can't be replicated any other way. People of faith care because the creation is one of the great gifts of the Almighty, and extinctions slowly destroy that gift. Schoolchildren, students, and others see extinctions as a clear sign that we are not properly taking care of the planet.

Because of these concerns, extinctions have a more formal role in international policy. Governments care about preventing extinctions because their citizens care. National laws and international agreements have been created to prevent extinctions. The Convention on Biological Diversity is an international treaty designed to prevent the loss of biodiversity, which means preventing extinctions. Many nations have created legislation for national parks and protected areas to give nature safe haven and to guard against extinction.

Most important in the field of climate change, extinctions are embodied in the international treaty on climate change. "Allowing ecosystems to adapt naturally" is one of the three benchmarks of the United Nations Framework Convention on Climate Change (the others being agricultural growth and sustainable development). Ecosystem adaptation is a more sensitive indicator than climate change: it can be impaired at levels of biological disruption far less severe than extinction. But extinction is an exclamation point, a red flag in that context. If species are going extinct due to climate change, then clearly something is very wrong. So from points of view from everyday life to the political, extinctions from climate change matter. The huge press attention to the initial research proved this interest. Virtually every story on the impacts of climate change now references extinction risk. The polar bear is the poster child for these extinctions, but species from the tropics to the poles are affected.

## Right for the Wrong Reasons?

The 2004 research was the first attempt to put numbers to climate change extinction risk. Did this first attempt to quantify this complex process get it right? Or was it simply a first straw man, to be torn down and replaced by more accurate estimates?

These questions are hotly debated. Many flaws have been found in the original research methods, some that could raise the estimate, some that could lower it. On the one hand, there is some evidence that the models used may overestimate range loss, and that there are substantial differences between modeling techniques. On the other hand, the climate scenarios used in Thomas et al. estimates were only for midcentury and didn't include interactions with habitat loss.

Climate models carry significant uncertainties, particularly with regard to precipitation. Precipitation change in one global climate model (GCM) will vary by region in ways very different from another GCM. Species distribution models rely on these GCM inputs to estimate changes in range sizes that can be used to estimate extinction risk. So where species range changes are sensitive to change in precipitation, very different results may emerge depending on what climate model is used. Species distribution models carry their own uncertainty, and it is not clear that the SAR can be used in the ways it was applied in the original research. The authors of this book explore these issues in part II.

Counterbalancing these possible sources of overestimation are climate change trajectories and land use interactions. The midcentury climate scenarios used in Thomas et al. are mild in comparison with likely overall change by the end of the century. Climate change is accelerating, and change in the latter half of the century is projected to be much greater than that in the first half century. Further, current emissions are above even the most extreme scenarios used by Thomas et al., so there is even more reason to think that change may be greater than that used to derive the million species estimate.

Conversion of land from natural habitat to human uses continues, and this wasn't factored into the original estimates. Because climate change causes species' ranges to shift, human land uses that block range shifts can dramatically increase likelihood of extinction. As ecosystems unravel and run into agricultural fields and expanding cities, critical interdependencies among species may begin to break down, ramping up extinction risk yet again.

Finally, marine species weren't considered at all in the million species estimate. Yet marine species are threatened not only by climate change, but by acidification caused when carbon dioxide dissolves in seawater. Thus, human $CO_2$ pollution threatens the oceans in two ways—directly through acidification of seawater and indirectly through climate change. There is little reason to think that the species that inhabit the oceans are less vulnerable than those that inhabit terrestrial environments, so there is a large group of potential marine extinctions to consider. There are many reasons to think the million species number is too low, and many important research questions to be pursued to reach final answers. These issues are examined in part V.

So flaws in the early methods that favor overestimation may be more than compensated by strong bias toward underestimation. The purpose of this book is to elaborate these biases, contribute evidence from other lines of inquiry, and let readers decide for themselves. The early estimates may well turn out to be right for the wrong reasons, or to be too low.

## How Can We Help?

The ultimate goal of this book is to suggest ways to stem a wave of extinctions due to climate change. By understanding the drivers and magnitude of change, policy makers and conservationists should gain critical insights into effective responses. It is certain that effective action will have two main foci: reduction of greenhouse gas emissions and improved conservation strategies.

Extinction risk helps identify acceptable and unacceptable levels of change relevant to global policy. Large numbers of extinctions are socially unacceptable, and make it impossible to achieve United Nations Framework Convention on Climate Change goal of allowing ecosystems to adapt naturally to climate change. We will need to understand extinction risk to help inform targets for limiting greenhouse gas pollution. The transition to a renewable energy economy is therefore the first ingredient in reducing the extinction risk from climate change.

But greenhouse gas levels are now unlikely to be tamed within the lower bounds safest for ecosystems and species. Thus, the second great challenge is to adapt our conservation strategies to cope with the stresses of climate change that can't be avoided. Given current emissions trajectories and the delays in international action, these stresses are likely to be large. Expanding protected areas, increasing connectivity, and creating ex situ safety nets for species will all be required. We hope that this book may also offer first insights into the magnitude and urgency of these needs.

The chapters that follow synthesize current research and suggest important avenues for advancing our understanding. They do not and cannot provide final answers. We hope that they speed the quest for answers and inform a wide range of readers deeply concerned about extinction risk from climate change.

## REFERENCES

Akçakaya, H. R., S. H. M. Butchart, G. M. Mace, S. N. Stuart, and C. Hilton-Taylor. 2006. "Use and misuse of the IUCN Red List Criteria in projecting climate change impacts on biodiversity." *Global Change Biology* 12 (11): 2037–2043.

Buckley, L. B., and J. Roughgarden. 2004. "Biodiversity conservation—Effects of changes in climate and land use." *Nature* 430 (6995).

Harte, J., A. Ostling, J. L. Green, and A. Kinzig. 2004. "Biodiversity conservation—Climate change and extinction risk." *Nature* 430 (6995).

Thomas, C. D., A. Cameron, R. E. Green, M. Bakkenes, L. J. Beaumont, Y. C. Collingham, B. F. N. Erasmus, et al. 2004. "Extinction risk from climate change." *Nature* 427 (6970): 145–148.

Thuiller, W., M. B. Araujo, R. G. Pearson, R. J. Whittaker, L. Brotons, and S. Lavorel. 2004. "Biodiversity conservation—Uncertainty in predictions of extinction risk." *Nature* 430 (6995).

# Chapter 2

# *First Estimates of Extinction Risk from Climate Change*

CHRIS D. THOMAS

This chapter reviews the first study that provided an international assessment of the risks to biodiversity associated with climate change. Rapid acceleration of information at the end of the twentieth century showed that the distributions of terrestrial species were responding to climate change (Parmesan et al., 1999; Pounds et al., 1999; Thomas and Lennon, 1999). Combined with the extreme El Niño event of 1998 that caused major bleaching damage to coral reefs, this work confirmed that climate variation and climate change were likely to have major impacts on biodiversity (Sala et al., 2000; IPCC, 2001; Walther et al., 2002; Parmesan and Yohe, 2003).

However, the question of whether climate change would be likely to cause many species to become extinct, as opposed to simply changing their distributions, remained unresolved. So Thomas et al. (2004a) decided to make a "first pass" estimate of what the level of threat might be. The authors accepted that there would be many uncertainties, but thought that preliminary estimates could still be useful in the context of policy development. It was also hoped that such an attempt would encourage scientific colleagues to develop improved estimates in the future.

## The General Approach

Thomas et al. (2004a) adopted a species distribution modeling (SDM) approach, alternatively termed niche or climate envelope models. The first step is to match the records of each species to geographic variation in the climate. This provides a description of the set of climatic conditions (e.g., temperatures at different times of year, precipitation, indices of drought) where the species has been found in recent decades. Although there are many different methods available, ultimately they all establish some form of correlation between the observed distribution of a species and a set of environmental (climatic) variables. It is then possible to use these models to predict where such climatic conditions might be found in the future, for a variety of climates that might be experienced in the future. These correlative models ignore all sorts of important things, especially the dynamics of birth, death, immigration, emigration, and the role of genetic variation within and among populations in determining responses. They also ignore the possibility that species will be able to live in areas where novel combinations of seasonal temperature and precipitation regimes will come into existence that do not currently exist anywhere on Earth under present-day conditions. In addition, the presence of other species (e.g., new invasive species), land use change, and other factors (e.g., nitrogen deposition, direct effects of carbon dioxide enrichment of the atmosphere) may cause some locations to be uninhabitable in the future, even though it might seem that they would be suitable, based on climate alone. So the SDM approach is very much a first approximation.

One then inserts the future climate variables into these models to evaluate where the climatic conditions favored by each species might be found in the future. For most species that are modeled, such projections generate (i) locations where both the recent and the future climate fall within the climatic conditions that are currently occupied (overlap—where conditions are assumed to remain suitable for the species), (ii) locations where the species currently occurs, but where the future climate will fall outside the set of conditions currently occupied (assumed to have declining suitability), and (iii) locations that currently lie outside the climatic conditions that are occupied, but that will lie within them in the future (assumed to have increasing suitability) (fig. 2-1A).

In some cases the geographic overlap zone was large—these species were not at risk—but in other cases there was no overlap zone at

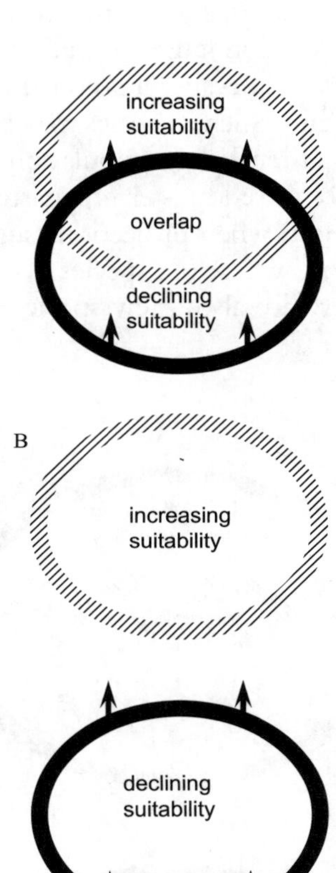


FIGURE 2-1. Schematic diagram of geographic range shifts under climate change. Solid-line circle represents the locations of the current distribution of a species, where climatic conditions were suitable for that species in the recent past. Hatched-line circles illustrate where similar climatic conditions might be found in the future. In some cases, past and future distributions partially overlap (A), whereas in other cases they do not (B).

all (fig. 2-1B). If such a species failed to colonize the region of "increasing suitability," it would potentially be at risk of extinction. Figure 2-1 could potentially indicate movements along a latitudinal or elevational gradient, or along a moisture gradient.

In other species, the area that is projected to remain climatically suitable is expected to be a geographic subset of the locations where it currently occurs (fig. 2-2A); a species that occupies the top half of a mountain is likely to retreat to an ever smaller subset of its former distribution as its lower elevation range boundary moves upward, and it is unable to expand upward because it is constrained by the maximum elevation of the mountain. Such a species may be at risk of extinction if, for example, the total remaining population size falls below some minimum required to ensure long-term persistence. However, much greater risk is experienced when projections suggest that the current set of climatic conditions where the species occurs may disappear entirely (fig. 2-2B). There are also a few species that show expanding

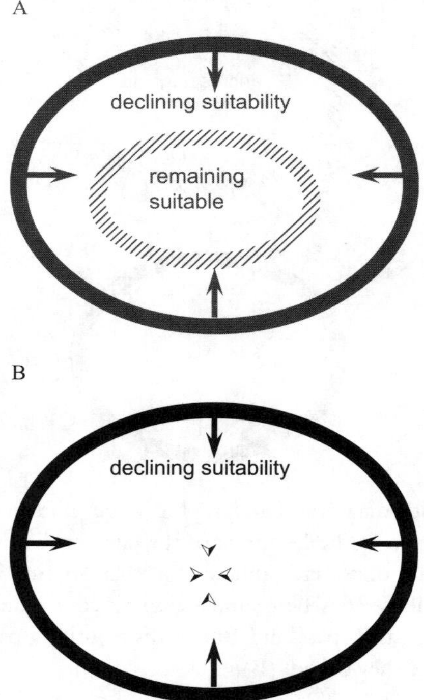


FIGURE 2-2. Schematic diagram of geographic range shrinkage under climate change. Solid-line circles represent the location of the current distributions of species, where climatic conditions were suitable for the species in the recent past. The hatched-line circle illustrates where these climatic conditions might be found in the future (A). Small arrows indicate the potential complete disappearance of such climatic conditions (B).

projected distributions (the reverse of the pattern shown in fig. 2-2A), and these are unlikely to be under any threat.

Future projected distributions are not explicit predictions of where a species will actually be at a given time. Species may survive for an unknown length of time in areas of declining climatic suitability, and they may or may not manage to colonize areas of increasing climatic suitability. If they do so, they may achieve these new distributions at variable rates. Empirical evidence indicates that the majority of species are shifting their distributions toward the poles (at "improving" range margins), but it has been suggested that some species may be doing so more slowly than the climate itself is shifting (e.g., Parmesan and Yohe, 2003; Parmesan, 2006; Menéndez et al., 2006). Many species with inadequate dispersal and rare habitats are apparently failing to expand into such areas at all (Warren et al., 2001; Menéndez et al., 2006). In the absence of adequate data on rates of range shifts in the species we examined, Thomas et al. (2004a, b) considered the extremes of complete dispersal (best-case scenario) and no dispersal at all (worst-case scenario), presuming that the real risks faced by these species would be intermediate.

Every species is affected by a wide range of factors, so SDMs can be thought of as a simple form of risk assessment, and not as a genuine prediction of how one might expect a given real species to behave. But averaged across many species, the results seem likely to be a fair representation of the risks that might be associated with climate change.

## The Basic Results

In the absence of an agreed methodology for converting projected range changes into extinction risk, Thomas et al. (2004a, b) examined the output of projections for 2050 in several ways. Although the estimates from the different methods varied, the "order of magnitude" answer was quite consistent. Around 10 or more percent of species, and not 1 percent, 0.1 percent, or 0.01 percent, appeared to be at risk. Thus, the risks from climate change appeared to be on a par with other major threats to global biodiversity.

Thomas et al. (2004b) estimated that, for midrange 2050 warming, the entire distributions of about 4 percent of the species analyzed would fall into the "climate disappearance" category (high risk of extinction; fig. 2-2B), and a further 8 percent of species would show a complete separation between their current ranges and where suitable climate conditions might be found in the future (as in fig. 2-1B). Thomas et al. (2004b) estimated that between 4 and 19 percent of species would potentially lose 100 percent of their modeled distributions, depending on which climate change scenario was considered and whether one assumed that they were able to disperse of not.

For every species that was projected to lose its entire climate space by 2050, at least as many again had lost more than 90 percent of their previous range area (i.e., the "overlap" zone in fig. 2-1A, or the "remaining suitable" area in fig. 2-2A, was less than 10 percent of the original range area). Species that are projected to lose 90 percent of their climate space by 2050 are likely to lose the remainder soon afterward, as will many species that are projected to lose 80–90 percent by 2050.

Even if the 2050 climatic conditions continued indefinitely, some proportion of the species projected to lose "only" 90 percent of their climate space by 2050 would still be at risk of extinction, either because they now fall below some population viability threshold, or because the area that remains climatically suitable no longer coincides so well with other habitat attributes that are important to the species. It was this realization that led Thomas et al. (2004a) to adopt the species-area approach to estimate the potential risk of extinction. It is widely known that the larger the area that is available of a particular habitat type, the more species are associated with it. If 90 percent of a particular habitat type is lost, approximately half of the species restricted to that habitat type are likely to be lost for two main reasons: because the remaining area no longer contains any resources (food, habitat, etc.) for that species (species that die out quickly); or because the remaining population size is no longer adequate for long-term persistence (species that die out more gradually). Thomas et al. (2004a) reasoned that loss of suitable climate is conceptually akin to habitat loss, and that the same general approach could be taken. This approach does not aim to identify exactly which species might become extinct, but to identify the proportion that might do so.

Thomas et al. (2004a) thus summed the projected range area losses to estimate that, on average, 17 percent of species might be at risk of extinction for midrange warming, under the optimistic assumption that species were able to colonize new areas that became climatically suitable for them; 31 percent might be at risk if dispersal was not possible. Using a variety of different methods of analysis (choice of analytical method, varying climate scenario, assuming full dispersal or none), the risk of extinction across all 1,103 species ranged from 9 percent to 52 percent. Given the several orders of magnitude of uncertainty that previously existed, the results helped identify that climate change was likely to cause high levels of extinction.

Finally, the maximum amount of climate warming expected by 2050 is quite similar to the minimum amount of warming expected by 2100. Depending on the analysis and assumptions, projections for 2050-maximum warming ranged between 13 and 52 percent potential extinction, a major threat.

Note that these are not projections of extinction *by* 2050, but estimates of the percentages of species that would, by 2050, be inhabiting climatic conditions outside the set of conditions they currently inhabit. The majority of these species might be expected to be declining by this time, potentially toward complete extinction, but the lifetime (extinction debt) of these dwindling species may often be long. The "extinction event" associated with anthropogenic climate change may take many centuries to be *fully* realized.

It is worth placing the estimated extinction in a geological context. Data are available for estimated extinction rates of marine genera; three of the recognized "Big Five" extinctions have taken place within the last 300 million years (Rohde & Muller, 2005; fig. 2-3). These estimates relate to the oceans, so quite how comparable they are to land species is questionable. Suppose that 13 percent of current terrestrial species are at risk of extinction from climate change (the likely minimum for 2100 warming), then somewhat less than 10 percent of genera might be expected to become extinct (lower dashed line, fig. 2-3), according to the genus-species conversion curve developed by Raup (1991). The upper estimates of species-level extinction from climate change run at around 50 percent of species, which, converted to generic extinctions, are shown by the higher dashed line in figure 2-3. Bear in mind that the geological extinction curve is the number of extinctions per million years, whereas projected extinctions from anthropogenic climate change are expected on a much shorter time scale, of tens of years to thousands of years.

Figure 2-3 suggests that climate change alone is unlikely to generate a mass extinction as large as one of the Big Five, although this is quite possible in combination with other factors. On the other hand, there is a high likelihood that climate change on its own could generate a level of extinction on a par with, or exceeding, the slightly "lesser"

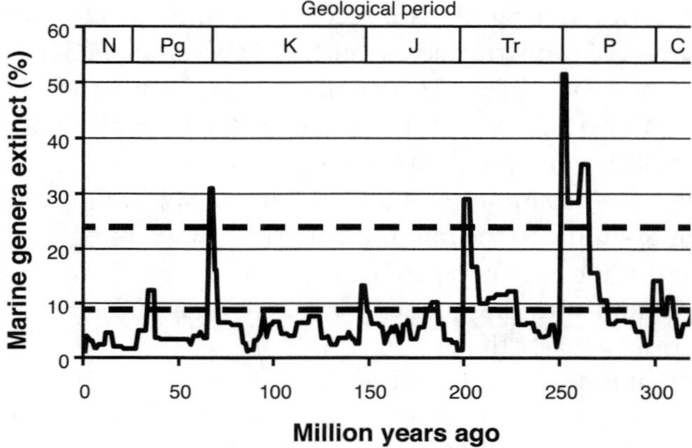


FIGURE 2-3. Percentage extinction for marine genera over the last 315 million years (data from Rohde & Muller, 2005), and the possible extent of the current, climate change–induced extinction threat (between the upper and lower dashed lines). The end-Permian (P), end-Triassic (Tr), and end-Cretaceous (K) events are the three Big Five extinctions to take place within this period. Other periods are the Neogene (N), which continues to the present day; Paleogene (Pg), which includes the Eocene-Oligocene extinction peak; Jurassic (J); and Carboniferous (C).

extinction events associated with the ends of the Carboniferous and Jurassic geological periods, and with the end of the Eocene epoch (within the Paleogene) (fig. 2-3). In this context, the potential impacts of climate change support the notion that we have recently entered the "Anthropocene" period (Crutzen, 2002).

## The Initial Response in the Media and Journals

The media response to the initial paper included front-page headlines in major newspapers across the world. The topic reached weekly news-quiz shows and generated questions in the UK parliament and US Congress. Incorporation within the Intergovernmental Panel on Climate Change (IPCC) (2007) conclusions and the Stern Review (Stern, 2006) cemented the influence of the work. Now the paper is widely cited in the literature as an emblematic illustration of the damage that may be wrought by climate change. With every retelling, the original uncertainties seem to fade further from sight!

Some commentary and criticism from colleagues (Ladle et al., 2004) stemmed from the press reports rather than from the original paper itself, where many caveats were included. The reporting was reasonably accurate, given the speed with which the story spread. The commonest problem was that almost all headlines reported that these species would be extinct "*by* 2050" when this was explicitly denied by the original paper, and personally by the authors when they spoke to journalists. These misleading headlines were printed even though most science journalists understood full well that the extinctions could take many decades or even centuries to achieve.

The second most common problem was the "up to" issue. Most journalists quote the highest of a range of values to generate maximum impact, so the largest estimates of extinction were the ones that were most widely quoted. Finally, objections were voiced to stating (in the press release) that a million species could be at risk of extinction from climate change. This still appears to be a reasonable conclusion, provided one accepts that the species modeled represent a sample of terrestrial biodiversity. Combining estimates of global diversity with per species risks from climate change suggests that a million species at risk could be conservative (Thomas, 2004).

Although the alarm raised by the media sometimes verged on exaggeration, policy makers and subsequent reports have tended to go back to the original sources and to the authors, and do not draw conclusions based on the clamor that was initially generated. Long-term harm does not seem to have arisen from reporting inaccuracies. Rather, it has alerted a wide constituency to the fact that there is a serious issue to address.

## The Major Concerns and Subsequent Estimates of Extinction

More than ten responses to the original paper were submitted to *Nature,* of which the journal published three. The first substantive point was that projections based on different distribution (climate envelope) models vary (Thuiller et al., 2004), to which Thomas et al. (2004b) replied by pointing out that the variation associated with this source of error was lower than the variation associated with other factors that were included in the original analysis (particularly different climate and dispersal scenarios, when applied to multispecies analyses). This was a good point, but did not affect the overall conclusions. The second key issue was raised by Harte et al. (2004), who noted that projections using SDMs assume that all populations of a species will be able to occupy all of the types of climate occupied by all other populations of that species, hence ignoring local adaptations. They argued that future conditions may fall outside the tolerances of local populations, causing them to decline or become extinct, even if there were some other (geographically remote) populations within the species that might be able to survive or even thrive within those new conditions. Harte et al. (2004) suggested that our projections might, therefore, underestimate the true level of threat. This is another good point. On the other hand, species may sometimes evolve to use novel conditions, so some species will survive "unexpectedly." It is uncertain whether these evolutionary considerations will increase or decrease overall expectations of species-level extinctions.

There are many other uncertainties and potential criticisms: one of the key motivations in writing the original paper was to stimulate this area of research. Rather than go through all of the pros and cons of each assumption and method used, the remainder of this chapter simply provides a very brief summary of how the Thomas et al. (2004a, b) estimates of extinction have fared since they were first published.

In SDMs, one generally assumes that the climate responses of species are directly linked to the climate, and these can generate high estimates of extinction risk (McClean et al., 2005). An alternative approach is to assume that climate affects the vegetation and that the impacts on biodiversity are mainly mediated through vegetation change. This was the approach taken by Malcolm et al. (2006). Malcolm et al. (2006) used output from dynamic global vegetation models (Scholze et al., 2006) to estimate the amount of vegetation change that would be expected to take place within the distributional ranges of species that are confined to global biodiversity hotspots—relatively small areas of the world that contain disproportionate numbers of rare species that do not occur outside these hotspot regions. They estimated that less than 1 percent to 43 percent of species (average 11.6 percent) are at risk of extinction for 2100 climate change, broadly the same range as found by Thomas et al. (2004a), although possibly somewhat lower. Jetz et al. (2007) took a similar approach, estimating that about 10 to 20 percent of the world's bird species could be threatened by 2100 from a combination of climate and land use changes. The Malcolm and Jetz approaches may somewhat underestimate the long-term threats from climate change because they consider only the changes associated with habitat/vegetation change that will have taken place by 2100, ignoring (i) subsequent vegetation changes that are already inevitable by then (because vegetation change lags behind the climate) and (ii) direct impacts of climate on species that are not mediated through vegetation change. In any event, Malcolm's, Jetz's, and Thomas's estimates are of comparable magnitude. The risk of extinction from climate change is high.

## Where Is the Risk Greatest?

Although the exact numbers of species at risk of extinction will remain contentious for the foreseeable future, it is worth evaluating whether a greater consensus might be reached on the locations where extinctions might be greatest. Jetz et al. (2007) identify that the risk per species may increase toward the poles, but that the concentration of species in the tropics may result in most extinctions taking place in equatorial regions. Many of these extinctions could be of species that have small geographic ranges, restricted to biodiversity hotspots (Malcolm et al., 2006).

Jack Williams et al. (2007) and Ohlemüller et al. (2006) found that there are areas of "disappearing" climate space—parts of the world where the type of climate that currently occurs there will cease to exist in the future, anywhere on Earth. Williams et al. (2007) showed that many of these areas are in tropical mountains, which are coincident with the distributions of many small-range species. Stephen Williams et al. (2003) found that endemic species that are completely restricted to high altitudes in Queensland, Australia, were particularly susceptible to climate change—they would be driven off the tops of the mountains. Ohlemüller et al. (2008) more generally found that small-range (endemic) species tend to be restricted to rare climates, that these climates tend to be cooler than surrounding areas, and that these rare climates are expected to shrink disproportionately with further climate change. Across these and other studies, it appears that species that already have small geographic ranges are likely to be most seriously threatened by climate change.

Over the last million years, the climate has fluctuated between cold glacial periods and warmer "interglacial" periods, such as the relatively warm Holocene period that we have experienced for the last approximately 10,000 years. During the extremes of these climatic fluctuations, many species become restricted periodically; warmth-loving species become restricted when the climate is generally very cold, and cool-adapted species become restricted when the climate is generally warm. At least some of the centers of endemism (areas that contain large numbers of species with small geographic ranges) we currently recognize are likely to be "interglacial refugia," mainly mountainous regions to which cool-adapted species have retreated during the warm Holocene. If these species are already restricted to unusually cold locations, further warming may eliminate them entirely, as suggested by Ohlemüller et al. (2008).

Using the original Thomas et al. (2004a) projections, predicted range declines (assuming perfect dispersal) are greatest in the Southern Hemisphere (negative latitudes in fig. 2-4). Range reductions are predicted to be especially high for species that are restricted to "centers of endemism." This initially seems odd because projected future warming is actually greater in the north. However, "natural" climate warming earlier in the Holocene, and potentially in previous interglacials, is likely to have resulted in greater warming at northern latitudes than in the south (Wright et al., 1993; Davis and Brewer, 2009). It is reasonable to suppose that heat-sensitive species would have been eliminated in the past and that the species that survive at high northern latitudes have shown the greatest capacity (e.g., dispersal, habitat range) to cope with massive climatic fluctuations there. In contrast, a smaller absolute amount of future warming in large parts of the tropics is likely to take local and regional climates outside the range of those experienced historically (Williams et al., 2007), threatening many tropical species with extinction.

Whether this historical interpretation turns out to be correct or not, analyses nonetheless suggest that the major risks, in terms of the extinction of species, lie predominantly in existing centers of endemism (fig. 2-4; Ohlemüller et al., 2008).

## The Way Forward

Very simple statistical models consistently suggest that there is a large extinction risk from climate change and that these risks will escalate with the level of warming that takes place. This is useful from the per-

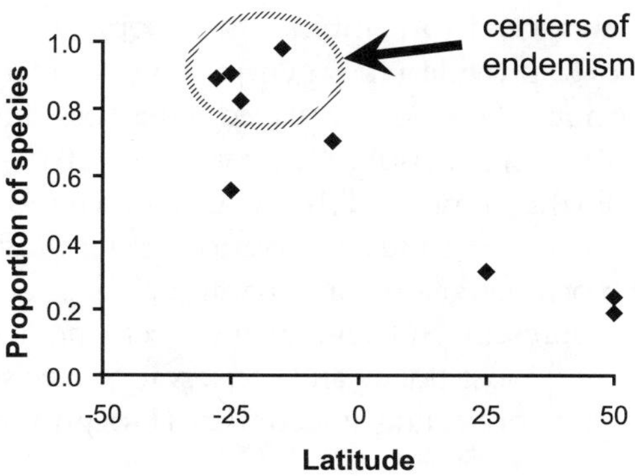


FIGURE 2-4. Proportion of species *projected* to lose more than 50 percent of their climate area by 2050, assuming full dispersal and the average of climate scenarios described by Thomas et al. (2004a). Each point represents a group of animals or plants in a given region. The *x*-axis represents the average latitude of each region.

spective of identifying that climate change represents a major extinction risk and hence supports the call for a need to "mitigate" or limit climate change. Many of the critiques of the climate SDM/climate envelope approach are from authors who are more interested in our capacity to "adapt" to climate change, for which much more detailed models may be required. More complex models are needed to assess: where a species might be at a particular date in the future; which areas need to be protected to enable a given species to survive climate change; and how we should design connected landscapes that will permit species to change their distributions. These important issues require dynamic models that incorporate the birth and death of individuals (or of populations), and the ability of individuals to colonize new regions.

For example, when the climate improves rapidly at the northern range boundary of a species (in the Northern Hemisphere), that species may no longer be constrained by the climate, but by its ability to colonize the new region. In such circumstances, range expansions can be simulated quite accurately by running colonization and extinction models across networks of suitable habitats (e.g., Thomas et al., 2001; Hill et al., 2001; Willis et al., 2009; Wilson et al., 2009). These models generally show that species expand most rapidly in areas with greatest habitat availability.

There is now a need to combine these population models with SDM approaches to provide a new generation of models in which the population dynamics of species are played out across climatically varying landscapes (Keith et al., 2008; Anderson et al., 2009). Such models should soon be forthcoming, and they will be much more appropriate for examining management and conservation strategy (adaptation) scenarios than are bioclimate-alone or dynamics-alone models. Although such models will represent a substantial improvement, many uncertainties will remain, and most policy and conservation decisions will have to be made despite continuing uncertainty (Hoegh-Guldberg et al., 2008).

## Conclusions

The Thomas et al. (2004a, b) analysis is far from perfect, but it nonetheless seems to have summarized the essence of the problem that species face from climate change. Large numbers of species live (or recently did so) in types of climates that will not exist in the future; other species may have difficulty keeping up with the shifting location of suitable climatic conditions. All of these species can be considered to be at risk, even if not all of them are at serious threat of extinction. On the other hand, additional species may be threatened by a combination of conditions not considered in such simplistic models. Climate change must now be incorporated routinely with national and international risk assessments and red-listing procedures in hopes of developing actions that will somewhat reduce the numbers of species that disappear entirely.

## Acknowledgments

I would particularly like to thank all of my coauthors who were involved in the work that contributed to the original 2004 papers: Alison Cameron, Rhys Green, Michel Bakkenes, L.J. Beaumont, Yvonne Collingham, Barend Erasmus, Marinez de Siqueira, Alan Grainger, Lee Hannah, Lesley Hughes, Brian Huntley, Albert van Jaarsveld, Guy Midgley, Lera Miles, Miguel Ortega-Huerta, Town Peterson, Oliver Phillips, and Steve Williams.

## REFERENCES

Anderson, B. J., H. R. Akçakaya, M. B. Araújo, D. A. Fordham, E. Martinez-Meyer, W. Thuiller, and B. W. Brook. 2009. "Dynamics of range margins for metapopulations under climate change." *Proceedings of the Royal Society B* 276: 1415–1420.

Crutzen, P. J. 2002. "Geology of mankind." *Nature* 415: 23.

Davis, B. A. S., and S. Brewer. 2009. "Orbital forcing and role of the latitudinal insolation/temperature gradient." *Climate Dynamics* 32: 143–165.

Harte, J., A. Ostling, J. L. Green, and A. Kinzig. 2004. "Biodiversity conservation—Climate change and extinction risk." *Nature* 430. doi:10.1038/nature02718.

Hill, J. K., Y. C. Collingham, C. D. Thomas, D. S. Blakeley, R. Fox, D. Moss, and B. Huntley. 2001. "Impacts of landscape structure on butterfly range expansion." *Ecology Letters* 4: 313–321.

Hoegh-Guldberg, O., L. Hughes, S. L. McIntyre, D. B. Lindenmayer, C. Parmesan, H. P. Possingham, and C. D. Thomas. 2008. "Assisted colonization and rapid climate change." *Science* 321: 345–346.

Intergovernmental Panel on Climate Change (IPCC). 2001. *Climate Change 2001: The Scientific Basis*. J. T. Houghton, Y. Ding, D. J. Griggs, M. Noguer, P. J. van der Linden, X. Da, K. Maskell, and C. A. Johnson, eds. Cambridge, UK: Cambridge University Press.

IPCC. 2007. *Climate Change 2007: Synthesis Report*. Core Writing Team, edited by R. K. Pachauri and A. Reisinger. Geneva, Switzerland: IPCC.

Jetz, W., D. S. Wilcove, and A. P. Dobson. 2007. "Projected impacts of climate and land-use change on the global diversity of birds." *PLoS Biology* 5: 1211–1219.

Keith, D. A., H. R. Akçakaya, W. Thuiller, G. F. Midgley, R. G. Pearson, S. J. Phillips, H. M. Regan, M. B. Araújo, and T. G. Rebelo. 2008. "Predicting extinction risks under climate change: Coupling stochastic population models with dynamic bioclimatic habitat models." *Biology Letters* 4: 560–563.

Ladle, R. J., P. Jepson, M. B. Araújo, and T. J. Whittaker. 2004. "Dangers of crying wolf over risk of extinctions." *Nature* 428: 799.

Malcolm, J. R., C. R. Liu, R. P. Neilson, L. Hansen, and L. Hannah. 2006. "Global warming and extinctions of endemic species from biodiversity hotspots." *Conservation Biology* 20: 538–548.

McClean, C. J., J. C. Lovett, W. Kuper, L. Hannah, J. H. Sommer, W. Barthlott, M. Termansen, G. E. Smith, S. Tokamine, and J. R. D. Taplin. 2005. "African plant diversity and climate change." *Annals of the Missouri Botanical Garden* 92: 139–152.

Menéndez, R., A. González-Megías, J. K. Hill, B. Braschler, S. G. Willis, Y. Collingham, R. Fox, D. B. Roy, and C. D. Thomas. 2006. "Species richness changes lag behind climate change." *Proceedings of the Royal Society B* 273: 1465–1470.

Ohlemüller, R., E. S. Gritti, M. T. Sykes, and C. D. Thomas. 2006. "Towards European climate risk surfaces: The extent and distribution of analogous and non-analogous climates 1931–2100." *Global Ecology and Biogeography* 15: 395–405.

Ohlemüller, R., B. J. Anderson, M. B. Araújo, S. H. M. Butchart, O. Kudrna, R. S. Ridgely, and C. D. Thomas. 2008. "The coincidence of climatic and species rarity: High risk to small-range species from climate change." *Biology Letters* 4: 568–572.

Parmesan, C. 2006. "Ecological and evolutionary responses to recent climate change." *Annual Review of Ecology, Evolution, and Systematics* 37: 637–669.

Parmesan, C., and G. Yohe. 2003. "A globally coherent fingerprint of climate change impacts across natural systems." *Nature* 421: 37–42.

Parmesan, C., N. Ryrholm, C. Stefanescu, J. K. Hill, C. D. Thomas, H. Descimon, B. Huntley, et al. 1999. "Poleward shifts in geographical ranges of butterfly species associated with regional warming." *Nature* 399: 579–583.

Pounds J. A., M. P. L. Fogden, and J. H. Campbell. 1999. "Biological response to climate change on a tropical mountain." *Nature* 398: 611–615.

Raup, D. M. 1991. *Extinction: Bad Genes or Bad Luck?* New York: W. W. Norton & Co.

Rohde, R. A., and R. A. Muller. 2005. "Cycles in fossil diversity." *Nature* 434: 208–210.

Sala, O. E., F. S. Chapin, J. J. Armesto, E. Berlow, J. Bloomfield, R. Dirzo, E. Huber-Sanwald, et al. 2000. "Biodiversity—Global biodiversity scenarios for the year 2100." *Science* 287: 1770–1774.

Scholze, M., W. Knorr, N. W. Arnell, and I. C. Prentice. 2006. A climate-change risk analysis for world ecosystems. *Proceedings of National Academy of Sciences, USA* 103: 13116–13120.

Stern, N. 2006. *The Economics of Climate Change: The Stern Review*. UK: Cabinet Office, HM Treasury.

Thomas, C. D. 2004. "Publish, publicise, and be damned." *Bulletin of the British Ecological Society* 35 (4): 14–16.

Thomas, C. D., and J. J. Lennon. 1999. "Birds extend their ranges northwards." *Nature* 399: 213.

Thomas, C. D., E. J. Bodsworth, R. J. Wilson, A. D. Simmons, Z. G. Davies, M. Musche, and L. Conradt. 2001. "Ecological and evolutionary processes at expanding range margins." *Nature* 411: 577–581.

Thomas, C. D., A. Cameron, R. E. Green, M. Bakkenes, L. J. Beaumont, Y. C. Collingham, B. F. N. Erasmus, et al. 2004a. "Extinction risk from climate change." *Nature* 427: 145–148.

Thomas, C. D., S. E. Williams, A. Cameron, R. E. Green, M. Bakkenes, L. J. Beaumont, Y. C. Collingham, et al. 2004b. "Biodiversity conservation: Uncertainty in predictions of extinction risk/Effects of changes in climate and land use/Climate change and extinction risk." *Nature* 430. doi:10.1038/nature02719.

Thuiller, W., M. B. Araújo, R. G. Pearson, R. J. Whittaker, L. Brotons, and S. Lavorel. 2004. "Biodiversity conservation—Uncertainty in predictions of extinction risk." *Nature* 430. doi:10.1038/nature02716.

Walther, G. R., E. Post, P. Convey, A. Menzel, C. Parmesan, T. J. C. Beebee, J. M. Fromentin, O. Hoegh-Guldberg, and F. Bairlein. 2002. "Ecological responses to recent climate change." *Nature* 416: 389–395.

Warren, M. S., J. K. Hill, J. A. Thomas, J. Asher, R. Fox, B. Huntley, D. B. Roy, et al. 2001. "Rapid responses of British butterflies to opposing forces of climate and habitat change." *Nature* 414: 65–69.

Williams, J. W., S. T. Jackson, and J. E. Kutzbach. 2007. "Projected distribution of novel and disappearing climates by 2100 AD." *Proceedings of National Academy of Sciences, USA* 104: 5738–5742.

Williams, S. E., E. E. Bolitho, and S. Fox. 2003. "Climate change in Australian tropical rainforests: An impending environmental catastrophe." *Proceedings of the Royal Society B* 270: 1887–1892.

Willis, S. G., C. D. Thomas, J. K. Hill, Y. C. Collingham, M. G. Telfer, R. Fox, and B. Huntley. 2009. "Dynamic distribution modelling: Predicting the present from the past." *Ecography* 32: 5–12.

Wilson, R. J., Z. G. Davies, and C. D. Thomas. 2009. "Modelling the effect of habitat fragmentation on range expansion in a butterfly." *Proceedings of the Royal Society B* 276: 1421–1427.

Wright, H. E., Jr., J. E. Kutzbach, T. Webb III, W. F. Ruddiman, F. A. Street-Perrott, and P. J. Bartlein, eds. 1993. *Global Climates since the Last Glacial Maximum*. Minneapolis: University of Minnesota Press.

# Chapter 3

## *Climate Change, Extinction Risk, and Public Policy*

JONATHAN MAWDSLEY, GUY MIDGLEY, AND LEE HANNAH

When initial estimates of the extinction risk from climate change appeared, there was immediate public and media interest. Authors of Thomas et al. appeared on CNN, BBC, and other major national and international television networks. Newspaper headlines, often front page, appeared on the day of the report's release. Magazine, radio, and other media treatments of the subject followed for weeks after. But to what extent was this media interest driven by policy relevance, and to what extent were the implications of the extinction risk estimates taken up in policy dialogue?

This chapter examines the policy relevance of extinction risk from climate change. It begins with examination of policy debates spurred by the 2004 estimates, moves to discussion of the incorporation of extinction risk into international policy instruments—from the United Nations climate change convention to international listing of threatened species—and concludes with a discussion of how additional research can help inform international policy debates.

### 2004 Policy Debates

The first major policy mention of extinction risk from climate change came only 4 days after the publication of the 2004 paper. On January

8, 2004, in the House of Commons of the United Kingdom, Margaret Beckett, the Secretary of State for Environment, Food, and Rural Affairs, referred to the Thomas et al. findings. Norman Baker challenged Secretary Beckett, referring to newspaper stories reporting the initial extinction risk estimates. He stated that those estimates added urgency to action to avert extinctions by acting on climate change. Beckett agreed that the extinction risk estimates were worrisome and went on to defend her government's efforts to reduce climate change–causing emissions.

Later that year, the US Senate held hearings specifically devoted to examining the consequences of extinction risk from climate change. The hearings were chaired by future presidential candidate John McCain, who opened the proceedings by saying "there are some who still deny that climate change is happening—despite their lying eyes." Witnesses at the hearing included authors of the Thomas et al. paper, representatives and scientists from conservation groups, and specialists in particular aspects of climate change. A significant focus of the hearing was on the marine impacts of climate change, which was noteworthy because the Thomas et al. estimates did not include marine extinctions.

It is impossible to say quantitatively how or whether the House of Commons and US Senate discussions on extinction risk affected international policy. However, qualitatively, it is clear that the extinction risk estimates captured policy makers' attention and were significant in contributing to concern about the possible impacts of climate change on wildlife. The United Kingdom subsequently went on to adopt more stringent greenhouse gas reduction targets, while the United States has struggled to address the issue for nearly a decade. Despite these differing short-term political outcomes, biological consequences of climate change are well embedded in international policy instruments, in particular the United Nations Framework Convention on Climate Change (UNFCCC).

## UNFCCC and "Dangerous Interference"

Probably the most prominent public policy arena where discussion of species extinction risks has occurred is the negotiations around the UNFCCC. Article 2 of the text of this convention states that the convention is intended to prevent "dangerous anthropogenic interference with the climate system" "within a time frame sufficient to allow ecosystems to adapt naturally to climate change" (UNFCCC, 1992). This language has fostered widespread interest in estimating the potential for "dangerous interference" in ecological systems (O'Neill and Oppenheimer, 2002), including the potential for species extinctions, under altered climate regimes (Thomas et al., 2004). Most nations in the world, including the United States, are signatories to the UNFCCC, so the language of the convention has major international significance. It establishes global response to climate change.

O'Neill and Oppenheimer (2002) were among the first to explore the implications of "dangerous interference" with regard to "unique and threatened systems," particularly coral reefs, where significant species loss from climate warming had already been projected (Hoegh-Guldberg, 1999). As evidence has mounted of impacts to species and ecosystems other than coral reefs—from conifer forests in western North America to amphibians in Central and South America—the difficulty in achieving the UNFCCC goal of allowing ecosystems to adapt naturally has become apparent. This has provided significant impetus toward lower greenhouse gas stabilization targets in international dialogue.

Estimates of possible increases in extinction risk in various ecosystems were summarized by the Intergovernmental Panel on Climate Change (IPCC) (2007) and have been used in UNFCCC communications (e.g., UNFCCC, 2007) as a key justification for adopting a comprehensive multinational approach to climate change that includes both mitigation and adaptation measures. The Convention on Biological Diversity has convened an Ad Hoc Technical Advisory Group to channel advice from convention participants to the UNFCCC on the possible impacts and interactions between climate change and biodiversity.

Estimates of extinction risk are of interest to a significant number of other intergovernmental entities that have been established to help implement particular international treaties related to biodiversity conservation. Some of the most important of these entities are the Convention on International Trade in Endangered Species, the Ramsar Convention (Convention on Wetlands of International Importance, Especially as Waterfowl Habitat), the Migratory Bird Treaty, and the Antarctic Treaty (USFWS, 2010).

## Climate Change and Threatened Species

The International Union for the Conservation of Nature (IUCN), a nongovernmental organization (NGO), maintains the international "Red Lists" of imperiled animal and plant species and is an important consumer of information on species extinction risks. The IUCN is currently reviewing information on climate vulnerabilities for threatened and imperiled species worldwide and has developed criteria to help identify species that may be susceptible to climate change (IUCN, 2008).

In addition to the IUCN, the IPCC, established by the United Nations Environment Programme and the World Meteorological Organization in 1989, is the pre-eminent authority of the science of climate change and its impacts. The IPCC has served as an important compiler and synthesizer of information on the possible effects of climate change on natural and managed ecosystems as well as the built human environment. Information about possible extinction risks associated with climate change is prominently featured in the Fourth Assessment Report of the IPCC (IPCC, 2007), particularly the Working Group II report "Impacts, Adaptation, and Vulnerability." Reports from the IPCC serve as an important resource body of knowledge and also a source of motivation for efforts to address climate change impacts through existing policy-making forums.

The IUCN was instrumental in drawing global attention to the importance of climate change in generating extinction risk. It convened the first meeting that brought together the authors of the Thomas et al. initial estimates of extinction risk from climate change. Despite this early involvement, the IUCN has struggled to incorporate climate change into its red-listing assessments of threatened species. The primary problem has been that the IUCN red-listing process is geared toward identifying species at immediate risk from pressing stressors such as habitat loss and pollution. The time lines in the IUCN Red List guidelines are therefore weighted toward timescales of decades or years. The criteria are poorly designed for dealing with stressors such as climate change that have effects decades or centuries in the future. After almost a decade of work, the IUCN is now poised to begin to systematically incorporate climate change as a threat in the red-listing assessment process (Foden et al., 2008).

Other important compilers of extinction risk information include the secretariat of the Convention on International Trade in Endangered Species, which reviews information and sets international trade restrictions on species deemed at risk for overexploitation through international trade (CITES, 2010).

## Extinction Risk and National Policy Debates

Most sovereign countries have some form of biodiversity conservation infrastructure, which usually includes legislation intended to protect biological diversity, and one or more governmental agencies that are specifically dedicated to nature protection or biodiversity conservation. Information about species extinction risk can be of great assistance to these conservation authorities in establishing priorities for conservation action and investment. National governments have a wide range of conservation tools available to prevent species extinctions.

Although these tools will undoubtedly be useful in managing the effects of climate change and preventing species extinctions, many existing conservation policies will need to be modified to incorporate new information about the effects of climate change (Mawdsley et al., 2009). Some of the most widely used conservation tools are "static" with respect to the landscape (e.g., a nature reserve with fixed boundaries) or are static with respect to species populations and distributions (e.g., an endangered species recovery plan that sets out a fixed desired population size in a certain fixed number of geographic areas) (Lovejoy, 2005; Scott and Lemieux, 2005; Lemieux and Scott, 2005; Hannah et al., 2005; Zacharias et al., 2006).

In addition to international agreements, many national governments have adopted conservation legislation that attempts to conserve individual species at high risk of extinction. One of the best known of these statutes is the Endangered Species Act of the United States (ESA, 1973). The act was originally adopted by the US Congress to protect those "species of fish, wildlife, and plants [that] have been so depleted in numbers that they are in danger of or threatened with extinction" (ESA, 1973). Because the language of the act was written at a time when the risks from anthropogenic climate change were not yet fully understood, some authors have questioned whether the act will be a useful tool for conserving species at risk from global climate change (Da Fonseca et al., 2005; Ruhl, 2007; Mawdsley et al., 2009).

The decision by the US Department of the Interior and the US Fish and Wildlife Service (2008) to list the polar bear (*Ursus maritimus*

Phipps, Ursidae) as a "threatened" taxon under the act provides an interesting example of the use of models of habitat loss and extinction risk in public policy debates. This decision also demonstrates that the act's listing mechanisms actually do have flexibility to accommodate new sources and types of information about extinction risks from climate change.

Although there are documented declines in some polar bear populations, other populations are stable or even increasing (IUCN–SSC PBSG, 2005). When this fact was noted in public comments on the formal listing decision, the US Department of the Interior and US Fish and Wildlife Service (2008) responded that "the polar bear is not currently in danger of extinction throughout all or a significant portion of [its] range, but [is] likely to become so within the 45-year 'foreseeable future' that has been established for this rule. This satisfies the definition of a threatened species under the [Endangered Species] Act; consequently listing the species as threatened is appropriate." According to this statement, the formal listing of the polar bear is based primarily on the potential for future decreases in populations of the species. Previous listings under the act have been based on evidence of actual declines in species populations (Goble et al., 2005). The record of decision (USDOI-USFWS, 2008) on the polar bear includes an extensive review of modeling studies suggesting that global climate change will have significant negative impacts on key sea ice features used by polar bears for hunting and feeding. Loss of these key habitat features is, in turn, expected to have negative impacts on polar bear populations by 2050. The listing of the polar bear thus depends on estimates of future habitat loss and inferred future population declines, rather than any significant observed decline in the species' populations. Acceptance of these modeled estimates by the US Department of Interior and the US Fish and Wildlife Service represents an important political as well as scientific validation of the methods for estimating extinction risk in polar bear populations.

## Extinction Risk and NGOs

NGOs are another set of important players who are involved in discussions and negotiations regarding climate change impacts at all levels (international, national, and subnational). Many of these organizations, such as World Wide Fund for Nature (WWF), African Wildlife Foundation, the Nature Conservancy (TNC), Conservation International (CI), Bird Life International, The Audubon Society, Flora and Fauna International, Wildlife Conservation Society, and National Wildlife Federation, have been established for the express purposes of promoting the conservation of wildlife and biological diversity. Some organizations, such as TNC, CI, and WWF, focus their efforts on biodiversity and ecosystem protection, while other organizations, such as IUCN and NatureServe, collect and disseminate information about plant and animal species at risk for extinction. Many of these organizations have already engaged in significant internal discussions regarding the potential impacts of climate change on their conservation work, and scientists at these organizations are frequently involved in assessments of the impacts of climate change on species and ecosystems.

NGOs have a number of important roles to play in policy debates around extinction risk and climate change:

- Advocate for international agreement on emissions reduction.
- Provide information on biological impacts and adaptation needs to decision makers at all levels.
- Help educate citizens and decision makers about extinction risk.
- Help connect concerned citizens with government decision makers.
- Facilitate interactions among government decision makers.
- Represent constituencies that might not otherwise be represented in public policy debates.

Ultimately, these organizations and agencies derive their power and authority from the strong interest of the general public in biodiversity conservation and the sustainable utilization of fish, animal, and plant resources. It is this concerned public that provides financial support for NGOs and pays taxes to support governmental conservation agencies. Given the strong public support for wildlife conservation activities in many countries (USFWS, 2006), it is likely that information about extinction risk will continue to play a significant role in public policy debates.

## Conclusions

Scientists who are modeling the potential effects of climate change on plant and animal species have the opportunity to contribute to a broad spectrum of public policy decisions. Information about extinction risk can be factored into decisions that are being made at national or subnational levels to protect individual species, as well as broad international agreements aimed at mitigating and adapting to global climate change. Further studies of extinction risks for individual species or suites of species will undoubtedly help guide these policy discussions and debates. We strongly encourage members of the scientific community to take the time to communicate the results of their studies to appropriate decision makers and policy-making bodies. Initial assessments of extinction risk and climate change (e.g., Thomas et al., 2004) have already painted a grim picture of future changes in global biodiversity. To paraphrase Charles Dickens, these shadows will remain unaltered in the future unless members of the global community, including biologists, can join together and decisively act to reduce the threat posed by anthropogenic climate change.

## REFERENCES


Convention on International Trade in Endangered Species of Wild Fauna and Flora (CITES). 2010. Accessed January 31, 2010. Available at http://www.cites.org

Da Fonseca, G. A. B., W. Sechrest, and J. Ogelthorpe. 2005. "Managing the matrix." In *Climate Change and Biodiversity,* edited by T. E. Lovejoy and L. Hannah, 346–358. New Haven: Yale University Press.

Endangered Species Act of the United States (ESA). 1973. Accessed January 31, 2010. http://www.fws.gov/Endangered/pdfs/esaall.pdf

Foden, W., G. Mace, J.-C. Vié, A. Angulo, S. Butchart, L. DeVantier, H. Dublin, A. Gutsche, S. Stuart, and E. Turak. 2008. "Species susceptibility to climate change impacts." In *The 2008 Review of The IUCN Red List of Threatened Species,* edited by J.-C. Vié, C. Hilton-Taylor, and S. N. Stuart. Gland, Switzerland: IUCN.

Goble, D. D., J. M. Scott, and F. W. Davis, eds. 2005. *The Endangered Species Act at 30: Volume 1: Renewing the Conservation Promise.* Washington, D.C.: Island Press.

Hannah, L., G. Midgely, G. Hughs, and B. Bomhard. 2005. "The view from the Cape: Extinction risk, protected areas, and climate change." *BioScience* 55 (3): 231–242.

Hoegh-Guldberg, O. 1999. "Coral bleaching, climate change and the future of the world's coral reefs." *Marine and Freshwater Research* 50: 839–866.

Intergovernmental Panel on Climate Change (IPCC). 2007. Fourth Assessment Report of the Intergovernmental Panel on Climate Change, 4 vols. Cambridge, UK: Cambridge University Press.

International Union for the Conservation of Nature (IUCN). 2008. Species susceptibility to climate change impacts. Accessed January 31, 2010. Available at http://cmsdata.iucn.org/downloads/climate_change_and_species.pdf

International Union for the Conservation of Nature–Species Survival Commission Polar Bear Specialist Group (IUCN–SSC PBSG). 2005. Summary of polar bear population status per 2005. Accessed January 31, 2010. Available at http://pbsg.npolar.no/en/status/status-table.html

Lemieux, C., and D. Scott. 2005. "Climate change, biodiversity conservation and protected area planning in Canada." *The Canadian Geographer* 49: 384–397.

Lovejoy, T. E. 2005. "Conservation with a changing climate." In *Climate Change and Biodiversity,* edited by T. E. Lovejoy and L. Hannah, 325–328. New Haven: Yale University Press.

Mawdsley, J. R., R. O'Malley, and D. Ojima. 2009. "A review of climate-change adaptation strategies for wildlife management and biodiversity conservation." *Conservation Biology* 23 (5): 1080–1089.

O'Neill, B. C., and M. Oppenheimer, 2002. "Dangerous climate impacts and the Kyoto Protocol." *Science* 296: 1971–1972.

Ruhl, J. L. 2007. "Climate change and the Endangered Species Act." Accessed January 31, 2010. Available at http://works.bepress.com/j_b_ruhl/1/

Scott, D., and C. Lemieux. 2005. "Climate change and protected area policy in Canada." *Forestry Chronicle* 81: 696–703.

Thomas, C. D., A. Cameron, R. E. Green, M. Bakkenes, L. J. Beaumont, Y. C. Collingham, B. F. N. Erasmus, et al. 2004. "Extinction risk from climate change." *Nature* 427: 145–148.

United Nations Framework Convention on Climate Change (UNFCCC). 1992. "United Nations Framework Convention on Climate Change." Accessed April 18, 2010. Available at http://unfccc.int/resource/docs/convkp/conveng.pdf

United Nations Framework Convention on Climate Change (UNFCCC). 2007. "Climate change: Impacts, vulnerabilities, and adaptation in developing countries." Accessed January 31, 2010. Available at http://unfccc.int/resource/docs/publications/impacts.pdf

US Department of the Interior and US Fish and Wildlife Service (USDOI and USFWS). 2008. "Endangered and Threatened Wildlife and Plants; Determination of threatened status for the Polar Bear (*Ursus maritimus*) throughout its range" (Final Rule). Federal Register: May 15, 2008 (Volume 73, Number 95). Accessed January 31, 2010. Available at http://frwebgate.access.gpo.gov/cgi-bin/getdoc.cgi?dbname=2008_register&docid=fr15my08-18

US Fish and Wildlife Service (USFWS). 2006. "2006 National survey of fishing, hunting, and wildlife-associated recreation." Accessed January 31, 2010. Available at http://library.fws.gov/pubs/nat_survey2006_final.pdf

US Fish and Wildlife Service (USFWS). 2010. "Digest of Federal Resource Laws of Interest to the U.S. Fish and Wildlife Service." Accessed January 31, 2010. http://www.fws.gov/laws/lawsdigest/treaty.html

Zacharias, M. A., L. R. Gerber, and D. K. Hyrenbach. 2006. "Review of the southern ocean sanctuary: Marine protected areas in the context of the International Whaling Commission sanctuary programme." *Journal of Cetacean Research and Management* 8: 1–12.

# PART II

## Refining First Estimates

Many studies have been published since 2004 focusing on extinction risk from climate change, looking at different taxa in different parts of the world, or looking at the methods used to derive the estimates. In this part of the book we examine these recent developments as a way of updating the findings that have been described in part I. We divide this discussion into two chapters. First Alison Cameron takes us through the multispecies modeling studies that have been published since the 2004 compendia. Cameron explores the similarities and differences and the gross numbers of studies before and after 2004. She finds that many more multispecies studies are now available, but that the methods used are seldom strictly comparable to the 2004 estimates, and therefore only qualitative confirmation of the 2004 results is available. In the subsequent chapter John Harte and Justin Kitzes look at the key species-area relationship methods that underlie extinction risk estimates using species distribution models. They find that results using the species-area relationship and the endemic-area relationships have limitations, but point to methodological improvements that show promise for the future. With the exploration of the 2004 results and their significance in part I, and the update on those results in the intervening years provided by this part of the book, we set the stage for later parts of the book that will explore the issue from a variety of disciplinary perspectives.

# Chapter 4

# *Refining Risk Estimates Using Models*

Alison Cameron

In 2004 nineteen scientists from fourteen institutions in seven countries collaborated in the landmark study described in chapter 2 (Thomas et al., 2004a). This chapter provides an overview of results of studies published subsequently and assesses how much, and why, new results differ from those of Thomas et al.

Some species distribution modeling (SDM) studies are directly comparable to the Thomas et al. estimates. Others using somewhat different methods nonetheless illuminate whether the original estimates were of the right order of magnitude. Climate similarity models (Williams et al., 2007; Williams and Jackson, 2007), biome, and vegetation dynamic models (Perry and Enright, 2006) have also been applied in the context of climate change, providing interesting opportunities for comparison and cross-validation with results from SDMs.

This chapter concludes with an assessment of whether the range of extinction risk estimates presented in 2004 can be narrowed, and whether the mean estimate should be revised upward or downward. To set the stage for these analyses, the chapter begins with brief reviews of advances in climate modeling and species modeling since 2004.

## Advances in Climate Change Projections

Thomas et al. used SDMs, which were based on climate models, which were in turn based on a range of emissions scenarios from the Intergovernmental Panel on Climate Change first (IPCC, 1990), second (IPCC, 1995), and third assessments (IPCC, 2001). The IPCC Fourth Assessment Report (2007a, b) has since revised the socioeconomic assumptions upon which the emissions scenarios are based, and global climate models (GCMs) have also developed significantly to couple ocean and atmosphere models and include aerosol forcing. As a result, mean global temperature increases are now predicted to range from 1.1 to 6.4 degrees Celsius by 2100 (IPCC, 2007a, b). Figure 4-1 shows the three categories from Thomas et al. over the warming scenarios published in the IPCC Fourth Assessment Report, illustrating the need to conduct SDM and extinction risk assessments for a temperature increase between 3 and 6.4 degrees Celsius (indicated as "new max" in fig. 4-1).

Through the four IPCC cycles, increasing numbers of climate modeling groups have produced GCMs, and SDM studies are increasingly attempting to compare, or assess the variation across, a range of GCMs (e.g., Peterson et al., 2008). SDM studies continue to focus on a range of times along the IPCC scenario trajectories, but the climate stabilization end points (beyond 2100) of each IPCC scenario are the most policy-relevant targets for SDM modeling and would provide even more extreme extinction risk scenarios. Relatively few studies have been published in the "new maximum" category (fig. 4-1), and even fewer studies (e.g., Bakkenes et al., 2006) have modeled species distributions for climate stabilization end points.

An additional and significant source of variation in climate change projections lies in the variability of outputs from separate runs of each GCM, known as realizations. Assessment of the effects of variation in realizations has become possible only with the Fourth Assessment Report (IPCC, 2007a, b), so few studies (e.g., Beaumont et al., 2007) have explored the effects on SDM outputs. Pierce et al. (2009) provided a convincing case for the use of ensembles of at least fourteen realizations, and concluded that averaging the outputs of multiple GCMs is more useful than choosing any single GCM based on model skill (how well the model performs against real data for the period of time over which climate observations have been made). However, to quantify the possible effects of the extreme predictions, sensitivity

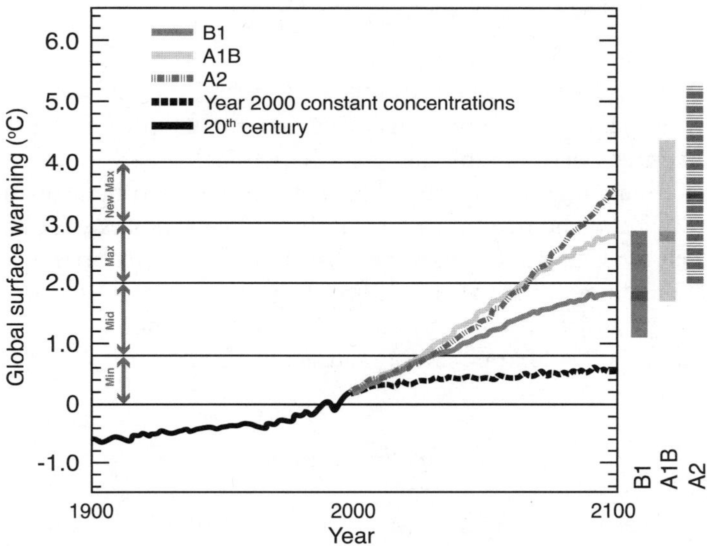


FIGURE 4-1. Potential global surface warming under three IPCC scenarios. Adapted from the IPCC Fourth Assessment Report, Summary for Policy Makers (2007a). Multimodel global averages of surface warming (relative to 1980–1999) for the scenarios A2, A1B, and B1 are shown as continuations of the twentieth-century simulations. The bars at right indicate the best estimate (solid line within each bar) and the likely range assessed for the three scenarios. The horizontal lines indicate the range of mean global warming that defined the three categories (min, mid, and max) included in the Thomas et al. (2004a) paper. The fourth category, "new max," was not included by Thomas et al. but has been used in table 4-1 later in this chapter.

analysis of the best- and worst-case GCM realizations should be considered, in addition to the ensemble or mean scenarios. To add to the intrinsic variation within and between GCMs, two main groups of methods (dynamic and statistical) are used to downscale the course grain outputs of GCMs to finer grain regional climate models. Downscaling incorporates fine-scale topographic and habitat heterogeneity and the effects of local weather patterns, which are agreed to influence the fine scale of species distributions (Luoto and Heikkinen, 2008; Ashcroft et al., 2009). However, review of the SDM literature reveals no comparisons of the effects of different downscaling methods on SDMs to date.

## Advances in SDM

SDMs use either machine learning or statistical methods to extrapolate from sparse field samples, using environmental predictors, such as habitat and climate variables, to create a predicted distribution map of the potential range of a species. The methods have strong foundations in ecological niche theory, the strengths and weaknesses were already relatively well understood over a decade ago (e.g., Morrison et al., 1987; Fielding and Haworth, 1995; Augustin et al., 1996), and they have been subjected to continuous review (Beaumont et al., 2008; Dormann, 2007; Hampe, 2004; Yates et al., 2009).

### *Multispecies SDM Studies*

Ninety-three multispecies SDM studies were published from 2001 to 2009. At the time of the 2004 extinction risk estimates, multispecies studies were rare. Thomas et al. drew on most that were available at that time. Since 2004, seventy-six multispecies studies have been published. Figure 4-2 categorizes the methods used in these studies, ranked in relation to their relative complexity, with climate envelopes being the simplest and multimodel products the most complex. Multimodel SDM products (best and ensemble solutions) have seen increasing application, but simpler methods have remained popular.

In general, confidence in SDMs increases as the realism of assumptions is improved, and increased realism usually requires increased complexity. In a review of sixteen SDM methods (Elith et al., 2006), the most common modeling methods used in Thomas et al.—GARP and BIOCLIM—performed relatively poorly in comparison to more recent methods from regression and machine learning (e.g., multivariate adaptive regression splines, boosted regression trees, and maximum entropy) and community models (generalized dissimilarity modeling). When modeling many species, as required to conduct a meaningful global extinction risk assessment, the trade-off between complexity and tractability imposes practical limitations. Fortunately significant advances in computational efficiency have been achieved in combination with the development of these new algorithms.

Although there have been significant efforts to compare and select "the best" SDM for individual species (e.g., Broennimann et al., 2006), many recent multispecies studies include only one complex

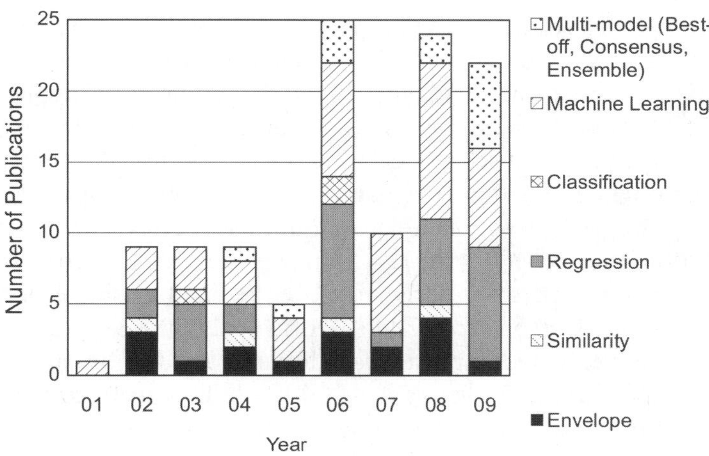


FIGURE 4-2. SDM methods published in ninety-three papers between 2001 and 2009 are categorized by algorithm class and rank ordered by approximate complexity. Climate envelopes are the simplest and multimodel products are the most complex SDM applications. Climate envelope methods include BIOCLIM, BOX, climate envelope, surface range envelope, and fuzzy climate envelope. Similarity methods include constrained Gower metric, Gower similarity, principal components analysis, climate matching, and Mahalanobis distance. Regression methods include generalized linear models, generalized additive models, logistic regression, stepwise logistic regression, locally weighted regression, and multivariate adaptive regression splines. Classification methods include classification and regression tree, classification tree analysis, and mixture discriminant analysis. Machine learning methods include genetic algorithms (including GARP), maximum entropy, boosted regression trees, random forests, and artificial neural network. Multimodel methods include best-off, model averaging, consensus, and multimodel ensembles.

(e.g., Fitzpatrick et al., 2008; Loarie et al., 2008; Jarvis et al., 2008) or simple SDM algorithm (e.g., Ritchie and Bolitho, 2008).

## *Dispersal Assumptions*

SDM provides an indication of where suitable conditions are likely to exist for a species, but not how likely a plant or animal is to be able to reach and colonize the suitable area. So plant dispersal and animal migration rates are widely regarded as the most significant uncertainties in predicting climate change impacts on biodiversity.

Most SDM studies have assessed one or both of two extreme scenarios—no-dispersal and unlimited dispersal—because these can be obtained directly from SDM outputs and don't require the application of additional sophisticated dispersal modeling methods. To progress toward realistic dispersal models requires an additional layer of dispersal modeling. A few papers have presented simple dispersal models that can easily be applied across multiple species (Fitzpatrick et al., 2008; Peterson et al., 2002). Some sophisticated mechanistic or process-based dispersal models have been applied to single or small sets of species (Iverson et al., 2005; Morin et al., 2008), but few studies have implemented sophisticated dispersal models on substantial suites of species (e.g., Williams et al., 2005; Phillips et al., 2008a; for a review of methods, see Phillips et al., 2008b). The most realistic dispersal models incorporate species-specific data on lifetime dispersal ability and account for variations in habitat suitability and matrix permeability. Dispersal parameters can be obtained from contemporary movement data, from genetic data, from paleontology data, or from expert estimation. However, such data are very limited in availability, and expert estimates are time-consuming to collate, so the number of species amenable to such complex modeling is currently very limited. Engler et al. (2009) found that simulations of realistic dispersal in alpine plants produced results closer to unlimited than no-dispersal assumptions. Although their results may not represent species from flatter environments, they are a hopeful sign that reality may lie nearer unlimited than no-dispersal scenarios.

## *Model Validation*

Probably the most significant advance in the last 5 years has been the increasing adoption of model validation procedures (e.g., area under the receiver operator curve, Cohen's kappa statistic), which provide reliability estimates of model performance. Models with poor validation results should be rejected and excluded from further analyses.

In the ninety-three multispecies SDM publications from 2001 to 2009, the frequency of published model validation results increased from 0 percent in 2001 to 78 percent in 2004, and has remained at about this level through 2009, when it was 79 percent (fig. 4-3).

The majority of models included by Thomas et al. (Bakkenes et al., 2002; Beaumont and Hughes, 2002; Huntley et al., 2004; Midgley et

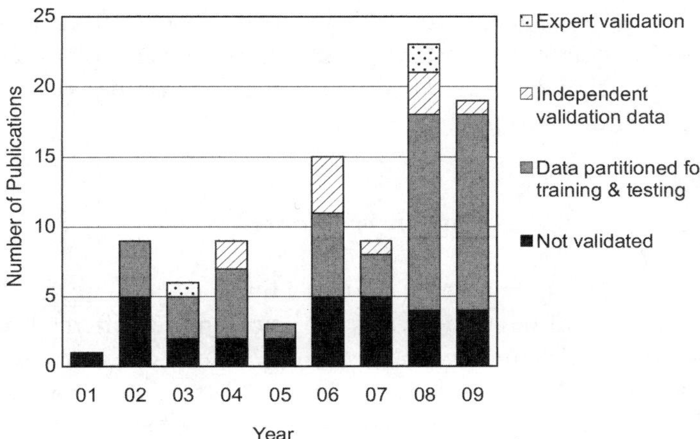


FIGURE 4-3. SDM methods published in ninety-three papers between 2001 and 2009 are categorized by their model evaluation procedures. Data partitioning methods include jack-knife, k-fold partitioning, and all other random partitioning methods. Independent validation includes data from other collections, other regions, or other time periods (hindcasting and forecasting cross-validations). Expert validation includes direct scrutiny by experts in comparison with published expert estimates of species distributions. Publications focusing on testing SDM algorithms were not included in this review, because it is intended to illustrate uptake of methods by modeling practitioners to inform species management or conservation policy.

al., 2002; Peterson et al., 2002; de Siqueira and Peterson, 2003) were not validated, and conducting model validation would likely result in a proportion of models being rejected. For example, in 2002 Bakkenes et al. accepted models for all 1,397 species, and in 2006 accepted only 856 of the same set after validation (a 39 percent rejection rate). Although Thomas et al. could not validate all models, to maximize model quality in the study, they included only endemics to each region because SDMs derived from incomplete range sampling are less reliable (e.g., Thuiller et al., 2004b). Thomas et al. (2004a) included only 197 of Bakkenes' European plant models.

Ideally SDM validations for climate change should incorporate temporal cross-validation methods whenever data are available (for forecasting, see Araûjo et al., 2005; for hindcasting, see Martinez-Meyer et al., 2004, Pearman et al., 2008, Green et al., 2008, Nogués-Bravo et al., 2008, Willis et al., 2009). The most rigorous validations would be to test against new and independent data from ground-truthing exercises (e.g., Costa et al., 2010; Rebello and Jones, 2010), and because climate change is now progressing at a detectable rate, models should be ground truthed through time.

### *Treatment of Uncertainty*

Much discussion (Araûjo et al., 2006; Hijmans and Graham, 2006; Pearson et al., 2006) has focused on variation among different modeling methods, and some very comprehensive studies of the relative performance of SDM methods (Elith et al., 2006; Tsoar et al., 2007; Thuiller et al., 2007) have been conducted since 2004.

The tendency of species to retain aspects of their fundamental niche over time is called niche conservatism, and the degree to which species' niches are conserved through time has received substantial attention among evolutionary biologists (Dormann et al., 2010; Kearney, 2006; Losos, 2008; Pearman et al., 2008; Wiens and Graham, 2005). It is therefore surprising that the range of assumptions that SDM algorithms make in relation to niche conservation and adaptation has not been discussed with regard to modeling future climate change, because it explains much of the variation observed among results. For example, from the six SDM methods included by Thomas et al., climate envelope (BIOCLIM) and similarity models (DOMAIN, Gower similarity, principal component analysis) do not make predictions beyond the current realized niche or into combinations of climate variables that do not currently exist. These methods assume that there will be no change in the realized niche or evolutionary adaptation of the fundamental niche within the time frame being modeled, and therefore produce conservative predictions of potential future distributions. In contrast, when model functions are fitted that extend beyond the range of climate conditions currently experienced by the species, as the regression (logistic regression, locally weighted regression) and machine learning (GARP) methods did, the underlying assumption is that the species niche may change and adapt, and the results are less conservative.

Proposals to deal with the variation among SDM methods have ranged from the selection of the "best" model (e.g., Broennimann et al., 2006), to compiling "consensus" distributions (Thuiller, 2004; Thuiller et al., 2005; Marmion et al., 2009), to more complex "ensemble" methods (Araûjo et al., 2006; Araûjo and New, 2007; Thuiller and Lafourcade, 2008; Thuiller et al., 2009; O'Hanley, 2009). Multimodel methods are certainly promising, but, particularly when applied in the context of climate change assessments, the SDM methods included in the ensemble should be restricted to methods with similar niche adaptation assumptions.

## Extinction Risk Estimation

Assuming that species distribution and dispersal models could produce predictions of potentially suitable habitat, a substantial challenge remains in interpreting the models to inform species action plans, conservation planning, and climate change policy. Predicted shifts and contractions in ranges do not readily translate into extinction risk probabilities for individual species and are even more problematic to summarize into overall extinction risk indices.

The main challenge arises from the increasing evidence that species may respond to climate change individualistically (e.g., Bakkenes et al., 2006; Williams and Jackson, 2007), both in terms of area change (expansion vs. contraction) and in terms of the geographical directions that their ranges may move in, so community composition is likely to be strongly affected. Biologists are faced with the challenge of applying the few accepted empirical relationships available for extinction risk prediction, which have been developed in relation to other extinction drivers such as habitat loss, or must develop new indices for extinction risk. Thomas et al. presented a suite of candidate extinction risk indices, three of which were based on the species-area relationship (SAR) and one on the International Union for Conservation of Nature (IUCN) Red List criteria (IUCN, 2001). These have generally been accepted in the spirit they were offered, provoking much discussion (Lewis, 2006; He and Hubbell, 2011), but as yet no resolution.

### *Comparison of Original and More Recent Extinction Risk Estimates*

Original results from studies included in Thomas et al. are compared with results from recent SDM studies, collated by literature review for this chapter, in table 4-1. The two sets of studies are grouped by taxa

TABLE 4-1. Comparisons among extinction risk indices from studies included in Thomas et al. (2004a) and more recent studies.

| | | | | *With Dispersal* | | | | *No Dispersal* | | | | |
|---|---|---|---|---|---|---|---|---|---|---|---|---|
| *Taxon* | *Region* | | *n* | *Min* | *Mid* | *Max* | *New Max* | *Min* | *Mid* | *Max* | *New Max* | *Citation* |
| **Mammals** | **Mexico** | | **96** | **0 (5)** | **0 (7)** | | | **3 (18)** | **3 (20)** | | | **Peterson et al., 2002** |
| | **Queensland** | | **11** | **0 (15)** | | **64 (80)** | | | | | | **Williams et al., 2003** |
| | Australia | ↓ | 5 | | 0 | 0 | 25 | | | | | Ritchie & Bolitho, 2008 |
| | **South Africa** | | **5** | | **20 (46)** | | | | **40 (59)** | | | **Erasmus et al., 2002** |
| | Africa | ↓ | 17 | | | | | | 6 (24) | | | Peterson & Martinez-Meyer, 2007 |
| | Africa | ↓ | 277 | | | 0 | | | | 4 | | Thuiller et al., 2006 |
| | Europe | | 40 | | | 3 | 3 | | | 5 | 15 | Levinsky et al., 2007 |
| **Birds** | **Mexico** | | **186** | **0 (3)** | **0 (4)** | | | **0 (8)** | **1 (8)** | | | **Peterson et al., 2002** |
| | Mexico | ↑ | 49 | | 8 (39) | | | | | | | Anciaes & Peterson, 2006 |
| | **Europe** | | **34** | | | **0 (6)** | | | | **18 (38)** | | **Huntley et al., 2004** |
| | Europe | ↓ | 40 | | | 0 (21) | 0 (27) | | | 4 (31) | 6 (38) | Huntley et al., 2008 |
| | **Queensland** | | **13** | **0 (10)** | | **39 (72)** | | | | | | **Williams et al., 2003** |
| | **South Africa** | | **5** | | **0 (32)** | | | | **0 (40)** | | | **Erasmus et al., 2002** |
| | South Africa | = | 6 | 0 | | | | | | | | Simmons et al., 2004 |
| | Northern Boreal | | 27 | | | | | | | 4 (37) | 7 (48) | Virkkala et al., 2008 |

| | | | | | | | | | | | | |
|---|---|---|---|---|---|---|---|---|---|---|---|---|
| **Frogs** | **Queensland** | | **23** | **4 (18)** | | **44 (67)** | | | | | | **Williams et al., 2003** |
| **Reptiles** | **Queensland** | | **18** | **0 (14)** | | **33 (64)** | | | | | | **Williams et al., 2003** |
| | **South Africa** | | **26** | | **4 (27)** | | | | **12 (45)** | | | **Erasmus et al., 2002** |
| **Butterflies** | **Mexico** | | **41** | **0 (4)** | **0 (5)** | | | **0 (11)** | **2 (15)** | | | **Peterson et al., 2002** |
| | **South Africa** | | **4** | | **0 (8)** | | | | **50 (70)** | | | **Erasmus et al., 2002** |
| | **Australia** | | **24** | **0 (7)** | **0 (16)** | **0 (26)** | | **0 (12)** | **0 (23)** | **4 (36)** | | **Beaumont & Hughes, 2002** |
| | Europe | | 100* | | | 1 (2) | | | | 1 (3) | | Luoto & Heikkinen, 2008 |
| **Other Invertebrates** | **South Africa** | | **10** | | **10 (24)** | | | | **70 (80)** | | | **Erasmus et al., 2002** |
| **Plants** | **Amazonia** | | **9** | | | **78 (79)** | | | | **100 (100)** | | **Miles et al., 2004** |
| | **Cerrado** | | **163** | | | | | **4 (45)** | **10 (57)** | | | **de Siqueira & Peterson, 2003** |
| | **Europe** | | **192** | **1 (5)** | **1 (6)** | **1 (6)** | | **3 (14)** | **3 (16)** | **4 (21)** | | **Bakkenes et al., 2002** |
| | Europe | ↓ | 856* | | 1 | | | | 1 (16) | | | Bakkenes et al., 2006 |
| | Europe | ↑ | 1200* | | | 5 | | | | 7 (16) | | Araújo et al., 2004 |
| | Europe | ↓ | 1350* | (4) | 1 (4) | 2 (5) | | (10) | 1 (11) | 4 (12) | | Thuiller et al., 2004a |
| | Europe | ↓ | 1350* | | | | | | 0 | 0 | 2 | Thuiller et al., 2005 |
| | Europe | ↓ | 84* | | | 0 | | | | 0 | | Normand et al., 2007 |

TABLE 4-1. Continued.

| Taxon | Region | | n | With Dispersal | | | | No Dispersal | | | | Citation |
|---|---|---|---|---|---|---|---|---|---|---|---|---|
| | | | | Min | Mid | Max | New Max | Min | Mid | Max | New Max | |
| | Europe | | 36* | | | 0 | 0 | | | 0 | 0 | Svenning & Skov, 2006 |
| | **South Africa** | | **243** | | **1 (27)** | | | | **7 (40)** | | | **Midgley et al., 2002** |
| | Africa | ↑ | 5197* | 15–36 | 19–49 | 25–57 | | 21–49 | 28–67 | 34–67 | | McClean et al., 2005 |
| | Australia | | 100 | | | 5 (16) | 24 (42) | | | 5 (18) | 17 (49) | Fitzpatrick et al., 2008 |
| | Mexico | | 32 | 0 (5) | 0 (8) | | | | | | | Gomez-Mendoza & Arriaga, 2007 |
| | Global | | 210 | | | 15 (41) | | | | 21 (57) | | Jarvis et al., 2008† |
| **All species** | **Global** | | | **4 (13)** | **4 (20)** | **13 (32)** | | **8 (31)** | **12 (37)** | **19 (52)** | | **Thomas et al., 2004** |

The columns indicate the climate change/dispersal scenarios used by Thomas et al. (2004a) with the addition of a new max category (categories: Min = 0.8°C–1.7°C, Med = 1.8°C–2°C, and Max = 2°C–3°C, expected global mean temperature increase). In each cell the value before the parentheses is the percentage of species predicted to have zero suitable climate space (after Thomas et al., 2004b), and the values inside parentheses are calculated using the third of the species-area relationships described in the first Thomas et al. paper (2004a).

Studies in bold were included in Thomas et al. (2004a). Values for "All species" were interpolated by three-way analysis of variance by Thomas et al. (2004a).

* Indicates studies that included nonendemic species.

↓, ↑, and =, represent decreases, increases, and no obvious change, respectively, in comparison to taxon/region studies published in Thomas et al. (2004a).

† Indicates an author who kindly made tables of their modeled area change data available for calculating the two indices.

and by region, divided by two dispersal assumptions (no dispersal; unlimited dispersal). Results are tabulated as the percentage of species for which suitable climate completely disappears, augmented by one species-area relationship–derived estimate.

The SAR is generally accepted as "one of ecology's few ironclad laws" (Pounds and Puschendorf, 2004). However, the simplicity of the SAR conceals a diversity of approximations, assumptions, and extrapolations that will be explored in the following chapter. In light of these limitations, table 4-1 emphasizes a more straightforward and conservative index, the proportion of species predicted to have zero suitable climate space in the future (after Thomas et al., 2004b). This conservative assumption counts up predictions of absolute extinction, ignoring the potential increase in extinction risk across the vast majority of species in each study that are predicted to suffer reductions in their distributions. In addition, for comparison with the original Thomas et al. results, their third SAR extinction risk index is also provided where possible.

The substantial efforts made by a range of authors to model European plants are noteworthy in the group of post-2004 studies. A comparative analysis of the Thomas et al. SAR extinction risk by Bakkenes et al. (2006), using the same SDM methods and climate data as their 2002 study, but including nonendemic species, found very little difference in the overall extinction risk. This result indicates that regional extinction risk from climate change may not be significantly different for endemic versus widespread species (note though that studies that do not model the entire current range of the species should not be included in estimating potential global extinction risk). Subsequent studies, of greater numbers of European plants, using the next IPCC scenarios (SRES scenarios) and a range of different SDM methods produced both higher (Araújo et al., 2004) and lower (Thuiller et al., 2004a; Thuiller et al., 2005) extinction risks.

Re-analysis of European bird distributions (Huntley et al., 2004 vs. Huntley et al., 2008) indicates lower extinction risk for a more severe climate change scenario than was included by Thomas et al., despite very few apparent differences in the SDM methods and data used.

Where studies are less directly comparable (i.e., for the same region and taxa but use different data or SDM methods), more recent assessments have also generally produced lower extinction risk indices (Williams et al., 2003 vs. Ritchie and Bolitho, 2008; Erasmus et al., 2002 vs. Peterson and Martinez-Meyer, 2007 vs. Thuiller et al., 2006).

However, at least two increases are indicated (Peterson et al., 2002 vs. Anciaes and Peterson 2006; Midgley et al., 2002 vs. McClean et al., 2005) and there are several disagreements between the two extinction risk indices, which complicates such comparisons.

Because the area change data were not available for the individual species for all new studies, it was impossible to generate the third SAR index for all studies, or to update the three-way analysis of variance across all species (bottom row of table 4-1) from Thomas et al. This makes it impossible to quantify the overall revision to the global extinction risk estimates that may arise from improved methods and expanding the taxon/region combinations included (i.e., it is impossible to revise the all-species result at the bottom of table 4-1).

## *IUCN Red List Criteria*

The general relationship between area loss and extinction risk is so widely accepted among biologists that the concept of increased extinction risk from range loss has been formalized within the IUCN Red List criteria (Mace et al., 2008). However, the climate change projections used by the majority of SDM studies range 50 to 100 years into the future, and for the vast majority of species the Red List criteria (IUCN, 2001) time frame of three generations for assigning species to threat categories fails to capture the consequences of slow-acting, but persistent, threats. This led Thomas et al. to adapt the time frames within which they applied the IUCN area change criteria.

In comparison to the SAR-based indices there has been a marked preference for applying Red List criteria to assess potential increases in threat from climate change (Bomhard et al., 2005; Levinsky et al., 2007; Normand et al., 2007; Thuiller et al., 2005; Thuiller et al., 2006), despite criticisms from Red List criteria experts (Akçakaya et al., 2006). However, most importantly, the IUCN has recently conducted its own global multi-taxon analysis (Foden et al., 2008) using trait-based methods to identify susceptibility to climate change. The IUCN found that 35 percent of all bird species possess traits that make them potentially susceptible to climate change and that 52 percent of all amphibian species are potentially susceptible to climate change (table 4-2). It should be noted that comparisons can not be made among, or summaries made across, taxa, but this method provides robust repeatable indices for comparison within each taxonomic group.

TABLE 4-2. The numbers and percentages of species assessed for "climate change susceptibility" in the 2008 IUCN Red List for birds and amphibians.

| | *Climate Change Susceptible* | *Threatened* | | |
|---|---|---|---|---|
| | | *Yes* | *No* | *Total* |
| Birds | Yes | 979 | 2,462 | 35% |
| | | 10% | 25% | |
| | No | 246 | 6,172 | 65% |
| | | 2% | 63% | |
| | | 15% | 88% | 9,856 |
| Amphibians | Yes | 1,488 | 1,729 | 52% |
| | | 24% | 28% | |
| | No | 503 | 2,502 | 48% |
| | | 8% | 40% | |
| | | 32% | 68% | 6,222 |

Species fall into four categories: (i) threatened and "climate change–susceptible"; (ii) threatened but not "climate change–susceptible"; (iii) not threatened but "climate change–susceptible"; and (iv) neither threatened nor "climate change–susceptible." From Foden et al. (2008).

The Thomas et al. analysis did not include amphibians, and included only 2 percent of the world's bird species. In comparison to the results of Foden et al. (2008) for birds the two area-based methods in table 4-1 indicate a wider range of risk, with the SAR-based method providing generally lower (0–39 percent across all studies, and both dispersal scenarios) and the absolute extinction method providing generally higher (3–72 percent across all studies, and both dispersal scenarios) estimates of extinction risk, than the trait-based method (25 percent) of Foden et al. (2008).

IUCN's trait-based susceptibility index incorporates more complex and realistic assumptions than the simple area change indices of Thomas et al. However, the method is data-intensive and accurate trait data are not readily available for many species, so expert estimates and extrapolations have to be substituted. Therefore, it would be advisable to repeat and compare these methods as data availability improves, because agreement between the two would be reassuring and differences in the results may suggest potential improvements in methods.

## Comparisons with Non-SDM Methods

Dynamic global vegetation models (DGVMs) simulate change in biomes based on processes, including biogeochemical fluxes, hydrological processes, and vegetation dynamics (such as establishment, productivity, and competition for resources; resource allocation; growth; disturbance; and mortality), to simulate the population dynamics of various plant functional types and produce global predictions of vegetation distribution. Many include benefit functions estimating the direct physiological benefits of increased carbon dioxide. However, they tend to be parameterized for particular regions or to represent a limited range of plant functional types, and a fundamental assumption of DGVMs is that the plant functional types are not dispersal-limited and can disperse at rates that match climatic shifts (Prentice et al., 2007).

Malcolm et al. (2006) used biome changes projected by an ensemble of DGVMs, applying species-area relationships to projected biome-area change to assess the effect of climate change (doubling of carbon dioxide) on global biodiversity hotspots. This study concluded that extinction risk will be between less than 1 percent and 43 percent, depending on species migration capabilities, the vegetation model used, the breadth of the biome definition, and how restricted species distributions are among biomes.

Malcolm et al. used a SAR exponent of 0.15, making their results most comparable to the results for $z = 0.15$ (for an explanation of $z$, see chapter 5) of Thomas et al. Their doubling of carbon dioxide scenario is comparable with the midrange scenario of Thomas et al. Malcolm et al.'s 2–26 percent and 3–43 percent extinction predictions, with dispersal and without dispersal, respectively, both have wider ranges than the 9–20 percent and 17–29 percent estimates for the same two dispersal assumptions provided by Thomas et al. Although the two studies are not based on the same regional boundaries or species data, it is encouraging that this DGVM approach generally agrees with the SDM method of Thomas et al. in predicting extinction risk.

Because species distributions and vegetation types are broadly limited by climate conditions, modeling the similarity of climate conditions through time is an elegant and informative approach. Williams et al. (2007) used IPCC Fourth Assessment Report scenarios to model change in climate across Earth's surface.

Figure 4-4, from Williams et al. (2007), shows the global distributions of climate changes for two climate change (A2 and B1) scenarios,

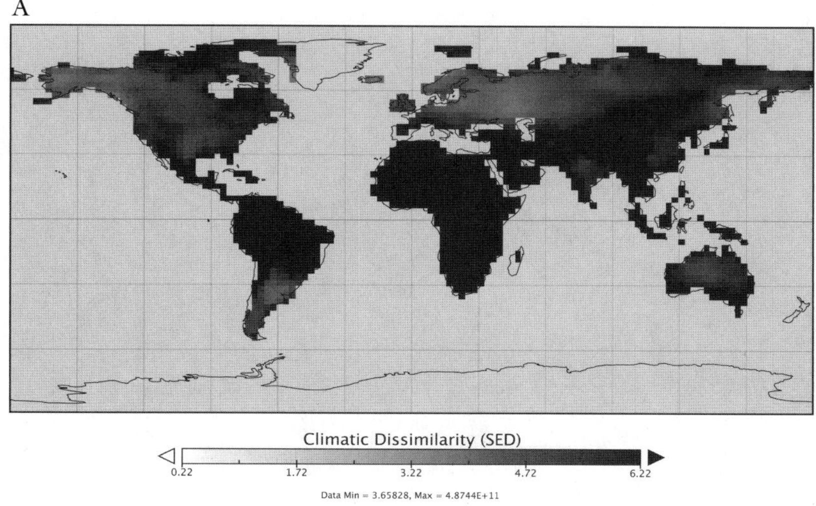


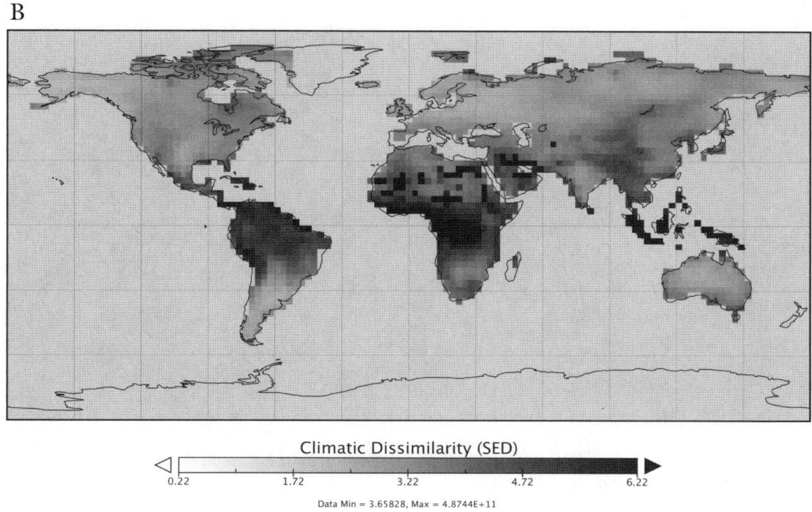


FIGURE 4-4. From Williams et al. (2007). A and B: Mapped indices of climate change risk for local climate change. C and D: Novel twenty-first-century climates. High dissimilarities indicate risk of novel climates. E and F: Disappearing twentieth-century climates. High dissimilarities indicate risk of disappearing climates. A, C, E: A2 Scenario. B, D, F: B1 Scenario.

C

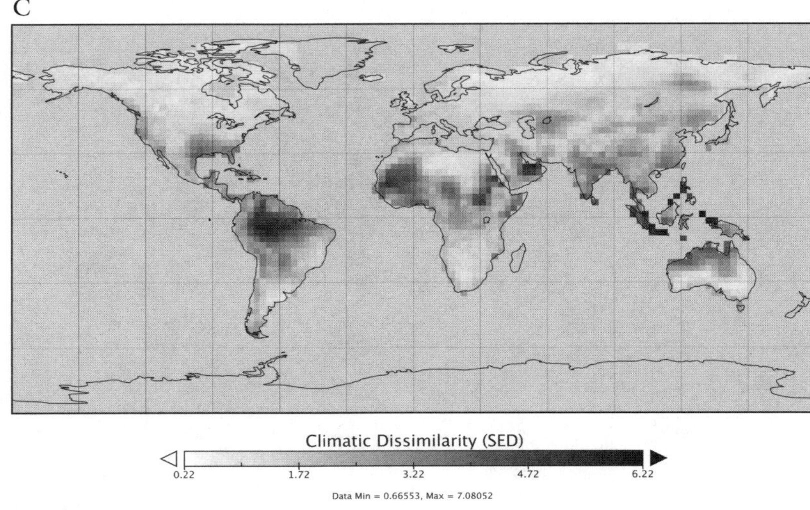


D

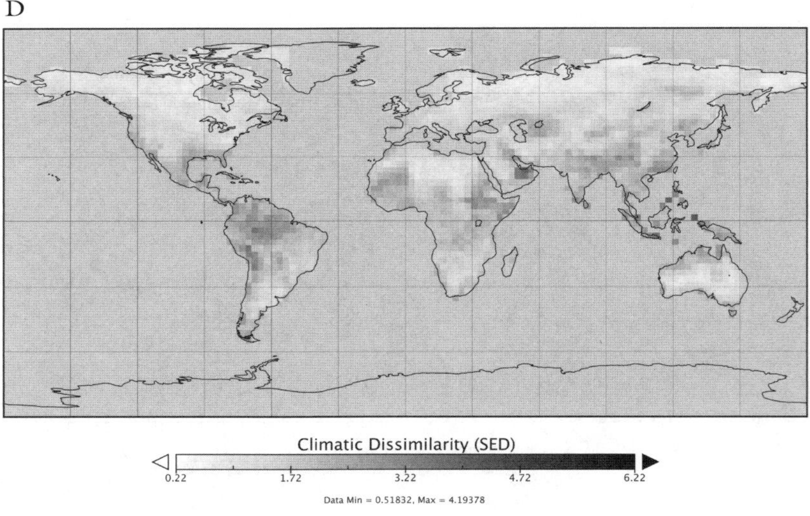


FIGURE 4-4. Continued.

highlighting the regions at most risk. Assuming unlimited dispersal, under the high-end A2 scenario (~3.6 degrees Celsius mean global temperature increase), 56–100 percent of global land area (Fig 4-4A) is predicted to experience biome-scale change, and application of the SAR ($z = 0.25$) to this scenario predicts extinction risk to be between

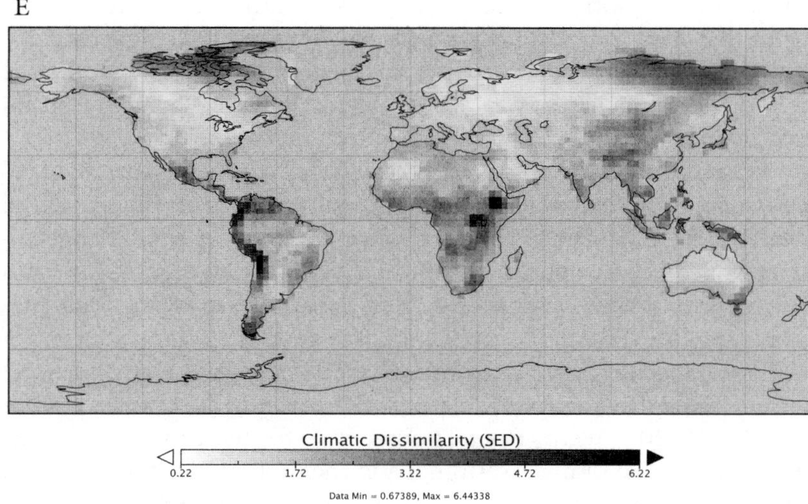


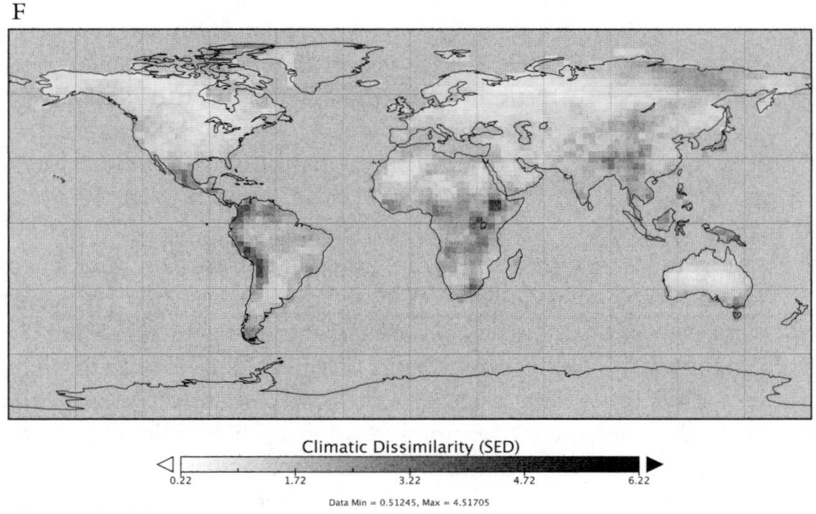


FIGURE 4-4. Continued.

18 and 100 percent. Within this, 12–39 percent of the global land area may experience combinations of climate variables that are not currently experienced anywhere else (novel future climates, fig. 4-4C), and the single assumption that species may fail to colonize novel climates results in 3–12 percent extinction risk. Also within this, 10–48

percent of the global land area (Fig 4-4E) currently experiences climates that may completely disappear by 2100 and the single assumption that species adapted to the disappearing climates would fail to adapt to other climates would result in 3–15 percent species extinction.

Corresponding extinction risk projections for the low-end B1 scenario (~1.8 degrees Celsius mean global temperature increase, fig. 4-4B) are 5–62 percent extinction from overall biome-scale change, 1–5 percent if only novel climates produce extinctions (fig. 4-4D), and 1–5 percent if only disappearing climates produce extinctions (fig. 4-4E).

Low-range extinction risk based on Williams et al. (2007) (5–62 percent) is higher than the all-species extinction risk (bottom row of table 4-1, 4–20 percent) from Thomas et al. for the equivalent warming scenario of midrange warming, and the high-range extinction risk (18–100 percent) overlaps with but ranges higher than the maximum-warming all-species extinction risk of Thomas et al. (19–52 percent).

## Synergies between Climate Change and Habitat Loss

Predictions of extinction risk are complicated by many potential interactions, including the direct effects of climate change on habitat transformation, as climate variables define suitable conditions for natural vegetation and agricultural crops, and interact with fire regimes.

Thomas et al. provided a crude comparison of relative extinction risk from climate change and land use change, and concluded that climate change is at least as important a threat as land use change. The studies included by Thomas et al. used only climate variables in the SDMs, so the assessment of extinction risk from habitat loss had to be made using separate habitat loss models and the traditional application of the species-area relationship. It would be better to directly include habitat variables in the SDMs, and to use climate and habitat scenarios within the models to allow direct comparison of the effects of climate change and habitat loss. Several studies have used habitat or land use as modeling variables to constrain potential species distributions to current vegetation distributions. For example, Thuiller et al. (2006) modeled 277 African mammals under future climate change and current land use scenarios, but not under future land use change scenarios. However, very few studies have attempted to provide assessments of the relative contributions of the two threats, or the possible synergies between them.

van Vuuren et al. (2006) estimated extinction risk from climate change (mean temperature increase 1.6–2.1 degrees Celsius), and land use change (using Millennium Ecosystem Assessment scenarios, 2005) using crop production and biome modeling methods. Applying the species-area relationship ($z = 0.34$) to the potential area changes between 2000 and 2050, they concluded that land use change will contribute more (7–13 percent) to global species diversity loss than climate change (2–4 percent). At 2–3 percent the extinction risk for the climate change–only scenario is lower than the 11–38 percent extinction risk predicted by Thomas et al. for $z = 0.35$, the equivalent (midrange) climate change scenario, and the assumption of unlimited dispersal.

Jetz et al. (2007) were the first to use the millennium ecosystem assessment scenarios to assess the effects of climate change on individual species, for an almost complete clade, across the whole world. Their results indicate that climate change and habitat loss will vary in importance at different latitudes, habitat loss in economically emerging tropical countries will continue to pose an even more direct and immediate threat to a greater number of bird species than climate change, and the combined effects of climate and habitat change will be greater than climate change on its own. Overall, Jetz et al. (2007) predicted 0.5–0.9 percent species extinctions (species with zero suitable habitat in future) for 1.6–3.28 degrees Celsius mean global temperature increase. These results are, not surprisingly, lower than those reported for birds by Thomas et al. on the basis of climate change only, with the same no-dispersal assumption, but are conservative in comparison to a similar analysis (also using a combination of climate scenarios and millennium assessment scenarios) by Sekercioglu et al. (2008) (fig. 4-5), who predicted that 4.7–6.4 percent of 8,500 land bird species will be committed to extinction by 2100, with a 2.8 degrees Celsius mean global temperature increase. The two studies (Jetz et al., 2007; Sekercioglu et al., 2008) are very similar in that they determine the future availability of suitable habitat within the species range using the millennium ecosystem assessment models of vegetation, and the difference in their results arises from their two assumptions of range stability. Sekercioglu et al. (2008) added the assumption that species ranges will be forced to move upslope at rates determined by climate scenarios in relation to the adiabatic lapse rate.

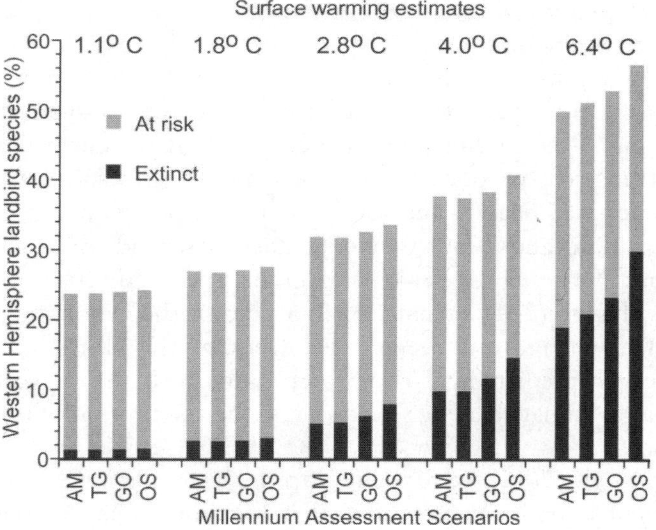


FIGURE 4-5. From Sekercioglu et al. (2008). Percentage of Western Hemisphere ($n$ = 3,349) land bird species projected to go extinct (current baseline = 0) by 2100 on the basis of estimates of various surface-warming estimates (IPCC 2007a), and millennium assessment habitat-change scenarios (Millennium Ecosystem Assessment, 2005; AM, adaptive mosaic; TG, technogarden; GO, global orchestration; OS, order from strength).

In total Sekercioglu et al. (2008) predicted that the number of bird species classed as threatened would increase by 19–30 percent by 2050 and 29–52 percent by 2100. The lower of these agrees with, and the higher exceeds, the 25 percent estimated by Foden et al. (2008) to be susceptible to climate change. However, although Foden's 25 percent susceptibility estimate is based purely on climate, Sekercioglu's extinction estimates incorporate both climate change and habitat loss.

## Conclusions

The new studies compared in this chapter (table 4-1) generally confirm the order of magnitude of extinction risk estimated by Thomas et al. Some of the newer studies produce somewhat lower extinction risk for similar amounts of change (i.e., less than 3 degrees Celsius), but emissions predictions continue to grow, making the midcentury scenarios used by Thomas et al. now more representative of change in the first third of the century. The results of Thomas et al. may now be thought of as representative of low- to midrange warming scenarios (i.e., 0–3 degrees Celsius mean global temperature increase). There is now a clear need to conduct SDM and related extinction risk assessments in the 3–6 degrees Celsius mean global temperature increase range that the IPCC Fourth Assessment Report predicts may occur by 2100 (see fig. 4-1). This would produce higher extinction risks for the studies presented in table 4-1.

Although none of the caveats listed by Thomas et al. and their critics can yet be completely eliminated, this chapter identifies a number of very significant advances in the contributing fields that can be expected to improve estimates of climate change extinction risk from SDM.

The following list of recommendations would improve SDM climate change assessments, would facilitate comparisons among studies, and would facilitate global multi-taxon analyses of extinction risk in the future.

- Species distribution modelers should use ensemble GCMs, including multiple realizations of multiple GCMs.
- A range of future time slices and climate stabilization end points for each IPCC scenario should be modeled.
- Publications should include the global mean temperature increase per IPCC scenario for the ensemble GCM at each of the time slices to facilitate comparison of results among studies.
- The effects of different GCM to regional climate model downscaling methods should be investigated.
- Models should be validated, at least using data-partitioning methods, and ideally complementing these with hindcasting or forecasting methods.
- Historical, current, and projected species distribution areas should be made widely available for further research.
- Modeling projects should be coupled with long-term monitoring in the field to assess their predictive utility over time. This could be done by identifying global and regional target species for collective monitoring, and, ideally, to stimulate rapid scientific advances, the data should be made widely and freely available for climate change research.

Finally, it is essential that rigorous comparisons among the range of modeling methods (DGVMs, species richness models, community models, climate similarity models) are conducted.

Although many studies employing these methods will be faced with high uncertainty, the number and quality of studies emerging helps reduce uncertainty, and Thomas et al. has demonstrated that even uncertain results can be very valuable in guiding public understanding and policy response.

## REFERENCES

Akçakaya, H. R., S. H. M. Butchart, G. M. Mace, S. N. Stuart, and C. Hilton-Taylor. 2006. "Use and misuse of the IUCN Red List Criteria in projecting climate change impacts on biodiversity." *Global Change Biology* 12: 2037–2043.

Anciaes, M., and A. T. Peterson. 2006. "Climate change effects on neotropical manakin diversity based on ecological niche modeling." *Condor* 108: 778–791.

Araújo, M. B., and M. New. 2007. "Ensemble forecasting of species distributions." *Trends in Ecology and Evolution* 22 (1): 42–47.

Araújo, M. B., M. Cabeza, W. Thuiller, L. Hannah, and P. H. Williams. 2004. "Would climate change drive species out of reserves? An assessment of existing reserve-selection methods." *Global Change Biology* 10: 1618–1626.

Araújo, M. B., R. G. Pearson, W. Thuiller, and M. Erhard. 2005. "Validation of species-climate impact models under climate change." *Global Change Biology* 11: 1504–1513.

Araújo, M. B., W. Thuiller, and R. G. Pearson. 2006. "Climate warming and the decline of amphibians and reptiles in Europe." *Journal of Biogeography* 33: 1712–1728.

Ashcroft, M. B., L. A. Chisholm, and K. O. French. 2009. "Climate change at the landscape scale: Predicting fine-grained spatial heterogeneity in warming and potential refugia for vegetation." *Global Change Biology* 15: 656–667.

Augustin, N. H., M. A. Mugglestone, and S. T. Buckland. 1996. "An autologistic model for the spatial distribution of wildlife." *Journal of Applied Ecology* 33: 339–347.

Bakkenes, M., J. R. M. Alkemade, F. Ihle, R. Leemansand, and J. B. Latour. 2002. "Assessing effects of forecasted climate change on the diversity and distribution of European higher plants for 2050." *Global Change Biology* 8: 390–407.

Bakkenes, M., B. Eickhout, and R. Alkemade. 2006. "Impacts of different climate stabilisation scenarios on plant species in Europe." *Global Environmental Change—Human and Policy Dimensions* 16: 19–28.

Beaumont, L. J., and L. Hughes. 2002. "Potential changes in the distributions of latitudinally restricted Australian butterfly species in response to climate change." *Global Change Biology* 8: 954–971.

Beaumont, L. J., A. J. Pitman, M. Poulsen, and L. Hughes. 2007. "Where will species go? Incorporating new advances in climate modelling into projections of species distributions." *Global Change Biology* 13: 1368–1385.

Beaumont, L. J., L. Hughes, and A. J. Pitman. 2008. "Why is the choice of future climate scenarios for species distribution modelling important?" *Ecology Letters* 11: 1135–1146.

Bomhard, B., D. M. Richardson, J. S. Donaldson, G. O. Hughes, G. F. Midgley, D. C. Raimondo, A. G. Rebelo, M. Rouget, and W. Thuiller. 2005. "Potential impacts of future land use and climate change on the Red List status of the Proteaceae in the Cape Floristic region, South Africa." *Global Change Biology* 11: 1452–1468.

Broennimann, O., W. Thuiller, G. Hughes, G. F. Midgley, J. R. M. Alkemade, and A. Guisan. 2006. "Do geographic distribution, niche property and life form explain plants' vulnerability to global change?" *Global Change Biology* 12: 1079–1093.

Costa, G. C., C. Nogueira, R. B. Machado, and G. R. Colli 2010. "Sampling bias and the use of ecological niche modeling in conservation planning: A field evaluation in a biodiversity hotspot." *Biodiversity Conservation* 19: 883-899.

Dormann, C. F. 2007. "Promising the future? Global change projections of species distributions." *Basic and Applied Ecology* 8: 387–397.

Dormann, C. F., B. Gruber, M. Winter, and D. Herrmann. 2010. "Evolution of climate niches in European mammals?" *Biology Letters* 6: 229–232.

Elith, J., C. H. Graham, R. P. Anderson, M. Dudik, S. Ferrier, A. Guisan, R. J. Hijmans, et al. 2006. "Novel methods improve prediction of species' distributions from occurrence data." *Ecography* 29: 129–151.

Engler, R., C. F. Randin, P. Vittoz, T. Czáka, M. Beniston, N. E. Zimmermann, and A. Guisan. 2009. "Predicting future distributions of mountain plants under climate change: Does dispersal capacity matter?" *Ecography* 32: 34–45.

Erasmus, B. F. N., A. S. van Jaarsveld, S. L. Chown, M. Kshatriya, and K. Wessels. 2002. "Vulnerability of South African animal taxa to climate change." *Global Change Biology* 8: 679–693.

Fielding, A. H., and P. F. Haworth. 1995. "Testing the generality of bird-habitat models." *Conservation Biology* 9: 1466–1481.

Fitzpatrick, M. C., A. D. Gove, N. J. Sanders, and R. R. Dunn. 2008. Climate change, plant migration, and range collapse in a global biodiversity hotspot: The Banksia (Proteaceae) of Western Australia. *Global Change Biology* 14: 1–16.

Foden, W., G. Mace, J.-C. Vié, A. Angulo, S. Butchart, L. DeVantier, H. Dublin, A. Gutsche, S. Stuart, and E. Turak. 2008. "Species susceptibility to climate change impacts." In *The 2008 Review of the IUCN Red List of Threatened*

*Species,* edited by J.-C. Vié, C. Hilton-Taylor, and S. N. Stuart. Gland, Switzerland: IUCN.

Gomez-Mendoza, L., and L. Arriaga. 2007. "Modeling the effect of climate change on the distribution of oak and pine species of Mexico." *Conservation Biology* 21 (6): 1545–1555.

Green, R. E., Y. C. Collingham, S. G. Willis, R. D. Gregory, K. W. Smith, and B. Huntley. 2008. "Performance of climate envelope models in retrodicting recent changes in bird population size from observed climatic change." *Biology Letters* 4: 599–602.

Hampe, A. 2004. "Bioclimate envelope models: What they detect and what they hide." *Global Ecology and Biogeography* 13: 469–476.

He, F., and S. P. Hubbell. 2011. "Species-area relationships always overestimate extinction rates from habitat loss." *Nature* 473: 368–371.

Hijmans, R. J., and C. H. Graham. 2006. "The ability of climate envelope models to predict the effect of climate change on species distributions." *Global Change Biology* 12: 2272-2281.

Huntley, B., R. E. Green, Y. C. Collingham, J. K. Hill, S. G. Willis, P. J. Bartlein, W. Cramer, W. J. M. Hagemeijer, and C. J. Thomas. 2004. "The performance of models relating species' geographical distributions to climate is independent of trophic level." *Ecology Letters* 7: 417–426.

Huntley, B., Y. C. Collingham, S. G. Willis, and R. E. Green. 2008. "Potential impacts of climatic change on European breeding birds." *PLoS ONE* 1: e1439.

Intergovernmental Panel on Climate Change (IPCC). 1990. *Climate Change: The IPCC Scientific Assessment*. Edited by J. T. Houghton, G. J. Jenkins, and J. J. Ephraums. New York: Cambridge University Press.

IPCC. 1995. *The Science of Climate Change*. Contribution of Working Group I to the Second Assessment of the Intergovernmental Panel on Climate Change. Edited by J. T. Houghton, L. G. Meira Filho, B. A. Callender, N. Harris, A. Kattenberg, and K. Maskell. UK: Cambridge University Press.

IPCC. 2001. *Climate Change 2001: The Scientific Basis*. Edited by J. T. Houghton, Y. Ding, D. J. Griggs, M. Noguer, P. J. van der Linden, and D. Xiaosu. Cambridge, UK: Cambridge University Press.

IPCC. 2007a. *Climate Change 2007: Synthesis Report*. Core Writing Team, Edited by R. K. Pachauri and A. Reisinger. Geneva, Switzerland: IPCC.

IPCC. 2007b. "Summary for policymakers." In *Climate Change 2007: The Physical Science Basis. Contribution of Working Group I to the Fourth Assessment Report of the Intergovernmental Panel on Climate Change,* edited by S. Solomon, D. Qin, M. Manning, Z. Chen, M. Marquis, K. B. Averyt, M. Tignor, and H. L. Miller. New York: Cambridge University Press.

International Union for Conservation of Nature (IUCN). 2001. *Red List Categories and Criteria, version 3.1*. Gland, Switzerland: IUCN Species Survival Commission.

Iverson, L. R., A. Prasad, and M. W. Schwartz. 2005. "Predicting potential changes in suitable habitat and distribution by 2100 for tree species of the eastern United States." *Journal of Agricultural Meteorology* 61 (1): 29–37.

Jarvis, A., A. Lane, and R. J. Hijmans. 2008. "The effect of climate change on crop wild relatives." *Agriculture Ecosystems & Environment* 126: 13–23.

Jetz, W., D. S. Wilcove, and A. P. Dobson. 2007. "Projected impacts of climate and land-use change on the global diversity of birds." *PLoS Biology* 5: e157.

Kearney, M. 2006. "Habitat, environment, and niche: What are we modelling?" *OIKOS* 115 (1): 186–191.

Levinsky, I., F. Skov, J. C. Svenning, and C. Rahbek. 2007. "Potential impacts of climate change on the distributions and diversity patterns of European mammals." *Biodiversity and Conservation* 16: 3803–3816.

Lewis, O. T. 2006. "Climate change, species-area curves and the extinction crisis." *Philosophical Transactions of the Royal Society* 361: 163-171.

Loarie, S. R., B. E. Carter, K. Hayhoe, S. McMahon, R. Moe, C. A. Knight, and D. D. Ackerly. 2008. "Climate change and the future of California's endemic flora." *PloS ONE* 3 (6): e2502.

Losos, J. B. 2008. "Phylogenetic niche conservatism, phylogenetic signal and the relationship between phylogenetic relatedness and ecological similarity among species." *Ecology Letters* 11: 995–1007.

Luoto, M., and R. K. Heikkinen. 2008. "Disregarding topographic heterogeneity biases species turnover assessments based on bioclimatic models." *Global Change Biology* 14: 483–494.

Mace, G. M., N. J. Collar, K. J. Gaston, C. Hilton-Taylor, R. H. Akçakaya, N. Leader-Williams, E. J. Milner-Gulland, and S. Stuart. 2008. "Quantification of extinction risk: IUCN's system for classifying threatened species." *Conservation Biology* 22 (6): 1424–1442.

Malcolm, J. R., C. Liu, R. P. Neilson, L. Hansen, and L. Hannah. 2006. "Global warming and extinctions of endemic species from biodiversity hotspots." *Conservation Biology* 20 (2): 538–548.

Marmion, M., M. Parviainen, M. Luoto, R. K. Heikkinen, and W. Thuiller. 2009. "Evaluation of consensus methods in predictive species distribution modelling." *Diversity and Distributions* 15: 59–69.

Martinez-Meyer, E., A. T. Peterson, and W. W. Hargrove. 2004. "Ecological niches as stable distributional constraints on mammal species, with implications for Pleistocene extinctions and climate change projections for biodiversity." *Global Ecology and Biogeography* 13: 305–314.

McClean, C. J., J. C. Lovett, W. Küper, L. Hannah, J. H. Sommer, W. Barthlott, M. Termansen, G. F. Smith, S. Tokumine, and J. R. D. Taplin. 2005. "African plant diversity and climate change." *Annals of the Missouri Botanical Garden* 92 (2): 139–152.

Midgley, G. F., L. Hannah, M. C. Rutherford, and L. W. Powrie. 2002. "Assessing the vulnerability of species richness to anthropogenic climate change in a biodiversity hotspot." *Global Ecology and Biogeography* 11: 445–451.

Miles, L., A. Grainger, and O. Phillips. 2004. "The impact of global climate change on tropical forest biodiversity in Amazonia." *Global Ecology and Biogeography* 13: 553–565.

Millennium Ecosystem Assessment. 2005. *Ecosystems and Human Well-being: Biodiversity Synthesis*. Washington, DC: World Resources Institute.

Morin, X., D. Viner, and I. Chuine. 2008. "Tree species range shifts at a continental scale: New predictive insights from a process-based model." *Journal of Ecology* 96 (4): 784–794.

Morrison, M. L., I. C. Timoss, and K. A. With. 1987. "Development and testing linear regression models predicting bird-habitat relationships." *Journal of Wildlife Management* 51: 247–253.

Nogués-Bravo, D., J. Rodríguez, J. Hortal, P. Batra, and M. B. Araújo. 2008. "Climate change, humans, and the extinction of the woolly mammoth." *PLoS Biology* 6 (4): e79.

Normand, S., J. C. Svenning, and F. Skov. 2007. "National and European perspectives on climate change sensitivity of the habitats directive characteristic plant species." *Journal for Nature Conservation* 15: 41–53.

O'Hanley, J. R. 2009. "Neural ensembles: A neural network based ensemble forecasting program for habitat and bioclimatic suitability analysis." *Ecography* 32: 89–93.

Pearman, P. B., A. Guisan, O. Broennimann, and C. F. Randin. 2007. "Niche dynamics in space and time." *Trends in Ecology and Evolution* 23 (3): 149–158.

Pearman, P. B., C. F. Randin, O. Broennimann, P. Vittoz, W. O. van der Knaap, R. Engler, G. L. Lay, N. E. Zimmermann, and A. Guisan. 2008. "Prediction of plant species distributions across six millennia." *Ecology Letters* 11: 357–369.

Pearson, R. G., W. Thuiller, M. B. Araújo, E. Martinez-Meyer, L. Brotons, C. McClean, L. Miles, P. Segurado, T. P. Dawson, and D. C. Lees. 2006. "Model-based uncertainty in species range prediction." *Journal of Biogeography* 33: 1704–1711.

Perry, G. L. W., and N. J. Enright. 2006. "Spatial modelling of vegetation change in dynamic landscapes: A review of methods and applications." *Physical Geography* 30 (1): 47–72.

Peterson, A. T., and E. Martinez-Meyer. 2007. "Geographic evaluation of conservation status of African forest squirrels (Sciuridae) considering land use change and climate change: The importance of point data." *Biodiversity and Conservation* 16: 3939–3950.

Peterson, A. T., M. A. Ortega-Huerta, J. Bartley, V. Sánchez-Corderos, J. Soberón, R. H. Buddemeier, and D. R. B. Stockwell. 2002. "Future projections for Mexican faunas under global climate change scenarios." *Nature* 416: 626–629.

Peterson, A. T., A. Stewart, K. L. Mohamed, and M. B. Araújo. 2008. "Shifting global invasive potential of European plants with climate change." *PLoS ONE* 3 (6): e2441.

Phillips, S. J., P. Williams, G. Midgley, and A. Archer. 2008a. "Optimizing dispersal corridors for the Cape Proteaceae using network flow." *Ecological Applications* 18 (5): 1200–1211.

Phillips, B. L., J. D. Chipperfield, and M. R. Kearney. 2008b. "The toad ahead: Challenges of modelling the range and spread of an invasive species." *Wildlife Research* 35 (3): 222–234.

Pierce, D. W., T. P. Barnett, B. D. Santer, and P. J. Gleckler. 2009. "Selecting global climate models for regional climate change studies." *Proceedings of the National Academy of Sciences, USA* 106 (2): 8441–8446.

Pounds, A., and R. Puschendorf. 2004. "Clouded futures." *Nature* 427: 107.

Prentice, I. C., A. Bondeau, W. Cramer, S. P. Harrison, T. Hickler, W. Lucht, S. Sitch, B. Smith, and M. T. Sykes. 2007. "Dynamic global vegetation modeling: Quantifying terrestrial ecosystem responses to large-scale environmental change." In *Terrestrial Ecosystems in a Changing World,* edited by J. G. Canadell, D. E. Pataki, and L. F. Pitelka, 175–192. Berlin: Springer.

Rebello, H., and G. Jones. 2010. "Ground validation of presence-only modelling with rare species: A case study on barbastelles *Barbastella barbastellus* (Chiroptera: Vespertilionidae)." *Journal of Applied Ecology.* doi:10.1111/j.1365-2664.2009.01765.x

Ritchie, E. G., and E. E. Bolitho. 2008. "Australia's savanna herbivores: Bioclimatic distributions and an assessment of the potential impact of regional climate change." *Physiological and Biochemical Zoology* 81 (6): 880–890.

Sekercioglu, C. H., S. H. Schneider, J. P. Fay, and S. R. Loarie. 2008. "Climate change, elevational range shifts, and bird extinctions." *Conservation Biology* 22 (1): 140–150.

Simmons, R. E., P. Barnard, W. R. J. Dean, G. F. Midgley, W. Thuiller, and G. Hughes. 2004. "Climate change and birds: Perspectives and prospects from southern Africa." *Ostrich* 75: 295–308.

de Siqueira, M. F., and A. T. Peterson. 2003. "Consequences of global climate change for geographic distributions of Cerrado tree species." *Bioneotropica* 3 (2): 1–14.

Svenning, J. C., and F. Skov. 2006. "Potential impact of climate change on the northern nemoral forest herb flora of Europe." *Biodiversity and Conservation* 15: 3341–3356.

Thomas, C. D., A. Cameron, R. E. Green, M. Bakkenes, L. J. Beaumont, Y. C. Collingham, B. F. N. Erasmus, et al. 2004a. "Extinction risk from climate change." *Nature* 427: 145–148.

Thomas, C. D., S. E. Williams, A. Cameron, R. E. Green, M. Bakkenes, L. J. Beaumont, Y. C. Collingham, et al. 2004b. "Biodiversity conservation: Uncertainty in predictions of extinction risk/Effects of changes in climate and land use/Climate change and extinction risk (reply)." *Nature* 430: Brief Communications Arising. doi:10.1038/nature02719.

Thuiller, W. 2004. "Patterns and uncertainties of species' range shifts under climate change." *Global Change Biology* 10 (12): 2020–2027.

Thuiller, W., and B. Lafourcade. 2008. MACIS deliverable 3.5. *Report on the results of the run of improved modelling to Europe*. Available at http://macis-project.net/

Thuiller, W., M. B. Araújo, R. G. Pearson, R. J. Whittaker, L. Brotons, and S. Lavorel. 2004a. "Uncertainty in predictions of extinction risk." *Nature* 430: Brief Communications. doi:10.1038/nature02716.

Thuiller, W., L. Brotons, M. B. Araújo, and S. Lavorel. 2004b. "Effects of restricting environmental range of data to project current and future species distributions." *Ecography* 27: 165–172.

Thuiller, W., S. Lavorel, M. B. Araújo, M. T. Sykes, and I. C. Prentice. 2005. "Climate change threats to plant diversity in Europe." *Proceedings of the National Academy of Sciences, USA* 102 (23): 8245–8250.

Thuiller, W., O. Broennimann, G. O. Hughes, J. R. M. Alkemade, G. F. Midgley, and F. Corsi. 2006. "Vulnerability of African mammals to anthropogenic climate change under conservative land transformation assumptions." *Global Change Biology* 12: 424–440.

Thuiller, W., C. Alberta, M. B. Araújo, P. M. Berry, M. Cabeza, A. Guisane, T. Hicklerf, et al. 2007. "Predicting global change impacts on plant species' distributions: Future challenges." *Perspectives in Plant Ecology, Evolution, and Systematics*. doi:10.1016/j.ppees.2007.09.004.

Thuiller, W., B. Lafourcade, R. Engler, and M. B. Araújo. 2009. "BIOMOD—A platform for ensemble forecasting of species distributions." *Ecography* 32: 369–373.

Tsoar, A., O. Allouche, O. Steinitz, D. Rotem, and R. A. Kadmon. 2007. "Comparative evaluation of presence-only methods for modelling species distribution." *Diversity and Distributions* 13 (4): 397–405.

Virkkala, R., R. K. Heikkinen, N. Leikola, and M. Luoto. 2008. "Projected large-scale range reductions of northern-boreal land bird species due to climate change." *Biological Conservation* 141: 1343–1353.

van Vuuren, D. P., O. E. Sala, and H. M. Pereira. 2006. "The future of vascular plant diversity under four global scenarios." *Ecology and Society* 11: 25.

Wiens, J. J., and C. H. Graham. 2005. "Niche conservatism: Integrating evolution, ecology, and conservation biology." *Annual Review of Ecology and Systematics* 36: 519–539.

Williams, J. W., and S. T. Jackson. 2007. "Novel climates, no-analog communities, and ecological surprises." *Frontiers in Ecology and the Environment* 5 (9): 475–482.

Williams, J. W., S. T. Jackson, and J. E. Kutzbach. 2007. "Projected distribution of novel and disappearing climates by 2100 AD." *Proceedings of the National Academy of Sciences, USA* 104: 5738–5742.

Williams, P., L. Hannah, S. Andelman, G. Midgley, M. B. Araújo, G. Hughes, L. Manne, E. Martinez-Meyer, and R. Pearson. 2005. "Planning for climate change: Identifying minimum-dispersal corridors for the Cape Proteaceae." *Conservation Biology* 19 (4): 1063–1074.

Williams, S. E., E. E. Bolitho, and S. Fox. 2003. "Climate change in Australian tropical rainforests: An impending environmental catastrophe." *Proceedings of the Royal Society B* 270: 1887–1892.

Willis, S. G., C. D. Thomas, J. K. Hill, Y. C. Collingham, M. G. Telfer, R. Fox, and B. Huntley. 2009. "Dynamic distribution modelling: Predicting the present from the past." *Ecography* 32: 5–12.

Yates C. J., J. Elith, A. M. Latimer, D. L. Maitre, G. F. Midgley, F. M. Schurr, and A. G. West. 2009. "Projecting climate change impacts on species distributions in megadiverse South African Cape and Southwest Australian Floristic Regions: Opportunities and challenges." *Austral Ecology*. doi:10.1111/j.1442-9993.2009.02044.x.

Chapter 5

# *The Use and Misuse of Species-Area Relationships in Predicting Climate-Driven Extinction*

JOHN HARTE AND JUSTIN KITZES

Thomas et al. (2004) pioneered the estimate of extinction risk due to climate change by coupling species range-loss simulations from species distribution models with species-loss estimates from the species-area relationships (SARs). Unfortunately, numerous conceptual and practical problems permeate this seemingly solid and straightforward approach. Chapter 4 explored developments in climate envelope modeling. Here we focus on the challenges associated with applying a SAR approach to climate-driven extinction estimates and propose a novel application of recent Maximum Entropy (MaxEnt) theory in ecology that may help to address some of them.

## Use of SARs to Estimate Climate-Driven Extinction

In general, there are several different fundamental forms of the SAR, and none is universally applicable to all climate change scenarios. More specifically, the SAR itself does not account for many important characteristics of the landscapes and species experiencing climate change.

## *General Considerations*

Overall, two broad categories of SARs can be distinguished. Nested SARs are constructed by averaging species richness across subplots of specified area, using presence-absence data for each species, at finer and finer scale within a larger plot. Island SARs are constructed from lists of species (within the taxonomic group of interest) found in spatially disjointed patches of habitat of differing area, such as actual islands.

These two SAR types tend to differ in shape. In particular, linear regressions of island SAR data, plotted on log-log axes, tend to exhibit different slopes than nested SARs (Rosenzweig, 1995; Drakare et al., 2006). Note that an "island" SAR can also be constructed from a single contiguous habitat by sampling disjointed patches of different areas (analogous to islands) within the habitat. This process, however, gives a subsample of a complete nested design, and there should be no fundamental difference between the nested and the disjointed island forms in this case.

Nested and island SARs are applicable to different types of climate change scenarios. Consider, first, the case in which a large biome such as Amazonia, of area $A_0$, containing $S_0$ species, shrinks under climate change to an area of $A$ located within $A_0$ (fig. 5-1, inset). If the species are plants with limited dispersal capability, for example, how many species would we expect to find remaining in area $A$? The better answer is clearly given by the nested SAR. On the other hand, consider the case in which the suitable patch of area $A$ is located completely outside the original area $A_0$, and all the individuals remaining following climate change must disperse from $A_0$ and colonize $A$. If the assumption is made that dispersal ability is not limiting and that a sufficient number of potential colonists are able to disperse from $A$, then an island SAR form is more appropriate.

Interestingly, in certain cases we can predict the slope of this "island" SAR form. In particular, if we assume that (i) total abundance in the new area $A$ is proportional to area, (ii) individuals found in $A$ are chosen by a random draw from all individuals in $A_0$, and (iii) the number of individuals of each species in $A_0$ follows a canonical lognormal abundance distribution, then the SAR takes the form of a species accumulation curve with a predicted slope of approximately 0.25 (Preston, 1962; May, 1975). Although a slope of 0.25 is often assumed for all forms of SARs, this theoretical justification for a SAR slope of approximately 0.25 does not apply to nested SARs and applies only to island SARs if the above conditions are assumed.

In practice, the nested SAR and the island SAR (constructed as a species accumulation curve as described above) approaches do give different estimates of the number of species expected to remain in a new, smaller habitat $A$. Figure 5-1 compares estimates of species remaining using both methods with data from the 50-hectare (123-acre) Barro Colorado Island tropical forest plot in Panama. The nested SAR approach consistently predicts higher levels of species loss than the island SAR approach.

In most cases of climate change–driven shifts in habitat, neither a pure island nor a nested SAR approach will be appropriate. First, dispersal capability will be neither unlimited nor completely absent. Second, newly suitable habitat is not likely to be only a subplot of the original habitat or only disjointed from it. Third, for the case of disjointed habitats, the individuals found in $A$ are not likely to be random draws from the species in $A_0$. As a result, it is impossible to choose a

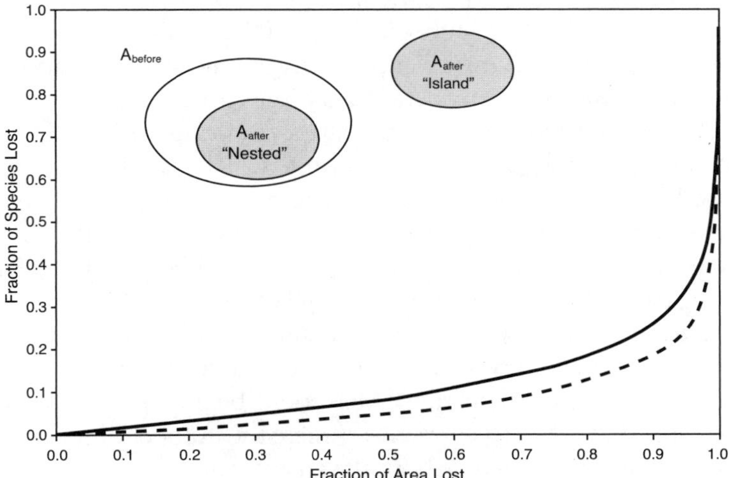


FIGURE 5-1. Differences between nested and island SAR approaches to estimating species remaining after habitat loss using data from the Barro Colorado Island forest plot. The nested SAR (solid line) predicts more extinction after habitat loss than the island SAR (dashed line). The island SAR curve is constructed as a species accumulation curve (see text), averaged over fifty trials, calculated by sampling individuals without replacement from the total individuals pool, assuming the number of individuals is proportional to area. Inset shows the geometry of climate envelope shifts for which nested and island forms of the SAR may be appropriate. If $A_{after}$ is contained within $A_{before}$, a nested form may be most appropriate, but if $A_{after}$ is outside of $A_{before}$, an island form may be most appropriate.

single SAR form, or exponent, to apply in all cases of habitat shifts under climate change.

As described above, the SAR can be used to determine how many species will be found in the remaining suitable habitat. Assuming that species cannot migrate to other ecosystems, the difference between the number of species in the original habitat and the number remaining after habitat contraction determines the number of species lost. Alternatively, however, the number of species that were found exclusively in the habitat area that was lost could be directly estimated. An endemics-area relationship (EAR) can be used for this second approach, as an EAR measures directly the number of species that are found uniquely in a subplot of area $A$ (Harte and Kinzig, 1997; Kinzig and Harte, 2000). For real landscapes, measuring the number of original species minus the number remaining (the SAR approach) and measuring the number of species found uniquely in the area that is lost (the EAR approach) will give an identical result. We note that a recently proposed analysis that claims to find that the EAR is always preferable for extinction estimation (He and Hubbell, 2011) violates this identity.

There is a difference, however, in which metric is most appropriate to use for projecting extinctions, and the choice hinges on geometry. Theoretical predictions for the shape and slope of the EAR and SAR often assume that the relationship will be applied to "nicely shaped" areas (i.e., approximately square or circular, with the longest dimension no more than perhaps two times the length of the shortest dimension) (Harte and Kinzig, 1997; Kinzig and Harte, 2000). Thus, if a long, skinny area is lost from the edge of a square, the SAR should be applied to the nearly square area remaining to correctly estimate the species remaining. If, however, habitat is lost such that only a long, skinny area remains, then the EAR should be applied to the nearly square area that is lost to correctly estimate the number of species lost. In more realistic cases where both the area lost and the area remaining are irregularly shaped, both methods will give only approximations. Thus, contrary to the findings of He and Hubbell (2011), for example, there is no universal answer to the question of which of the two metrics will best estimate extinction.

## *Specific Critiques*

In addition to the general concerns described above, SARs can fail to provide a clear or complete picture of climate-driven extinctions for several other specific reasons. We briefly mention seven of these below.

### DISPERSAL IS POORLY UNDERSTOOD.

The appropriate species richness metric to use for extinction projections hinges critically on assumptions about species dispersal, as described above. In practice, existing climate-extinction estimates have generally assumed two extreme cases, zero dispersal and unlimited dispersal, which can provide a wide range of estimates for proportion of species driven extinct (e.g., approximately 20 percent difference in Thomas et al., 2004). The reality, of course, lies somewhere in between and is difficult to determine.

### POWER-LAW SARS ARE RARE TO NONEXISTENT OVER LARGE SPATIAL INTERVALS.

Despite their widespread use in estimating extinctions at the regional to continental scale, the empirical validity of the power-law SAR model is questionable over large spatial intervals. The usual test for power-law behaviors is to graph $log(S)$ versus $log(A)$ and examine the $R^2$ value of the straight-line fit, the slope of which is an estimate of the power law exponent, $z$. When SAR data spanning a spatial interval of several orders of magnitude in area are plotted on log-log axes, the data nearly always exhibit distinct curvature (e.g., Rosenzweig, 1995; Drakare et al., 2006; Harte et al., 2008) even when the $R^2$ values of the linear regressions exceed 0.98, as is typical in SAR analyses. This curvature can be very significant—the observed $z$-values can range from less than 0.1 to greater than 0.5 across a single site (Harte et al., 2009).

In the next section of this chapter, we suggest a method, based on recent theory, for incorporating the typical non-power-law behavior of SARs into species-loss estimates.

### SPECIES RICHNESS DEPENDS ON THE SHAPE, AS WELL AS SIZE, OF HABITAT PATCHES.

Even if accurate scaling rules could be provided that answer the points raised above, the number of species found in a habitat patch depends not only on the patch area but also on the shape of the patch (Kunin, 1997; Harte et al., 1999). With nonrandomly placed populations, irregularly shaped patches tend to hold more species than square patches of the same area. This is because patches with a long dimension, such as a skinny rectangle, are more likely to include species turnover across the increasingly varied habitats often encountered over larger and larger distance intervals (i.e., beta diversity).

More generally, as habitat is lost or disturbed in various spatial configurations, the slope and characteristic shape of the SAR can change dramatically (Ney-Nifle and Mangel, 2000). Therefore, knowing the SAR for an undisturbed system does not necessarily allow predictions of species loss once the system becomes disturbed. This poses particular problems in the case of climate change, because climatic boundaries on species' ranges are often irregular in shape.

### POPULATIONS MAY BE MORE APPROPRIATE UNITS FOR EXTINCTION ESTIMATION THAN SPECIES.

The standard application of SARs takes species as the unit for extinction estimation. This is appropriate if all individuals within a species can be assumed to have identical climate tolerances. If, however, local climate adaptation has occurred, then the use of a simple SAR approach could underestimate, perhaps substantially, the amount of species loss.

Take, for example, the case where a population at the warmer edge of the species' range is adapted to a warmer climate than one in the middle or at the colder edge of the range. This can have a substantial effect on the likelihood of the species surviving climate change if individuals are limited in their dispersability (Harte et al., 2004). Figure 5-2 illustrates this for an extreme case in which no dispersal is possible. In this example, with no population-level adaptation in climate preferences, only the population inhabiting the warmer end of the range would be unable to survive the warming resulting from climate change (fig. 5-2B). In contrast, with population-level adaptation, no individuals in any of the populations could survive that change (fig. 5-2C). The effect of this population-level adaptation will matter less, of course, as dispersability increases.

The result shown in the figure may seem unintuitive, as intraspecies variability is normally associated with survivability, but the issue is one of time frames. The scenario shown in figure 5-2B, in which all populations exhibit variation in temperature tolerance (i.e., all populations have some members that could survive in all present-day temperature regimes), demonstrates the usefulness of variability, which can help the species under climate change scenarios. However, as shown in figure 5-2C, if variability does not exist within individual populations, and adaptation to local climate occurred slowly in the past and cannot keep pace with the speed of temperature change, then the results can be disastrous for the species. A recent review (Donoghue, 2008) suggests that species may more easily disperse to suitable climates than adapt to locally and rapidly changing conditions.

## KNOWLEDGE OF SPECIES TURNOVER IS NEEDED IN FRAGMENTED HABITATS.

All of the applications of the SAR described so far apply to estimates of extinctions within a single habitat patch. We may more frequently,

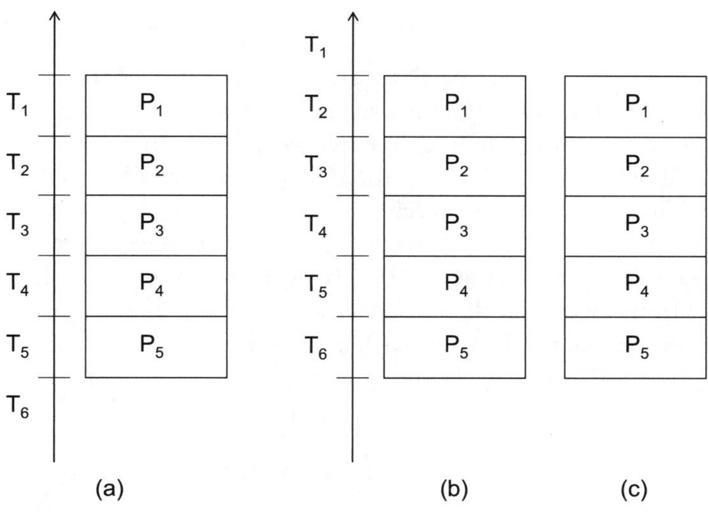


FIGURE 5-2. Consequences of population-level adaptation to climate, assuming no dispersal. (A) Before climate change. A species with five adjacent populations has a defined range before climate change, with population $P_1$ found in a range with temperature $T_1$, $P_2$ found in a range with temperature $T_2$, and so forth. If there is no population-level adaptation to temperature, then populations $P_1$ through $P_5$ can all persist in any of temperatures $T_1$ through $T_5$. If there is population-level adaptation to temperature, then population $P_1$ can only persist in $T_1$, $P_2$ can only persist in $T_2$, and so forth. (B) After climate change. If there is no population-level adaptation to temperature, then only $P_1$ finds itself in an unsuitable temperature and is lost. (C) After climate change. If there is population-level adaptation to temperature, then every population finds itself in an unsuitable temperature and all populations are lost.

however, encounter landscapes in which the same collection of species is found in two or more locations, each of which shrinks in area. The problem then becomes that the SAR extinction calculations do not define which particular species are lost from each of the shrunken areas, and, as a result, the overlap in the list of lost species cannot be predicted. The number of species lost from a set of two shrinking patches could range anywhere from zero to the sum of the calculated losses in each patch individually. The problem of commonality between habitat patches is only compounded if species are able to migrate or disperse among patches.

#### "MINIMUM VIABLE POPULATION" EFFECTS ARE IGNORED.

The use of the SAR or EAR to estimate species loss under habitat degradation does not take into account the possibility that even if a species is not immediately eliminated because of inadequate area, it may still become extinct over time if its population size is sufficiently small. This process has been referred to as realizing an extinction debt (Tilman et al., 1994; Kuussaari et al., 2009) or species relaxation (Diamond, 1975; Soule et al., 1979). In addition to the risks that small populations face from demographic and environmental stochasticity, additional causal mechanisms such as genetic bottlenecks and Allee effects (Soule, 1986) may also be important.

#### WEB CASCADES MAY RESULT IN LARGER LOSSES.

Because the species that make up present-day communities do not all individually have the same limits of climate tolerance, climate change will result in novel communities. A species may migrate to a new location that no longer supports its food species, mutualists, pollinators, or seed dispersers. In these cases, there may be cascades of secondary (and beyond) effects, resulting in larger losses than predicted simply from SARs (Dunne et al., 2002; Srinivasan et al., 2007).

## New Directions

Estimating species loss under climate change by applying SARs clearly faces both practical and conceptual problems. Given all of the uncertainties in the use of the SAR, can we say anything about the direction in which existing estimates are likely to be wrong?

We argue that, overall, our knowledge gaps are optimistically lopsided, with our models generally projecting less extinction than is likely to actually occur. Examining the list of critiques above, we find only one critique that, if corrected, would decrease extinction estimates (i.e., the decreasing slope of SARs at large scales). Conversely, we find several critiques that, if corrected, would likely increase extinction estimates (i.e., assumption of "nicely shaped" habitat loss, neglect of population-level differences in climate adaptation, neglect of minimum population effects, neglect of species interactions). Several others (i.e., choosing the appropriate SAR type, dispersal, the use of EARs versus SARs, species commonality) are of unclear direction.

To improve our estimates of species loss, we believe that the following are particularly important research tasks:

- Develop better prediction methods for the number of species shared among sets of disjointed habitat patches.
- Enrich understanding of the shapes and slopes of SARs at large spatial scales.
- Enrich understanding of secondary species losses due to trophic web–induced and other interaction-induced cascades.

The second item on this list warrants special attention. Recent theoretical investigation of the SAR based on application of the MaxEnt principle from information theory has led to new insights into the value of the SAR slope parameter $z = d(log(S)) / d(log(A))$ over large ranges of spatial scales within a biogeographical region (Harte et al., 2008, 2009). The theory predicts that $z$ at any scale, $A$, within a biome is a universal, decreasing function of the ratio $N(A) / S(A)$, where $N(A)$ is the summed abundance of all the species at scale $A$, and $S(A)$ is the species richness at that scale. While more testing is needed, particularly with animal data, the theoretical prediction is in good agreement with observations for a wide range of habitats and spatial scales ranging from plots on the order of several square meters to the 60,000-square-kilometer Western Ghats preserve in India.

Moreover, MaxEnt theory may provide a means of improving a novel method of estimating extinction that was suggested by Thomas et al. (2004). In their landmark paper (see also chapter 2, this volume), they introduced a "Species-Area Method 3," which posited that each

species obeys a scaling formula for the probability of its persistence when the area of that species' suitable habitat shrinks. The formula they suggested reads:

$$P = P_0 \left( \frac{A}{A_0} \right)^z \qquad (1)$$

where $P$ and $P_0$ are the probability of survival in the reduced habitat of area $A$ and the probability of survival in the original habitat of area $A_0$, respectively. Further assuming that $P_0 = 1$, and using common values of $z = 0.15, 0.25$, and $0.35$ for all species, Thomas et al. estimated species losses under climate scenarios. They refer to the range for the parameter $z$ of $0.25$ to $0.35$ as most likely, and the $0.15$ is taken as a "conservative" case in the sense that it leads to the smallest extinction estimates. Importantly, $z$ is assumed to be the same for all species and constant across spatial scale.

In MaxEnt theory (Harte et al., 2008, 2009), a formula that can accomplish the same goal as eq. 1 can be derived, but the effective $z$-value for each species is now a known function of the abundance of the species in $A_0$ and of the ratio $A_0 / A$. To see this, we first note that the probability of any species with initial abundance $n_0$ in $A_0$ having abundance $n$ in $A$ can be written as (Harte et al., 2008)

$$P(n, A \mid n_0, A_0) = ce^{-\lambda n} \qquad (2)$$

where $c$ is a normalization constant and $\lambda$ is a "Lagrange multiplier" that is a function of $n_0$ and the ratio $A_0 / A$. For $n_0$ and $A_0 / A > 1$, $c$ and $\lambda$ are given to a good approximation by

$$c \cong \frac{1}{\frac{An_0}{A_0} + 1} \qquad (3)$$

$$\lambda \cong \ln\left(1 + \frac{A_0}{An_0}\right) \qquad (4)$$

More complicated expressions are obtained when the inequalities are not satisfied.

To proceed, we assume some functional dependence of the probability of survival of a species, $P$, in $A$ on either $n$, the species' abundance remaining in $A$, or ratio $n / n_0$. A simple example of such an assumption, and one that is in keeping with the spirit of Thomas et al., is to assume that the probability of survival is equal to the probability that $n / n_0$ exceeds a critical threshold, $r_c$. Alternatively, we could employ a critical value of $n_c$ that relates to a minimum viable population size across all species, although we do not treat this case here.

The probability of a species reaching the critical threshold $r_c$ can be calculated by integration (Kitzes and Harte, in preparation):

$$P\left(\frac{n}{n_0} > r_c\right) = \int_{r_c n_0}^{n_0} ce^{-\lambda n}\,dn \tag{5}$$

The integration yields

$$P\left(\frac{n}{n_0} > r_c\right) = \frac{[n_0\beta / (1+n_0\beta)]^{r_c n_0} - [n_0\beta / (1+n_0\beta)]^{n_0}}{(1+n_0\beta)\ln(1+1/n_0\beta)} \tag{6}$$

where $\beta \equiv A / A_0$. Table 5-1 compares some key results from the Thomas et al. and the MaxEnt approaches.

TABLE 5-1. Comparison of Thomas et al. (2004) and MaxEnt estimates of species extinction risk following range contraction.

| | | *Probability of Extinction* | | |
|---|---|---|---|---|
| *Method* | | *Remaining habitat* $\beta = 0.5$ | *Remaining habitat* $\beta = 0.1$ | *Remaining habitat* $\beta = 0.05$ |
| Thomas et al. (eq. 2) | $z = 0.35$* | 0.22 | 0.55 | 0.65 |
| | $z = 0.25$ | 0.16 | 0.44 | 0.53 |
| | $z = 0.15$ | 0.10 | 0.29 | 0.36 |
| MaxEnt (eq. 6) | $r_c = 0.1$ | 0.10 | 0.63 | 0.86 |
| | $r_c = 0.01$ | 0.01 | 0.09 | 0.18 |
| | $r_c = 0.001$ | 0.002 | 0.01 | 0.02 |

*$z$ is the slope of the power-law species-area relationship in eq. 2, $r_c$ is the critical ratio of remaining to original individuals that indicates extinction, and $\beta$ is the fraction of a species range remaining following climate change. In MaxEnt calculations, $n_0 = 1{,}000$ (so, for example, the values in the $r_c = 0.1$ row are thus the probability of less than 100 individuals remaining). We note that MaxEnt gives probabilities of extinction that are within +/- 0.01 for species with abundances up to $n_0 = 10{,}000$. Compared to the Thomas et al. approach, MaxEnt gives lower estimates of the probability of species extinction at low levels of clearing and an overlapping range of predictions at high levels of clearing for this set of $r_c$. MaxEnt results are from the exact equations that are the analogs to eq. 6 (Harte et al., 2008).

If the necessary criteria for the approximations in eqs. 3 and 4 are not met, a more complicated but still tractable equation results. In either case, the expression for the survival probability can be arranged into the form of the Thomas et al. expression, eq. 1, but with $z$ now an explicit function of $n_0$ and $\beta$. Limited knowledge of $n_0$ and the critical value $r_c$ for many species may make this method difficult to apply, although eq. 2, rewritten as $P(n > 0) = 1 - P(n = 0) = 1 - c$, where $c$ represents true absences in a gridded presence-absence map at any arbitrary resolution, provides a method for estimating $n_0$ for any species based on presence-absence data.

It may also be useful for ecologists to seek new methods, not necessarily involving the traditional combination of climate envelopes and SAR analysis, to develop more insight into future species losses. There may be an opportunity to combine existing insights from studies of interannual climate and population variability, from observations along climate transects, from warming experiments, and from paleoclimatic analyses of species compositional changes.

## Conclusions

Whether or not we can improve our ability to forecast the magnitude of impending species losses, one generalization from available knowledge seems robust. All estimates using available methods point to a huge impending extinction episode from unmitigated anthropogenic climate warming and suggest that the uncertainty in these methods is likely to underestimate the extent of extinctions in the coming centuries. The data suggest that life on Earth, as we have known it, is going to change dramatically over the coming century.

After years of investigation and debate, the academic and public discussion on climate change has begun to move beyond whether climate change is occurring and on to how to stop it or adapt to it. We argue that the same approach is now warranted for the biodiversity crisis. Making further incremental gains in our ability to predict the magnitude of coming extinctions should not slow our efforts to respond with appropriate policies on the basis of the information we now have in hand. We urge our colleagues to begin to move beyond asking whether we are in the midst of the next great mass extinction and focus more attention on how to reduce species losses by mitigating global warming and reducing other anthropogenic threats to species.

## Acknowledgments

Funding from the National Science Foundation (DEB 0516161 to JH and Graduate Research Fellowship to JK) and conversations with Adam Smith, Danielle Svehla Christianson, Stacy Jackson, and Alison Cameron are gratefully acknowledged.

### REFERENCES

Diamond, Jared. 1975. "The island dilemma: Lessons of modern biogeographic studies for the design of natural reserves." *Biological Conservation* 7: 129–146.

Donoghue, Michael J. 2008. "A phylogenetic perspective on the distribution of plant diversity." *Proceedings of the National Academy of Sciences, USA* 105 (Supplement 1): 11549–11555.

Drakare, Stina, Jack J. Lennon, and Helmut Hillebrand. 2006. "The imprint of the geographical, evolutionary and ecological context on species-area relationships." *Ecology Letters* 9 (2): 215–227.

Dunne, Jennifer A., Richard J. Williams, and Neo D. Martinez. 2002. "Network structure and biodiversity loss in food webs: Robustness increases with connectance." *Ecology Letters* 5 (4): 558–567.

Harte, John, and Ann P. Kinzig. 1997. "On the implications of species-area relationships for endemism, spatial turnover, and food web patterns." *Oikos* 80 (3): 417–427.

Harte, John, Sarah McCarthy, Kevin Taylor, Ann Kinzig, and Marc L. Fischer. 1999. "Estimating species-area relationships from plot to landscape scale using species spatial-turnover data." *Oikos* 86 (1): 45–54.

Harte, John, Annette Ostling, Jessica L. Green, and Ann Kinzig. 2004. "Biodiversity conservation—Climate change and extinction risk." *Nature* 430 (6995).

Harte, J., T. Zillio, E. Conlisk, and A. B. Smith. 2008. "Maximum entropy and the state-variable approach to macroecology." *Ecology* 89 (10): 2700–2711.

Harte, John, Adam B. Smith, and David Storch. 2009. "Biodiversity scales from plots to biomes with a universal species-area curve." *Ecology Letters* 12 (8): 789–797.

He, F., and P. Legendre. 2002. "Species diversity patterns derived from species-area models." *Ecology* 83 (5): 1185–1198.

He, F., and S. P. Hubbell. 2011. "Species-area relationships always overestimate extinction rates from habitat loss." *Nature* 473: 368–371.

Hubbell, Stephen P. 2001. *The Unified Neutral Theory of Biodiversity and Biogeography*. Princeton, NJ: Princeton University Press.

Kinzig, Ann P., and John Harte. 2000. "Implications of endemics-area relationships for estimates of species extinctions." *Ecology* 81 (12): 3305–3311.

Kunin, William E. 1997. "Sample shape, spatial scale and species counts: Implications for reserve design." *Biological Conservation* 82 (3): 369–377.

Kuussaari, Mikko, Riccardo Bommarco, Risto K. Heikkinen, Aveliina Helm, Jochen Krauss, Regina Lindborg, Erik Ockinger, et al. 2009. "Extinction debt: A challenge for biodiversity conservation." *Trends in Ecology & Evolution* 24 (10): 564–571.

Lomolino, Mark V. 2001. "The species-area relationship: New challenges for an old pattern." *Progress in Physical Geography* 25 (1–21): 1.

May, Robert M. 1975. "Patterns of species abundance and diversity." In *Ecology and Evolution of Communities*. Edited by Martin L. Cody and Jared M. Diamond. Cambridge, MA: Belknap Press of Harvard University Press.

Ney-Nifle, Muriel, and Marc Mangel. 2000. "Habitat loss and changes in the species-area relationship." *Conservation Biology* 14 (3): 893–898.

Plotkin, Joshua B., Matthew D. Potts, Douglas W. Yu, Sarayudh Bunyavejchewin, Richard Condit, Robin Foster, Stephen Hubbell, et al. 2000a. "Predicting species diversity in tropical forests." *Proceedings of the National Academy of Sciences, USA* 97 (20): 10850–10854.

Plotkin, Joshua B., Matthew D. Potts, Nandi Leslie, N. Manokaran, James LaFrankie, and Peter S. Ashton. 2000b. "Species-area curves, spatial aggregation, and habitat specialization in tropical forests." *Journal of Theoretical Biology* 207 (1): 81–99.

Preston, Frank W. 1962. "The canonical distribution of commonness and rarity: Part I." *Ecology* 43 (2): 185–215.

Rosenzweig, Michael L. 1995. *Species Diversity in Space and Time*. Cambridge: Cambridge University Press.

Soule, Michael E. 1986. *Conservation Biology: The Science of Scarcity and Diversity*. Sunderland, MA: Sinauer Associates.

Soule, Michael E., Bruce A. Wilcox, and Claire Holtby. 1979. "Benign neglect: A model of faunal collapse in the game reserves of East Africa." *Biological Conservation* 15 (4): 259–272.

Srinivasan, U. Thara, Jennifer A. Dunne, John Harte, and Neo D. Martinez. 2007. "Response of complex food webs to realistic extinction sequences." *Ecology* 88 (3): 671–682.

Thomas, Chris D., Alison Cameron, Rhys E. Green, Michel Bakkenes, Linda J. Beaumont, Yvonne C. Collingham, Barend F. N. Erasmus, et al. 2004. "Extinction risk from climate change." *Nature* 427 (6970): 145–148.

Tilman, David, Robert M. May, Clarence L. Lehman, and Martin A. Nowak. 1994. "Habitat destruction and the extinction debt." *Nature* 371 (6492): 65–66.

# PART III

# Current Extinctions

Having examined the results of the 2004 research and refined it in light of more recent research in the first two parts of this book, we are now ready to begin a more multidisciplinary exploration of extinction risk from climate change. We start with the present. What extinctions have been recorded in response to the nearly half century of climate change that the world has already experienced?

Sarah McMenamin and Peter Glynn explore this question for the terrestrial and marine realms, respectively, in the first two chapters of this part. Both find that contemporary extinctions from climate change may be controversial, both because climate change is controversial, but also because it is difficult to establish extinction and causality without the benefit of long-term historical perspective. This difficulty in establishing contemporary extinctions adds to the difficulty of assessing extinction risk from climate change. The relatively few extinctions that have been observed thus far may indicate that long lag times are involved in extinctions from climate change, that extinctions are difficult to identify given our still limited information about most taxa on the planet, or that modeling techniques overestimate extinction risk.

In the final chapter of this part, Eric Post and Jedediah Brodie look at the region of the world that has experienced the most physical change due to climate change thus far—the polar regions. One might expect that the high levels of physical change in this part of the world would have led to extinctions from climate change, or at least downward trends in demography that might easily be expected to lead to extinction. Post and Brodie find some support for this expectation, but the view is not as straightforward and simple as one might expect. Change in Antarctica is more complex than change over the northern polar oceans, and even in the north, some species, such as polar bears, show both increasing and decreasing populations. The contemporary record tells us that answers about extinction risk from climate change will not be simple. Some positive policy responses are emerging, such as listing the polar bear as threatened under the US Endangered Species Act. The current policy challenge is arriving at appropriate responses to extinction events that may be decades in the future, and not waiting to act until the problem is unresolvable.

# Chapter 6

## *First Extinctions on Land*

Sarah K. McMenamin and Lee Hannah

The golden toad (*Bufo periglenes*) disappeared from Costa Rica in 1989 and became the first terrestrial extinction to be linked to climate change. Like the first marine extinction attributed to climate change (see chapter 7), the extinction of the golden toad was linked to El Niño events. The marine extinction is irrefutably linked to coral bleaching, but the causes of the golden toad extinction are far more controversial. Golden toad sightings have been reported in Guatemala since the 1980s, but these sightings have never been confirmed. Although there is some hope that residual populations still survive, *Bufo periglenes* is currently listed as extinct in the International Union for Conservation of Nature (IUCN) Red List, and the cause of the extinction is hotly debated.

In this chapter, we review the evidence that climate change was indeed the critical causal factor in the golden toad extinction. We detail several ongoing debates about climate change causality and examine several other amphibian extinctions that have followed a pattern similar to that of the golden toad. Finally, the chapter concludes with an exploration of the lessons of the golden toad and some of the considerations in attributing other contemporary extinctions to climate change and its associated effects.

## Catastrophe in Monteverde

In the early 1980s, a young researcher named Alan Pounds went to the Monteverde cloud forest of Costa Rica to study tropical herpetofauna populations. Among the most charismatic of the Monteverde lizards and frogs was the golden toad, a bright orange frog with an enigmatic life history. Golden toads were featured on tourism posters and conservation brochures for Costa Rica, but little was known of their biology. They were believed to spend much of their lives underground, emerging for spectacular (and photogenic) annual mass breedings.

As Pounds's field studies continued, he watched the lizard and frog populations of Monteverde crash catastrophically. In 1987 the spectacular breeding aggregation of the golden toad suddenly ceased. A single individual was observed in each of the following two years, and after 1989, researchers never encountered another golden toad at Montverde. After a decade of fruitless searching, in 2004 the species was pronounced extinct on the IUCN Red List.

Pounds knew that there must be a common environmental factor driving these profound biological changes and was determined to identify the culprit. He and colleague Martha Crump noted that the extinctions of the golden toad and the harlequin frog (*Atelopus varius*) had occurred immediately following several warm, dry years caused by the El Niño Southern Oscillation (Pounds and Crump, 1994). Further analysis showed that unusually dry years in 1983 (the previous El Niño event), 1987, 1994, and 1998 were all associated with demonstrable changes in reptile, amphibian, and bird populations. Reptile and amphibian populations crashed in the dry years, while bird species inhabiting lowlands moved up in elevation. During these years, more than a dozen lowland bird species were newly observed at 1,500 meters (4,921 feet), and bird species previously abundant at 1,500 meters were crowded out by these new arrivals. The declines in midelevation birds, lizards, and frogs were all associated with declines in mist frequency in the dry years. Mist frequency declined for extended periods during these years, causing dramatic changes in ecological conditions and favoring some species (e.g., premontane birds) while harming others (midelevation birds, lizards, and frogs). Pounds published this data linking climate fluctuations with the extinctions, declines, and demographic changes of Monteverde in *Nature* (Pounds et al., 1999). During the same year, Chris Still, Pru Foster, and Steve Schneider used mathematical simulations to show that climate trends associated with human-induced climate change could lead to elevated cloud bases and drying in cloud forests such as Monteverde (Still et al., 1999). Together, these studies contributed robust evidence that the first extinction linked to climate change had occurred.

## An Amphibian Pandemic

Still, Pounds knew that the story was not complete and hypothesized that climate may have enhanced a proximate cause of the declines, speculating that the proximate cause may have been infectious disease (Pounds et al., 2006; Pounds and Crump, 1994). A proximate factor was indeed identified, and researchers realized that a fungal skin disease, chytridiomycosis, had caused the extinction of the golden toad and was now ravaging amphibian populations worldwide (Berger et al., 1998; Rachowicz et al., 2006). Pounds updated his original climate-linked extinction hypothesis to include this pathogen, theorizing that climate interacted with the epidemiology of chytridiomycosis to enhance the spread and lethality of the infection. He believed that climate conditions acted as the causal factor allowing chytridomycosis to destroy amphibian populations, while climate was directly damaging to other herpetofaunal populations, including anoline lizards, which were not susceptible to chytridiomycosis but had also crashed during warm periods.

Amphibian populations continued to decline catastrophically throughout Latin America, with *Atelopus* frogs especially devastated by chytridiomycosis. Within two decades, nearly 70 percent of more than one hundred species in the genus had been annihilated. With this newly enlarged data set, Pounds and a large team of collaborators linked many of the extinctions to dry periods (Pounds et al., 2006; see fig. 6-1). They demonstrated that most extinct *Atelopus* species were last observed immediately before significant dry periods, and that dry periods were highly correlated with large numbers of chytridiomycosis-linked extinctions. In a stroke, this paper increased the number of extinctions attributed to climate change from a single disappearance to dozens of extinctions.

Chytridiomycosis is caused by the pathogenic chytrid fungus *Batrachochytrium dendrobatidis* (Bd), and is now unequivocally the largest proximate cause of the current global amphibian extinction crisis (Skerratt et al., 2007; Lips et al., 2006). Hundreds of frog extinctions

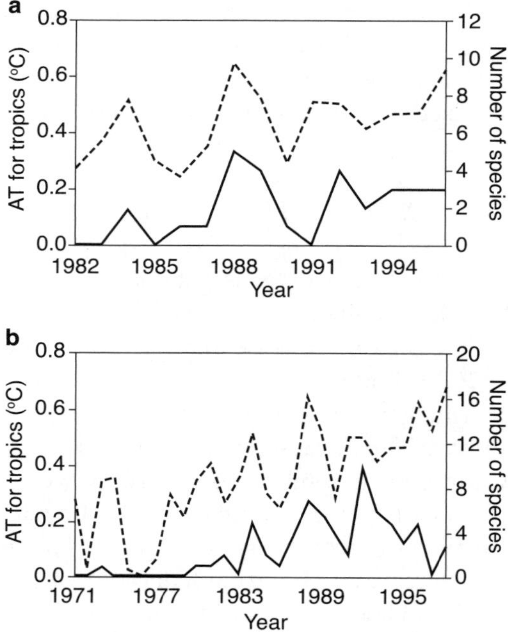


FIGURE 6-1. Number of new world *Atelopus* species observed for the last time versus air temperature. The number of species observed for the last time (solid line) is plotted with the annual deviation in air temperatures (AT) above a long-term normal. *Atelopus* extinctions coincided with warm, dry periods, perhaps because these promote conditions in which *Atelopus* are unable to thermally regulate to minimize virulence of chytridiomycosis infection. From Pounds 2006.

have been attributed to Bd infection, and each year more papers and reports demonstrate the destructive global presence of the fungus. More than two hundred papers were published during 2010 alone confirming the distribution and deadly effects of Bd, which is now present on every inhabited continent.

A link between warmer temperatures and chytridiomycosis lethality is immediately counterintuitive, because warm temperatures actually kill Bd fungus. Growing in vitro in laboratory conditions, Bd fungus is found to flourish at 23 degrees Celsius and dies when exposed to warmer temperatures above 30 degrees Celsius (Longcore et al., 1999; Piotrowski et al., 2004). Amphibians infected with Bd actively increase their body temperatures to ward off the disease (Richards-

Zawacki, 2009), and some studies indicate that Bd mortalities are highest during cold seasons and in the coolest areas of distribution (Berger et al., 1998; Lips, 1999; Bradley et al., 2002). A recent study tested the lethality of Bd infection under different controlled conditions and found that infected frogs actually lived significantly longer at 23 degrees Celsius than at 17 degrees Celsius (Bustamante, 2010).

Pounds and colleagues explained this apparent inconsistency with the chytrid-thermal-optimum hypothesis, which posited that environmental changes associated with warming primed amphibian populations for infection and enhanced the growth of the fungus. Changes in tropical cloud cover generate warmer nights and cooler days, possibly shifting conditions toward those favored by Bd. During Bd infection, amphibians actively seek warm microhabitats to increase body temperature (Richards-Zawacki, 2009). If a lifting cloud base and increased cloud cover destroy warm refugia, Pounds reasoned that this behavioral fever response might be curtailed. Further, the general physiological stress of warmer, drier, and less misty conditions might broadly predispose amphibians to illness. Indeed, frogs infected with Bd can live for years and show little mortality (Murray et al., 2009), and environmental conditions may cause the disease to become lethal. Climate change and its associated effects may well serve as the trigger allowing Bd to eradicate entire amphibian populations.

## Chytrid-Climate Controversy

Publication of the hypothesis linking Bd extinctions and climate change generated immediate and intense controversy (Alford et al., 2007; Lips et al., 2008; Di Rosa et al., 2007). Karen Lips and colleagues rejected the chytrid-thermal-optimum hypothesis, explaining the timing of Central and South American *Atelopus* extinctions through the simple spatiotemporal propagation of the disease itself (Lips et al., 2008). Analyses of declines in other geographic areas similarly show little correlation with interannual climate or regional temperature variation (Retallick et al., 2004; Kriger, 2009; Walker et al., 2010). Other studies argue that although the Central and South American population crashes were indeed correlated with climate fluctuation, the correlation is a statistical coincidence and causation cannot be confidently assigned (Rohr et al., 2008).

Still other researchers assert that the climatic changes Pounds observed may be due to deforestation (Lawton et al., 2001) or were merely effects of El Niño (Anchukaitis and Evans, 2010), rather than being caused by anthropogenic climate warming, as Pounds initially suggested. Whatever the causes of the actual climatic variation, numerous statistical analyses link amphibian declines with climate fluctuations. Although causation remains unclear, Bd outbreaks have been correlated with periods of unusually warm temperature in Venezuela (Lampo et al., 2006), as well as Australia (Laurance, 2008; Drew et al., 2006), Spain (Bosch et al., 2007), and Italy (Di Rosa et al., 2007). Pounds and others maintain that the spatiotemporal-epidemic-spread hypothesis is insufficient to explain patterns of Bd outbreaks, and invoke climate as a significant factor in Bd-linked extinctions (Pounds et al., 2007; Rohr et al., 2010 ).

Nonetheless, this circumstantial evidence leaves room for much debate and thus far allows only speculation into potential mechanisms by which climate conditions might allow Bd to become lethal. Attempts to determine whether the spread of Bd is influenced by climate, and indeed whether amphibian declines in general are enhanced by climate, are confounded by the unequivocal fact that the earth is warming at the same time that amphibian declines and extinctions are intensifying worldwide. Because these two trends are temporally correlated, causation can be extraordinarily difficult to assign (Rohr et al., 2008). The chytrid-thermal-optimum hypothesis has been vigorously debated, and rebuttals, defenses, and new evidence on either side of the controversy go to press every year. Two decades after the first recent amphibian extinctions occurred, we are still debating the factors that contributed to the first of what proved to be a devastating onslaught of amphibian eradication.

## Population-Level Extinctions

The scientific community is still deciding whether amphibian chytridiomycosis extinctions were indeed due in part to anthropogenic climate change. In the meantime, groups other than amphibians are clearly threatened by warming, both by the proximate effects of rising temperature and the secondary effects of climate and environmental change. Scientists are observing numerous population-level extinctions attributed to the effects of climate change. Lifting cloud bases over rain forest areas and cloud cover changes similar to those in Costa Rica have occurred in the Australian North Queensland forest and directly threaten endemic biodiversity. Rising temperatures have dissipated low-lying fog, causing changes that will potentially exterminate more than fifty endemic vertebrates (Williams, 2003; Shoo et al., 2005). The record high temperatures during the summer of 2005 were devastating to the lemuroid ringtail possum (*Hemibelideus lemuroids*). This arboreal species relies on precipitation and collection of water in the canopy for hydration and is extremely sensitive to increases in temperature, and although it is too soon to declare the species entirely extinct, it is clear from recent surveys that the population has been severely harmed.

Climate modification has had a major impact on the timing of seasonal changes, with spring temperatures arriving earlier in the year. Phenological transitions for terrestrial species are triggered by environmental cues, and when cues become decoupled (i.e., daylight hours are not consistent with temperature signals), transitions may be inappropriately expressed and interdependent populations can fall out of sync. This phenomenon has been observed in invertebrates, and decoupling between butterflies and seasonal precipitation caused the extirpation of several populations of *Euphydryas editha bayensis* when emergence of preferred food plants failed (McLaughlin et al., 2002). Changes in phenology have also resulted in population-level declines in several avian species, including the great tit (*Parus major*). Earlier maturation of juveniles now occurs at a period of the year that no longer corresponds to seasonal periods that provide sufficient food resources, and populations are suffering as a result (Both et al., 2006; Sanz, 2002; Visser et al., 1998).

Climate changes can decouple other species relationships as well. Historically, populations of bark beetles (Scolytidae) were limited in their geographic range by temperature. In the last decade, warming trends have allowed populations of bark beetles to move up in elevation and north in range, and warmer winters allow them to complete multiple life cycles in a single season. Released from these environmental checks, the beetles are now destroying vast areas of Rocky Mountain pine forest. Due to the unchecked growth of the pest, hundreds of millions of lodgepole pines and thousands of square kilometers of high-elevation five-needle pines such as the whitebark pine (*Pinus albicaulis*) have been damaged and destroyed (Bale et al., 2002; Logan et al., 2003; Raffa et al., 2008). Temperature changes at high elevations have harmed mammalian populations as well. Pika and marmot extirpations in the Rockies are attributed to rising temperatures exceeding thermal tolerances of these high-elevation mammals (Parmesan, 2006).

In Mexico, warming has led to the extinction of lizard populations (Sinervo et al., 2010). These population extinctions have been linked to climate change by models that demonstrate that populations are lost where thermal optima for foraging disappear. Sinervo et al. (2010) have used these same models to project global extinctions in lizards that are on the same order of magnitude as those estimated by Thomas et al. (2004).

## Conclusions

The warming of the last century has produced far-reaching impacts on the world's biota, apparent on every continent, at every trophic level, and in most major taxonomic groups (Ceballos and Ehrlich, 2002; Parmesan and Yohe, 2003; Thomas et al., 2004; Parmesan, 2006). Climate-induced environmental changes precipitate complex ecosystem effects, disrupting the timing of life events, perturbing equilibria with pathogens, and destroying microhabitats that served as refugia. Climate is linked to population-level declines and extirpations in terrestrial mammals, birds, invertebrates, and forest conifers, and likely contributed to the species-level extinction of about eighty tropical amphibians in Central and South America.

Nonetheless, considering the myriad ways in which climate can disrupt biological systems and the 0.74 degree Celsius mean rise in global temperatures during the last century (IPCC, 2007), we might have anticipated more extinctions to be unequivocally attributable to the documented changes in climate. Several factors may account for this lack of attribution. First, because climate is often merely one of numerous interacting factors contributing to the decline of a species, it can be difficult to explicitly link extinctions to warming (Singer and Parmesan, 2010). Most of the decline mechanisms we have presented involve secondary changes in abiotic conditions (i.e., cloud cover, precipitation), organismal behavior (e.g., migration, phenology, defensive behavior), and crucial interactions among numerous species. These changes frequently involve the breakdown of complex ecological relationships that may be difficult to characterize or quantify. Evidence for climate linkage is situational or observational, and thus attributions are sometimes highly controversial in the scientific community. Second, many extinctions are likely to be under way in species that have yet to be discovered (Smith et al., 1993; Stork, 1993). This sobering scenario is especially likely in tropical regions, where in particular dozens of amphibian species are disappearing before they are ever described (Wake and Vredenburg, 2008). Third, many of the easily characterized species that were most vulnerable to warming may have already succumbed to warming-related extinction during the Quaternary megafaunal extinction (see chapter 11, this volume).

Global temperature will increase between 2 and 4 degrees Celsius in the coming century (IPCC, 2007), making Earth's climate warmer than it has been in 3 million years, and thus warmer than most terrestrial vertebrate species have experienced in the time since they evolved (Hadly and Barnosky, 2009). Scientists have documented signs of severe climate-related distress in contemporary populations, and we are witnessing the first bellwether species falling to warming. In the absence of mitigation policies, climate change will continue to interact with other environmental stressors to break down ecological relationships, push species toward extinction, and drive us to a climatically novel and biologically impoverished Earth.

## Acknowledgments

Many thanks to Elizabeth Hadly, Alan Pounds, Terry Root, and Rebecca Terry.

### REFERENCES

Anchukaitis, K. J., and M. N. Evans. 2010. "Tropical cloud forest climate variability and the demise of the Monteverde golden toad." *Proceedings of the National Academy of Sciences, USA* 107 (11): 5036–5040.

Alford, R. A., K. S. Bradfield, and S. J. Richards. 2007. "Global warming and amphibian losses." *Nature* 447: E3–E4.

Bale, J. S., G. J. Masters, I. D. Hodkinson, C. Awmack, T. M. Bezemer, V. K. Brown, J. Butterfield, et al. 2002. "Herbivory in global climate change research: Direct effects of rising temperature on insect herbivores." *Global Change Biology* 8 (1): 1–16.

Barnosky, A. D., P. L. Koch, R. S. Feranec, S. L. Wing, and A. B. Shabel. 2004.

"Assessing the causes of Late Pleistocene extinctions on the continents." *Science* 306 (5693): 70–75.

Berger, L., R. Speare, P. Daszak, D. E. Green, A. A. Cunningham, C. L. Goggin, R. Slocombe, et al. 1998. "Chytridiomycosis causes amphibian mortality associated with population declines in the rain forests of Australia and Central America." *Proceedings of the National Academy of Sciences, USA* 95 (15): 9031–9036.

Bosch, J., L. Carrascal, L. Durn, S. Walker, and M. Fisher. 2007. "Climate change and outbreaks of amphibian chytridiomycosis in a montane area of Central Spain; Is there a link?" *Proceedings of the Royal Society B: Biological Sciences* 274 (1607): 253.

Both, C., S. Bouwhuis, C. Lessells, and M. Visser. 2006. "Climate change and population declines in a long-distance migratory bird." *Nature* 441: 81–83.

Bradley, G., P. Rosen, M. Sredl, T. Jones, and J. Longcore. 2002. "Chytridiomycosis in native Arizona frogs." *Journal of Wildlife Diseases* 38 (1): 206.

Bustamante, H. M., L. J. Livo, and C. Carey. 2010. "Effects of temperature and hydric environment on survival of the Panamanian golden frog infected with a pathogenic chytrid fungus." *Integrative Zoology* 5 (2): 143–153.

Ceballos, G., and P. Ehrlich. 2002. "Mammal population losses and the extinction crisis." *Science* 296 (5569): 904.

Di Rosa, I., F. Simoncelli, A. Fagotti, and R. Pascolini. 2007. "Ecology: The proximate cause of frog declines?" *Nature* 447 (7144): E4–E5.

Drew, A., E. Allen, and L. Allen 2006. "Analysis of climatic and geographic factors affecting the presence of chytridiomycosis in Australia." *Diseases of Aquatic Organisms* 68 (3): 245.

Hadly, E. A., and A. D. Barnosky. 2009. "Vertebrate fossils and the future of conservation biology." *Conservation Paleobiology: Using the Past to Manage for the Future*, Paleontological Society Short Course. G. P. Dietl and K. W. Flessa, The Paleontological Society Papers. 15.

IPCC. 2007. *Fourth Assessment Report of the Intergovernmental Panel on Climate Change*. New York: Cambridge University Press.

Kriger, K. 2009. "Lack of evidence for the drought-linked chytridiomycosis hypothesis." *Journal of Wildlife Diseases* 45 (2): 537.

Lampo, M., A. Rodriguez-Contreras, E. La Marca, and P. Daszak. 2006. "A chytridiomycosis epidemic and a severe dry season precede the disappearance of *Atelopus* species from the Venezuelan Andes." *The Herpetological Journal* 16 (4): 395–402.

Laurance, W. 2008. "Global warming and amphibian extinctions in eastern Australia." *Austral Ecology* 33 (1): 1–9.

Lawton, R. O., U. S. Nair, R. A. Pielke, and R. M. Welch. 2001. "Climatic impact of tropical lowland deforestation on nearby montane cloud forests." *Science* 294 (5542): 584–587.

Lips, K. 1999. "Mass mortality and population declines of anurans at an upland site in western Panama." *Conservation Biology* 13: 117–125.

Lips, K. R., F. Brem, R. Brenes, J. D. Reeve, R. A. Alford, J. Voyles, C. Carey, et al. 2006. "Emerging infectious disease and the loss of biodiversity in a Neotropical amphibian community." *Proceedings of the National Academy of Sciences, USA* 103 (9): 3165–3170.

Lips, K. R., J. Diffendorfer, J. R. I. Mendelson, and M. W. Sears. 2008. "Riding the wave: Reconciling the roles of disease and climate change in amphibian declines." *PLoS Biology* 6 (3): e72.

Logan, J., J. Regniere, and J. Powell. 2003. "Assessing the impacts of global warming on forest pest dynamics." *Frontiers in Ecology & Environment* (1): 130–137.

Longcore, J., A. Pessier, and D. Nichols. 1999. "*Batrachochytrium dendrobatidis* gen. et sp. nov., a chytrid pathogenic to amphibians." *Mycologia* 91: 219–227.

McLaughlin, J. F., J. J. Hellmann, C. L. Boggs, and P. R. Ehrlich. 2002. "Climate change hastens population extinctions." *Proceedings of the National Academy of Sciences, USA*. doi:10.1073/pnas.052131199.

Murray, K. A., L. F. Skerratt, R. Speare, and H. McCallum. 2009. "Impact and dynamics of disease in species threatened by the amphibian chytrid fungus, *Batrachochytrium dendrobatidis*." *Conservation Biology* 23 (5): 1242–1252.

Parmesan, C. 2006. "Ecological and evolutionary responses to recent climate change." *Annual Review of Ecology, Evolution, and Systematics* 37: 637–669.

Parmesan, C., and M. C. Singer. 2008. "Amphibian extinctions: Disease not the whole story." *PLoS Biology* (28 March).

Parmesan, C., and G. Yohe. 2003. "A globally coherent fingerprint of climate change impacts across natural systems." *Nature* 421: 37–42.

Piotrowski, J., S. Annis, and J. Longcore. 2004. "Physiology of *Batrachochytrium dendrobatidis,* a chytrid pathogen of amphibians." *Mycologia* 96 (1): 9.

Pounds, J. A., and M. L. Crump. 1994. "Amphibian declines and climate disturbance: The case of the golden toad and the harlequin frog." *Conservation Biology* 8 (1): 72–85.

Pounds, J. A., M. P. L. Fogden, and J. H. Campbell. 1999. "Biological response to climate change on a tropical mountain." *Nature* 398 (6728): 611–616.

Pounds, J. A., M. R. Bustamante, L. A. Coloma, J. A. Consuegra, M. P. L. Fogden, P. N. Foster, E. La Marca, et al. 2006. "Widespread amphibian extinctions from endemic disease driven by global warming." *Nature* 439 (12): 161–167.

Pounds, J. A., M. R. Bustamante, L. A. Coloma, J. A. Consuegra, M. P. L. Fogden, P. N. Foster, E. La Marca, et al. 2007. "Global warming and amphibian losses; The proximate cause of frog declines? (Reply)." *Nature* 447 (7144): E5–E6.

Rachowicz, L. J., R. A. Knapp, J. A. T. Morgan, M. J. Stice, V. T. Vredenburg, J. M. Parker, and C. J. Briggs. 2006. "Emerging infectious disease as a proximate cause of amphibian mass mortality." *Ecology* 87 (7): 1671–1683.

Raffa, K., B. Aukema, B. Bentz, A. Carroll, J. Hicke, M. Turner, and W. Romme. 2008. "Cross-scale drivers of natural disturbances prone to anthropogenic amplification: The dynamics of bark beetle eruptions." *Bioscience* 58 (6): 501–517.

Retallick, R., H. McCallum, and R. Speare. 2004. "Endemic infection of the amphibian chytrid fungus in a frog community post-decline." *PLoS Biology* 2 (11): e351.

Richards-Zawacki, C. 2009. "Thermoregulatory behaviour affects prevalence of chytrid fungal infection in a wild population of Panamanian golden frogs." *Proceedings of the Royal Society B: Biological Sciences* 227: 519–528.

Rohr, J. R., T. R. Raffel, J. M. Romansic, H. McCallum, and P. J. Hudson. 2008. "Evaluating the links between climate, disease spread, and amphibian declines." *Proceedings of the National Academy of Sciences, USA* 105 (45): 17436.

Sanz, J. J. 2002. "Climate change and breeding parameters of great and blue tits throughout the western Palaearctic." *Global Change Biology* 8 (5): 409–422.

Shoo, L., S. Williams, and J. Hero. 2005. "Climate warming and the rainforest birds of the Australian Wet Tropics: Using abundance data as a sensitive predictor of change in total population size." *Biological Conservation* 125 (3): 335–343.

Sinervo, B., F. Mendez-de-la-Cruz, D. B. Miles, B. Heulin, E. Bastiaans, M. V. S. Cruz, R. Lara-Resendiz, et al. 2010. "Erosion of lizard diversity by climate change and altered thermal niches." *Science* 328 (5980): 894–899.

Singer, M. C., and C. Parmesan. 2010. "Phenological asynchrony between herbivorous insects and their hosts: Signal of climate change or pre-existing adaptive strategy?" *Philosophical Transactions of the Royal Society B: Biological Sciences* 365 (1555): 3161–3176.

Skerratt, L. F., L. Berger, R. Speare, S. Cashins, K. R. McDonald, A. D. Phillott, H. B. Hines, et al. 2007. "Spread of chytridiomycosis has caused the rapid global decline and extinction of frogs." *EcoHealth* 4 (2): 125–134.

Smith, F. D. M., R. M. May, R. Pellew, T. H. Johnson, and K. R. Walter. 1993. "How much do we know about the current extinction rate?" *Trends in Ecology & Evolution* 8 (10): 375–378.

Still, C., P. Foster, and S. H. Schneider. 1999. "Simulating the effects of climate change on tropical montane cloud forests." *Nature* (398): 608–610.

Stork, N. 1993. "How many species are there?" *Biodiversity and Conservation* 2 (3): 215–232.

Thomas, C. D., A. Cameron, R. E. Green, M. Bakkenes, L. J. Beaumont, Y. C. Collingham, B. F. N. Erasmus, et al. 2004. "Extinction risk from climate change." *Nature* 427 (6970): 145–148.

Visser, M., A. Van Noordwijk, J. Tinbergen, and C. Lessells. 1998. "Warmer springs lead to mistimed reproduction in great tits (*Parus major*)." *Proceedings of the Royal Society B: Biological Sciences* 265 (1408): 1867–1870.

Wake, D. B., and V. T. Vredenburg. 2008. "Are we in the midst of the sixth mass extinction? A view from the world of amphibians." *Proceedings of the National Academy of Sciences, USA* 105 (Supplement 1): 11466–11473.

Walker, S. F., J. Bosch, V. Gomez, T. W. J. Garner, A. A. Cunningham, D. S. Schmeller, M. Ninyerola, et al. 2010. "Factors driving pathogenicity vs. prevalence of amphibian panzootic chytridiomycosis in Iberia." *Ecology Letters* 13 (3): 372–382.

Williams, S., E. Bolitho, and S. Fox. 2003. "Climate change in Australian tropical rainforests: An impending environmental catastrophe." *Proceedings of the Royal Society B: Biological Sciences* 270 (1527): 1887–1892.

Chapter 7

# *Global Warming and Widespread Coral Mortality: Evidence of First Coral Reef Extinctions*

PETER W. GLYNN

Coral reefs cover 255,000 square kilometers of the earth's surface (Spalding and Grenfell, 1997) and likely harbor more than a million species globally, perhaps as many as 3 million (Reaka-Kudla, 1997; Small et al., 1998). Coral reefs benefit humankind in numerous ways. They provide ecosystem services and advantages to tropical human communities, including coastal protection, nurseries and sources of nutrition for fisheries, tourism, and great stores of genetic material and species (biodiversity). In addition, a less tangible benefit relates to the esthetics of coral reefs—the sheer beauty and wonders of these diverse ecosystems offer inspiration to lay persons and scientific investigators alike.

Human-induced climate change is warming sea surface waters, causing coral bleaching and widespread coral mortality. Bleaching occurs when corals expel the symbiotic algae that live within their cells. Most reef-building or zooxanthellate corals (i.e., cnidarians engaged in an obligate symbiotic relationship with photoautotrophic dinoflagellates in the genus *Symbiodinium*) occupy habitats whose temperature conditions are perilously close to their upper thermal tolerance limits (Coles and Brown, 2003; Jokiel and Brown, 2004; McWilliams et al., 2005; Hoegh-Guldberg et al., 2007). A slight elevation in temperature (1 to 1.5 degrees Celsius above the climatological thermal mean), often in combination with increased duration, can cause coral bleaching and mortality. When corals die and reef structures are eroded, many of the species taking shelter in reefs also disappear. Coral bleaching episodes are rapidly increasing around the world, and have led to widespread coral mortality in all oceans, especially after severe El Niño events.

Coral reefs are therefore the marine ecosystem in which extinctions due to climate change might first be expected. This chapter explores the evidence for first climate change–linked reef extinctions. In addition to the various difficulties in documenting extinctions in terrestrial ecosystems, validating marine extinctions on coral reefs has been hampered by (i) a meager research effort directed toward reefs until only the past few decades, and (ii) the difficulty in locating and monitoring subtidal populations. Here, examples of severe declines and purported extinctions of reef-building coral species are examined first, followed by reef-associated invertebrates. These results show widespread losses of coral species, resulting in local and regional extinctions. Loss of one species after a particularly severe El Niño event was initially interpreted as a global extinction, but subsequent discoveries have led to it being reclassed as a regional extinction. It is believed that no global coral extinctions have yet occurred due to climate change, but the observed local and regional extinctions following coral bleaching events indicate that global extinctions may be expected in the future if climate change isn't constrained.

## First Apparent Extinction and Rediscovery

The first surveys of the Uva Island coral reef on the Pacific coast of Panamá in March 1970 resulted in an exhilarating and exceptional discovery. Numerous large colonies of a hydrocoral (*Millepora*) species, never before reported from the eastern Pacific region, were found inhabiting a broad expanse of the Uva Island reef flat. Continuing surveys of deeper reef zones, the forereef slope and reef base, revealed two additional species of *Millepora* (Glynn, 1972; Glynn et al., 1972; Porter, 1972). The disappearances of these three species in the aftermath of the severe 1982–83 El Niño mortality event are among the best documented regional extinctions of corals. These zooxanthellate species were *Millepora intricata* (fig. 7-1A), *Millepora platyphylla,* and a recently described species, *Millepora boschmai* (fig. 7-1B), all narrowly

FIGURE 7-1. Hydrocoral species severely affected during the 1982–83 El Niño-Southern Oscillation bleaching event in Panamá, Gulf of Chiriquí, eastern tropical Pacific. (A) *Millepora intricata* experienced a temporary (3–4 year) decline, disappearing entirely from shallow reef zones. Secas Islands reef, 4 meters (13 feet) depth, March 18, 1990. (B) *Millepora boschmai,* now likely regionally extinct. Lazarus Cove, Uva Island, 6 meters (20 feet) depth, February 22, 1992. Photograph (B) courtesy of J. S. Feingold.

restricted to the Gulf of Chiriquí in the tropical Panamic Pacific Province.

When first observed in Panamá in 1970, *M. boschmai* was considered a possible new species; this was verified from material sent to the hydrocoral systematist Hilbrandt Boschma at the Leiden Museum in the Netherlands. Boschma died before describing the new species. To avoid a "centinelan extinction," or the loss of a species before it is known to science (Wilson, 1992), a concerted effort was undertaken to search for *Millepora* spp. over their former ranges. After not finding live colonies of *M. platyphylla* or *M. boschmai* after extensive surveys over an 8-year period (1983–1990), with a total search effort of 204 diver hours, it was concluded that these two species experienced regional and global extinctions, respectively.

Only 7 months after the publication of the extinction of *M. boschmai* (Glynn and de Weerdt, 1991), five live colonies were discovered at a cove on the north shore of Uva Island, Gulf of Chiriquí (Glynn and Feingold, 1992). Based on the sizes of these recently discovered colonies, and their estimated growth rates, it is probable that they recruited to this site after 1983. A total of eight live colonies of *M. boschmai* were found at Uva and Coiba Islands, but have not been seen alive since the last surveys at these sites in 2007 and 2001, respectively (Brenes et al., 1993; Maté, 2003).

A second surprising rediscovery occurred when Razak and Hoeksema (2003) recognized *M. boschmai* in collections from Indonesia, based on skeletal morphological characters. The occurrence of five colonies at south Sulawesi and Sumba, ~17,600 kilometers (10,900 miles) west of Panamá, indicates that *M. boschmai* can no longer be considered an eastern Pacific endemic.

The fates of the other eastern Pacific hydrocorals discovered in 1970 are mixed. *M. platyphylla* has not been seen alive in the eastern Pacific since 1983 (i.e., for 28 years). *M. intricata* was again abundant in shallow reef zones at Uva Island and other reefs 14 years after the 1982–83 mortality event. Living populations were present in deeper water (12–25 meters [39–82 feet]) at several reef sites after 1983, and it is likely these served as source populations for re-establishment in shallow reef zones (Glynn et al., 2001). All of the shallow reef populations of *M. intricata* that had recovered since 1983 again bleached and died during the 1997–98 El Niño event. As in 1983, deeper populations at 12–20 meters (39–66 feet) depth did not bleach or experience any marked increase in mortality. Shallow reef areas again were colonized by *M. intricata* (~ 2 centimeter [1 inch]-high colonies) at two sites as early as 2000, and by 2002, several colonies (one 21 centimeters [8 inches] in height) were present at 2–3 meters (7–10 feet) deep.

In summary, *M. boschmai* and *M. platyphylla* are best considered regionally extinct species because they both are known to occur elsewhere in the Indo-Pacific region. *M. intricata* is also a wide-ranging Indo-Pacific species that is confined to a single gulf (Gulf of Chiriquí, Panamá) in the eastern Pacific. Because it is capable of repopulating shallow reefs from deepwater refuges following El Niño disturbances, it is not at present considered regionally or globally endangered.

## Observed Declines in Other Corals and Associated Reef Species

Four scleractinian coral species experienced extreme reductions in population size in the eastern Pacific as a consequence of the 1982–83 El Niño bleaching event (Glynn, 1997). These species and their respective localities were *Pocillopora capitata,* Costa Rica; *Porites panamensis,* Costa Rica; *Psammocora stellata,* Panamá and Galápagos Islands; *Gardineroseris planulata,* Costa Rica and Galápagos Islands. In addition, two species (*Acropora valida* and *Porites rus*) disappeared from the eastern Pacific during the same period or slightly later, but the cause(s) of their extirpations is less certain. Since these species were discovered only in 1983, it is unclear what their status was prior to the El Niño event or why they disappeared. *Acropora valida* was discovered in 1983 at Gorgona Island, Colombia (Zapata and Vargas-Ángel, 2003), and *Porites rus* in 1983 at Samaná, Costa Rica (Cortés and Jiménez, 2003). Because these species range widely throughout the Indo-Pacific, their losses from the eastern Pacific represented regional extinctions.

The only two western Atlantic acroporid species, *Acropora palmata* and *Acropora cervicornis,* have undergone major regionwide declines since the early 1980s. These losses are primarily a result of disease-related mortality, elevated temperature-induced bleaching, and physical damage from hurricanes. Based on observed high rates of population decline through their ranges, the National Marine Fisheries Service of the National Oceanographic and Atmospheric Administration listed these acroporids as "threatened" species in 2006 under the Endangered Species Act (*Acropora* Biological Review Team, 2005). More recently, since early 2000, both *Acropora* species have shown signs of recovering at several Caribbean localities, in the Bahamas, along southeastern Florida, and at the Flower Gardens site in the Gulf of Mexico (Vargas-Ángel et al., 2003; Precht and Aronson, 2006).

Whether these will serve as source populations to ensure survival of these species and eventually renewed reef framework construction over the wider Caribbean remains uncertain.

Among reef-associated invertebrates, the three mollusc (*Stiliger vossi, Hippopus hippopus,* and *Tridacna gigas*) and single sea urchin (*Tripneustes gratilla*) species listed by Dulvy et al. (2003) represent severe population reductions that could result in local extinctions. It is difficult to judge the status of *Stiliger vossi,* a minute (ca. 1 millimeter in length) ascoglossan described in 1960. Clark (1994) noted that it is a very rare species, but long-term monitoring at its type locality has not been performed. The two tridacnid giant clams, *Hippopus hippopus* and *Tridacna gigas,* were overfished and may best be regarded as local, commercial extinctions. *Tripneustes gratilla,* a sea urchin harvested for its gonads at Bolinao in the Philippines, experienced severe declines in the 1990s, but has since demonstrated a degree of recovery from grow-out culture efforts and management intervention (Juinio-Meñez et al., 2008). *Diadema antillarum* population declines during the early 1980s were regionwide in the Caribbean, Bahamas, and Gulf of Mexico, but the species has experienced recovery in Jamaica (Edmunds and Carpenter, 2001) and some other areas of the western Atlantic during the past decade.

## Functional Extinctions

Although there is no evidence of global extinction in individual coral reef species, many reefs are experiencing functional extinctions. Functional extinctions refer to reductions in species populations that lead to important changes in their ecological roles, with cascading effects on community structure and/or function. The degradation of coral reef frameworks due to bleaching is leading to marked declines in topographic complexity. This loss of structure has caused at least two levels of functional extinctions: loss of essential habitats for coral-dependent animals, and changes in the physical structure of reef formations (Williams et al., 1999; Alvarez-Filip et al., 2009).

Thus, coral reefs have experienced four of the five types of extinctions that have been recognized in the marine environment (Carlton et al., 1999)—local, regional, functional, and commercial. The fourth type, commercial extinction, occurs when a commercially exploited species no longer provides profitable yields. If continued overexploitation of reef-associated species results in additional commercial extinctions, an increase in functional extinctions can be expected. Coupled with increasing climate change stressors, future coral reef extinctions would likely lead to coral reef ecosystem dysfunction.

## What Is at Stake

Coral reefs support the highest concentration of phyla of all ecosystems on planet Earth. Of the thirty-three presently recognized animal phyla, no fewer than thirty-one contain species associated with coral reefs (table 7-1). Many thousands of described species belong to the ten prominent reef phyla listed. All phyla with reef-associated species contain numerous cryptic and symbiotic members that reside within different kinds of reef cavities or with living host organisms, respectively. This concealed fauna is phyletically rich and speciose, contributing significantly to the biodiversity and ecological function of coral reefs. The remaining animal taxa, many of which are visible, occur in open surface and water column habitats (plankton, nekton).

Species contributing to reef-building—that is, construction of reef frameworks, structural integrity, consolidation, and calcification—belong to the phyla Porifera, Cnidaria, Annelida, Mollusca, Ectoprocta, Echinodermata, and Chordata. Equally important are invertebrates engaged in destructive reef processes such as corallivory and bioerosion, the consumption of tissues or the erosion of skeletons of reef-building organisms, respectively. Phyla with species involved in these activities are the Porifera, Platyhelminthes, Sipuncula, Annelida, Arthropoda, Mollusca, and Echinodermata.

## Projected Trends, Hope, and Despair

Based on contemporary responses of reef communities to sea warming episodes, some hypotheses of the fate of coral reef ecosystems into the next century are suggested. There are uncertainties attendant to such predictions, especially considering the wide range of projected values of the critical climate change variables (Kleypas, 2007). For example, under low-emission (B1) and high-emission (A2) scenarios, mean global temperature rise is projected to increase by 1.8 degrees Celsius and 3.4 degrees Celsius, respectively, by the end of this century

TABLE 7-1. Ten of thirty-one major animal phyla whose members contribute to the community composition of coral reefs.

| *Phylum*[1] | *Number of Described Species* | *Ecological Functions*[2] |
|---|---|---|
| Porifera—sponges | 5,500 | reef building (m)[3], consolidation, erosion |
| Cnidaria—coelenterates | 9,795 | reef building (s)[3], consolidation |
| Platyhelminthes—flat worms | 15,000 | corallivores |
| Nemata—round worms | 12,000 | n/a[2] |
| Annelida—segmented worms | 12,148 | reef building (m)[3], consolidation, corallivores, erosion |
| Arthropoda—arthropods | 47,217 | corallivores, erosion |
| Mollusca—molluscs | 52,525 | reef building (m)[3], consolidation, corallivores, erosion |
| Ectoprocta—moss animals | 5,700 | reef building (w)[3], consolidation |
| Echinodermata—echinoderms | 7,000 | reef building (w)[3], corallivores, erosion |
| Chordata—chordates | 4,932 | reef building (w)[3], corallivores, erosion |
| Miscellaneous (n = 21) (comb jellies, ribbon worms, *inter alia*)[4] | 4,561 | Functions phylum-dependent |

[1]An additional twenty-one phyla listed under "Miscellaneous" contain several members that are also closely associated with coral reefs (see Glynn, 2011 for more information on these taxa). Noted for each phylum is the known number of global marine species (after Bouchet, 2006). The actual number present on reefs is likely an order of magnitude higher.
[2]Principal ecological roles of respective species in terms of direct effects on reef growth, persistence, and decline.
[3]Reef-building potential of some member species: (s) strong, (m) moderate, (w) weak
[4]Dominantly indirect effects

(IPCC, 2007). Temperature increases in the tropics during past interglacial periods are thought to have been within this range. Although reef-building corals survived the elevated temperatures associated with past glacial to interglacial transitions, the magnitude and rate of increase are expected to be greater this century. If true, such a warming event would exceed the tolerance limits of those species that do not have the capacity to acclimate or adapt to these changes (Hoegh-Guldberg et al., 2007; Baker et al., 2008). In addition, atmospheric carbon dioxide concentration alone, with projected values of 555–825 parts per million volume, would lead to ocean acidification and reduced calcification in coral reef ecosytems.

Under such conditions the survival of reef-building corals, the essential architects of coral reef structures, is at risk. Their survival and condition will determine the continued presence or demise of the multitude of associated coral reef metazoans that live as integral members of reef ecosystems. The risk of these functional and global extinctions may be moderated by resistance and resilience, as well as by the possible survival of corals inhabiting refugia (benign environments) that could repopulate affected reef areas.

### *Resistance and Resilience in Reef-Building Corals*

Certain coral genera, especially those with species exhibiting nonbranching colony morphologies, have been observed to survive severe bleaching events (table 7-2). Coles and Brown (2003) and Baird et al.

TABLE 7-2. Some of the generally hardy coral genera that often contribute importantly to reef building and postbleaching recovery, listed by family and with the authors who offered evidence for resistance or resilience.

| *Family* | *Genera* | *Authority* |
|---|---|---|
| Acroporidae | *Astreopora* | McClanahan, 2000 |
| Agariciidae | *Pavona* | Hueerkamp et al., 2001; McClanahan et al., 2004 |
| Oculinidae | *Galaxea* | McClanahan et al., 2004 |
| Siderastreidae | *Siderastrea* | Gates and Edmunds, 1999 |
| Faviidae | *Cyphastrea* | McClanahan et al., 2004 |
| | *Diploastrea* | Schuhmacher et al., 2005 |
| | *Favia* | Loya et al., 2001; LaJeunesse et al., 2003 |
| | *Favites* | Loya et al., 2001; LaJeunesse et al., 2003 |
| | *Goniastrea* | Loya et al., 2001; Brown et al., 2002 |
| | *Leptastrea* | Loya et al., 2001; LaJeunesse et al., 2003 |
| | *Platygyra* | Loya et al., 2001; LaJeunesse et al., 2003 |
| Poritidae | *Porites* | Hoegh-Guldberg and Salvat, 1995; Gates and Edmunds, 1999; Hueerkamp et al., 2001; Loya et al., 2001 |
| | *Goniopora* | McClanahan et al., 2004 |

(2009) have offered reviews of the known and potentially effective physiological mechanisms that could play a role in the acclimatization and adaptation of coral animals to elevated temperature and irradiance stress. Some coral hosts can minimize stressful solar radiation flux to their symbionts by producing fluorescent pigments and sequestering mycosporine-like amino acids. Fluorescent pigments have been shown to absorb, scatter, and dissipate high-energy radiation, thus reducing photodamage. Mycosporine-like amino acids also absorb and dissipate ultraviolet energy, thus limiting the formation of toxic intermediate by-products of photosynthesis. Two additional mechanisms that could reduce bleaching damage are antioxidant systems and heat shock proteins, both observed in corals. Also, coral species that can supplement their energy needs by heterotrophic feeding during stressful periods have been shown to survive experimental bleaching better than nonfeeding corals (Grottoli et al., 2006). Finally, Loya et al. (2001) hypothesized that the survival of nonbranching corals during a major bleaching event on Okinawa was related to (i) the possession of thick tissues, and (ii) a colony morphology that facilitates a high mass transfer of reactive oxygen species produced during photosynthesis.

The survival of brooding coral species, in contrast to broadcast-spawning corals, in the Caribbean during the 1980s and 1990s, has prompted some workers to consider the brooding reproductive mode to be advantageous in a global warming scenario (Knowlton, 2001). In the western Atlantic, brooding poritid, agariciid, and siderastreid corals have demonstrated relatively high survivorship during disturbance events (Hughes, 1994; Aronson and Precht, 2001; Kikuchi et al., 2003) and increasing relative abundances of poritids (Green et al., 2008), but broadcast-spawning corals in the *Montastraea* complex seem to have fared better than brooders in the US Virgin Islands (Rogers et al., 2008). The broadcast-spawning eastern Pacific scleractinian coral fauna has survived multiple recent El Niño-Southern Oscillation bleaching events (Glynn and Colley, 2008). It is premature to conclude which, if any, of these reproductive traits might be advantageous.

Since the pioneering study of Rowan et al. (1997), who demonstrated the existence of diverse, multicladal communities of zooxanthellae in the Caribbean *Montastraea* species complex and their differing sensitivities to increasing temperature and solar radiation, similar responses have been observed in several additional coral species. Evidence is mounting that the potential for different *Symbiodinium* variants (e.g., based on the internal transcribed spacers 1 and 2 of a multicopy ribosomal gene family, ITS-1 and ITS-2 rDNA) to enhance the physiological tolerance of at least some reef coral species to climate change stressors may be effective into the next century, providing greenhouse gas emissions are moderated (Baskett et al., 2009).

## *Refugia and Recovery*

Some habitats with relatively benign environmental conditions ("refugia") provide a degree of protection to corals during major bleaching events. Observations on coral survival and recovery in the Bahamas, Caribbean, northern Red Sea, and southeastern Africa during the 1998 bleaching event demonstrated a strong correlation with local upwelling and medium depth sites (Riegl and Piller, 2003). Based on an analysis of sea surface temperature time series in Madagascar, McClanahan et al. (2009) found a close relationship between the condition of coral communities and long-term trends in site-specific thermal environments. Coral communities in northern Madagascar with high coral cover and relatively high numbers of coral genera occurred in thermally benign environments.

High coral survivorship on offshore banks, compared with insular fringing reefs, was also monitored and modeled in the western Caribbean during the 1998 bleaching event (Riegl et al., 2008). The high survivorship of *Acropora cervicornis* at offshore locations was attributed to vigorous flushing of reef waters and reduced runoff. In the eastern Pacific (Panamá), no coral bleaching was observed in an upwelling center during 1998, but corals were severely affected in non-upwelling areas during the height of the warming event (Glynn et al., 2001).

Deep-living zooxanthellate hydrocorals that experienced tidally forced pulses of cool water also survived the 1998 bleaching event. Broadcast spawning of *Montastraea* spp. between 33 meters and 45 meters (108 feet and 148 feet) depth at the Flower Garden Banks in the Gulf of Mexico (Vize, 2006) demonstrates the reproductive viability of mesophotic populations and the possibility that they could figure effectively in repopulating depleted shallow reef zones. The role that deeper living corals will play in ameliorating the risk of extinction will depend on their connectivity with shallow-occurring populations and their ability to maintain relative resistance to mortality from regional thermal extremes.

### *Coral Community Structure and Reef Building*

With the extirpation of thermally sensitive corals, coral community diversity would decline both locally and regionally. It is also likely that some of the 200+ species at an elevated risk of extinction could become globally extinct (Carpenter et al., 2008). These would be major impacts on already threatened corals. Carlton et al. (1999) have argued that numerous reef species could already have become extinct from nonclimate causes, based on theoretical estimations of species extinctions from species richness and area relationships (Reaka-Kudla, 1997). If global coral reefs support more than 1 million species total (known and unknown), and if 5 percent of the world's reef area has been irreparably degraded, it is possible that a loss of 50,000–60,000 species has already occurred. Species losses would amount to 300,000–400,000 if 30 percent of global reefs were destroyed by bleaching. In Wilkinson's (2008) assessment of reef condition, it is estimated that 19 percent of the world's coral reefs have been effectively destroyed and show no immediate prospects of recovery. In addition, it is predicted that 35 percent of global reefs are under risk of collapse due to the interaction of multiple factors, including climate change, during the next 20–40 years.

## Conclusions

Coral bleaching and mortality caused by episodes of elevated sea temperature have resulted in local, regional, functional, and commercial extinctions, but no documented global extinctions. In the near future although reef survival in some state may be possible, it is likely that the structure and function of coral reef ecosystems would differ substantially from those of today. In a worst-case scenario, if coral reef–bleaching disturbances become more severe and remain in a chronic state, it is probable that several coral species will become globally extinct by the end of the century. With the demise of many corals, we could expect the loss of tens of thousands of known and unknown coral associates. Remaining metazoan corallivores and bioeroders would continue to deplete coral populations and erode reef structures. These are not encouraging prospects, but unfortunately remain squarely within the realm of possibilities.

## Acknowledgments

I thank D. Holstein for technical assistance in producing photographs. Acquisition of pertinent literature was greatly facilitated by L. McManus, A. Campbell, and D. Holstein. Discussions with V. W. Brandtneris, S. D. Cairns, I. C. Enochs, C. Langdon, A. M. S. Correa, and C. Wilkinson improved the focus of this contribution. Finally, L. Hannah's interest in addressing coral responses to climate change was largely responsible for this contribution. Research results of PWG in the eastern Pacific were funded by NSF grant OCE-0526361 and earlier awards.

## REFERENCES

*Acropora* Biological Review Team. 2005. "*Atlantic* Acropora *Status Review Document*." Report to National Marine Fisheries Service, Southeast Regional Office. March 3, 2005.

Alvarez-Filip, L., N. K. Dulvy, J. A. Gill, I. M. Côté, and A. R. Watkinson. 2009. "Flattening of Caribbean coral reefs: Region-wide declines in architectural complexity." *Proceedings of the Royal Society B*. doi:10.1098/rspb.2009.0339.

Aronson, R. B., and W. F. Precht. 2001. "Evolutionary paleoecology of Caribbean coral reefs." In *Evolutionary Paleoecology: The Ecological Context of Macroevolutionary Change*, edited by W. D. Allman and D. J. Bottjer, 171–233. New York: Columbia University Press.

Baird, A. H., R. Bhagooli, P. J. Ralph, and S. Takahashi. 2009. "Coral bleaching: The role of the host." *Trends in Ecology and Evolution*, 24: 16–20.

Baker, A. C., P. W. Glynn, and B. Riegl. 2008. "Climate change and coral reef bleaching: An ecological assessment of long-term impacts, recovery trends, and future outlook." *Estuarine, Coastal and Shelf Science*, 80: 435–471.

Baskett, M. L., S. D. Gaines, and R. M. Nisbet. 2009. "Symbiont diversity may help coral reefs survive moderate climate change." *Ecological Applications*, 19: 3–17.

Bouchet, P. 2006. "The magnitude of marine biodiversity." In *The Exploration of Marine Biodiversity, Scientific and Technological Challenges*, edited by C. M. Duarte, 21–64. Bilbao: Fundación BBVA.

Brenes, R., J. Cuadras, M. Durbán, A. Fernández-López, L. M. González, and A. Miranda. 1993. "Plan de Manejo del Parque Nacional Coiba (1ª fase)." Informe inédito, *AECI, ICONA, INRENARE*. Panamá.

Brown, B. E., C. A. Downs, R. P. Dunne, and S. W. Gibb. 2002. "Exploring the basis of thermotolerance in the reef coral *Goniastrea aspera*." *Marine Ecology Progress Series* 242: 119–129.

Carlton, J. T., J. B. Geller, M. L. Reaka-Kudla, and E. A. Norse. 1999. "Historical extinctions in the sea." *Annual Review of Ecology and Systematics,* 30: 515–538.

Carpenter, K. E., M. Abrar, G. Aeby, R. B. Aronson, S. Banks, A. Bruckner, Á. Chiriboga, et al. 2008. "One-third of reef-building corals face elevated extinction risk from climate change and local impacts." *Science,* 321: 560–563.

Clark, K. B. 1994. "Ascoglossan ( = Sacoglossa) molluscs in the Florida Keys: Rare marine invertebrates at special risk." *Bulletin of Marine Science,* 54: 900–916.

Coles, S. L., and B. E. Brown. 2003. "Coral bleaching—Capacity for acclimatization and adaptation." *Advances in Marine Biology,* 46: 183–223.

Cortés, J., and C. Jiménez. 2003. "Corals and coral reefs of the Pacific of Costa Rica: History, research and status." In *Latin American Coral Reefs,* edited by J. Cortés, 361–385. Amsterdam: Elsevier.

Dulvy, N. K., Y. Sadovy, and J. D. Reynolds. 2003. "Extinction vulnerability in marine populations." *Fish and Fisheries,* 4: 25–64.

Edmunds, P. J., and R. C. Carpenter. 2001. "Recovery of *Diadema antillarum* reduces macroalgal cover and increases abundance of juvenile corals on a Caribbean reef." *Proceedings of the National Academy of Sciences, USA,* 98: 5067–5071.

Gates, R. D., and P. J. Edmunds. 1999. "The physiological mechanisms of acclimatization in tropical reef corals." *American Zoologist* 39: 30–43.

Glynn, P. W. 1972. "Observations on the ecology of the Caribbean and Pacific coasts of Panamá." In *The Panamic Biota: Some Observations Prior to a Sea-Level Canal,* edited by M. L. Jones, 13–30. *Bulletin of the Biological Society of Washington*, no. 2.

Glynn, P. W. 1997. "Eastern Pacific reef coral biogeography and faunal flux: Durhams's dilemma revisited." *Proceedings of the 8th International Coral Reef Symposium, Panamá,* 1: 371–378.

Glynn, P. W. 2011. "In tandem reef coral and cryptic metazoan declines and extinctions." *Bulletin of Marine Science,* 87. http://dx.doi.org/10.5343/bms.2010.1025

Glynn, P. W., and S. B. Colley. 2008. "Survival of brooding and broadcasting reef corals following large scale disturbances: Is there any hope for broadcasting species during global warming?" *Proceedings of the 11th International Coral Reef Symposium*. Dania, FL: Nova Southeastern University/NCRI, 11: 361–365.

Glynn, P. W., and W. H. de Weerdt. 1991. "Elimination of two reef-building hydrocorals following the 1982–83 El Niño warming event." *Science,* 253: 69–71.

Glynn, P. W., and J. S. Feingold. 1992. "Hydrocoral species not extinct." *Science,* 257: 1845.

Glynn, P. W., R. H. Stewart, and J. E. McCosker. 1972. "Pacific coral reefs of Panamá: Structure, distribution and predators." *Geologische Rundschau,* Stuttgart 61: 483–519.

Glynn, P. W., J. L. Maté, A. C. Baker, and M. O. Calderón. 2001. "Coral bleaching and mortality in Panamá and Ecuador during the 1997–1998 El Niño-Southern Oscillation event: Spatial/temporal patterns and comparisons with the 1982–1983 event." *Bulletin of Marine Science,* 69: 79–101.

Green, D. H., P. J. Edmunds, and R. C. Carpenter. 2008. "Increasing relative abundance of *Porites astreoides* on Caribbean reefs mediated by an overall decline in coral cover." *Marine Ecology Progress Series,* 359: 1–10.

Grottoli, A. G., L. J. Rodrigues, and J. E. Palardy. 2006. "Heterotrophic plasticity and resilience in bleached corals." *Nature,* 440: 1186–1189.

Hoegh-Guldberg, O., and B. Salvat. 1995. "Periodic mass-bleaching and elevated sea temperatures: Bleaching of outer reef slope communities in Moorea, French Polynesia." *Marine Ecology Progress Series* 121: 181–190.

Hoegh-Guldberg, O., P. J. Mumby, A. J. Hooten, R. S. Steneck, P. Greenfield, E. Gomez, C. D. Harvell, et al. 2007. "Coral reefs under rapid climate change and ocean acidification." *Science* 318: 1737–1742.

Hueerkamp, C., P. W. Glynn, L. D'Croz, J. L. Maté, and S. B. Colley. 2001. "Bleaching and recovery of five eastern Pacific corals in an El Niño-related temperature experiment". *Bulletin of Marine Science*, 69: 215–236.

Hughes, T. P. 1994. "Catastrophes, phase shifts, and large-scale degradation of a Caribbean coral reef." *Science*, 265: 1547–1551.

IPCC, 2007. *The physical science basis. Contribution of working group 1 to the fourth assessment report of the Intergovernmental Panel on Climate Change,* edited by S. Solomon, D. Qin, M. Manning, Z. Chen, M. Marquis, K. B. Averyt, M. Tignor, and H. L. Miller. Cambridge: Cambridge University Press.

Jokiel, P. L., and E. K. Brown. 2004. "Global warming, regional trends and inshore environmental conditions influence coral bleaching in Hawaii." *Global Change Biology,* 10: 1627–1641.

Juínio-Meñez, M. A., H. G. Bangi, M. C. Malay, and D. Pastor. 2008. "Enhancing the recovery of depleted *Tripneustes gratilla* stocks through out culture and restocking." *Reviews in Fisheries Science*, 16: 35–43.

Kikuchi, R. K. P., Z. M. A. N. Leão, V. Testa, L. X. C. Dutra, and S. Spanó. 2003. "Rapid assessment of the Abrolhos Reefs, eastern Brazil (Part 1: Stony corals and algae)." In *Status of Coral Reefs in the Western Atlantic: Results of Initial Surveys, Atlantic and Gulf Rapid Reef Assessment (AGRRA) Program. Atoll Research Bulletin* 496, edited by J. C. Lang, 172–187.

Kleypas, J. A. 2007. "Constraints on predicting coral reef response to climate change." In *Geological Approaches to Coral Reef Ecology,* edited by R. B. Aronson, 386–424. New York: Springer.

Knowlton, N. 2001. "The future of coral reefs." *Proceedings of the National Academy of Sciences, USA* 98: 5419–5425.

Loya, Y., K. Sakai, K. Yamazato, Y. Nakano, H. Sambali, and R. van Woesik. 2001. "Coral bleaching: The winners and the losers." *Ecological Letters,* 4: 122–131.

LaJeunesse, T. C., W. K. W. Loh, R. van Woesik, O. Hoegh-Guldberg, G. W. Schmidt, and W. K. Fitt. 2003. "Low symbiont diversity in southern Great Barrier Reef corals, relative to those of the Caribbean." *Limnology and Oceanography* 48: 2046–2054.

Maté, J. L. 2003. "Corals and coral reefs of the Pacific coast of Panamá." In *Latin American Coral Reefs,* edited by J. Cortés, 387–417. Amsterdam: Elsevier.

McClanahan, T. R. 2000. "Bleaching damage and recovery potential of Maldivian coral reefs." *Marine Pollution Bulletin* 40: 587–597.

McClanahan, T. R., A. H. Baird, P. A. Marshall, and M. A. Toscano. 2004. "Comparing bleaching and mortality responses of hard corals between southern Kenya and the Great Barrier Reef, Australia." *Marine Pollution Bulletin* 48: 327–335.

McClanahan, T. R., M. Ateweberhan, J. Omukoto, and L. Pearson. 2009. "Recent seawater temperature histories, status, and predictions for Madagascar's coral reefs." *Marine Ecology Progress Series,* 380: 117–128.

McWilliams, J. P., I. M. Côté, J. A. Gill, W. J. Sutherland, and A. R. Watkinson. 2005. "Accelerating impacts of temperature-induced coral bleaching in the Caribbean." *Ecology,* 86: 2055–2060.

Porter, J. W. 1972. "Ecology and species diversity of coral reefs on opposite sides of the Isthmus of Panamá." In *The Panamic Biota: Some Observations Prior to a Sea-Level Canal,* edited by M. L. Jones, 89–116. *Bulletin of the Biological Society of Washington*, no. 2.

Precht, W. F., and R. B. Aronson. 2006. "Death and resurrection of Caribbean coral reefs: A paleoecological perspective." In *Coral Reef Conservation, Zoological Society of London,* edited by I. Côté and J. D. Reynolds, 40–77. Cambridge: Cambridge University Press.

Razak, T. B., and B. W. Hoeksema. 2003. "The hydrocoral genus *Millepora* (Hydrozoa: Capitata: Milleporidae) in Indonesia." *Zool. Verh. Leiden,* 345: 313–336.

Reaka-Kudla, M. L. 1997. "The global biodiversity of coral reefs: A comparison with rain forests." In *Biodiversity II: Understanding and Protecting Our Biological Resources,* edited by M. L. Reaka-Kudla, D. E. Wilson, and E. O. Wilson, 83–108. Washington, D.C.: Joseph Henry Press.

Riegl, B., and W. E. Piller. 2003. "Possible refugia for reefs in times of environmental stress." *International Journal of Earth Sciences,* 92: 520–531.

Riegl, B., S. J. Purkis, J. Keck, and G. P. Rowlands. 2008. "Monitored and modeled coral population dynamics and the refuge concept." *Marine Pollution Bulletin,* 58, 24–38.

Rogers, C. S., J. Miller, E. M. Muller, P. Edmunds, R. S. Nemeth, T. B. Smith, R. Boulon, et al. 2008. "Ecology of coral reefs in the US Virgin Islands." In

*Coral Reefs of the USA,* edited by B. M. Riegl and R. E. Dodge, 303–373. Berlin: Springer.

Rowan, R., N. Knowlton, A. Baker, and J. Jara. 1997. "Landscape ecology of algal symbionts creates variation in episodes of coral bleaching." *Nature,* 388: 265–269.

Schuhmacher, H. K., K. Loch, W. Loch, and W. R. See. 2005. "The aftermath of coral bleaching on a Maldivian reef—A quantitative study." *Facies*, 51: 80–92.

Small, A. M., W. H. Adey, and D. Spoon. 1998. "Are current estimates of coral reef biodiversity too low? The view through the window of a microcosm." *Atoll Research Bulletin,* 458: 1–20.

Spalding, M. D., and A. M. Grenfell. 1997. "New estimates of global and regional coral reef areas." *Coral Reefs,* 16: 225–230.

Vargas-Ángel, B., J. D. Thomas, and S. M. Hoke. 2003. "High-latitude *Acropora cervicornis* thickets off Fort Lauderdale, Florida, USA." *Coral Reefs,* 22: 465–473.

Vize, P. D. 2006. "Deepwater broadcast spawning by *Montastraea cavernosa, Montastraea franski* and *Diploria strigosa* at the Flower Garden Banks, Gulf of Mexico." *Coral Reefs,* 25: 169–171.

Wilkinson, C. R. 2008. *Status of Coral Reefs of the World: 2008*. Townsville, Australia: Global Coral Reef Monitoring Network and Reef and Rainforest Research Centre.

Williams, E. H., Jr., P. J. Bartels, and L. Bunkley-Williams. 1999. "Predicted disappearance of coral-reef ramparts: A direct result of major ecological disturbances." *Global Change Biology,* 5: 839–845.

Wilson, E. O. 1992. *The Diversity of Life*. New York: W. W. Norton & Company.

Zapata, F. A., and B. Vargas-Ángel. 2003. "Corals and coral reefs of the Pacific coast of Colombia." In *Latin American Coral Reefs,* edited by J. Cortés, 419–447. Amsterdam: Elsevier.

# Chapter 8

## *Extinction Risk at High Latitudes*

ERIC POST AND JEDEDIAH BRODIE

Of the abiotic changes associated with the current phase of warming occurring on Earth, the loss of sea ice, snow cover, and glaciers are among the most apparent, rapid, and potentially ecologically devastating. These physical effects make polar regions the most likely places to experience first extinctions due to climate change. Have extinctions already been recorded in high latitude species, or do population trends suggest that extinctions are imminent? This chapter answers these questions.

Since 1979, the annual extent of sea ice in the Arctic Ocean during winter and summer has diminished steadily (fig. 8-1). Although large sections of the Antarctic ice cap continue to melt into the ocean, the rate and trend in changes in Antarctic sea ice are less clear. Environmental changes throughout the Arctic resulting from climate change have coincided with a number of clear downward trends in populations of Arctic species (Post et al., 2009), while Antarctic species have exhibited a more mixed response, with some populations increasing and some populations declining, depending on species and region. For this reason, climate change appears to pose greater risks of extinction in the Arctic, and we focus mostly on Arctic species in this chapter.

Many species living at high latitudes are adapted to or dependent upon snow and ice cover either seasonally or permanently for

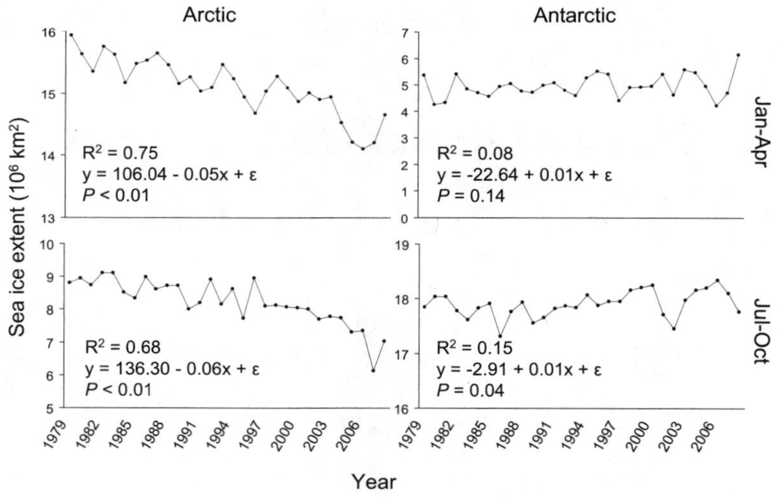


Sea ice extent>"N by season" & "S by season"

FIGURE 8-1. Trends and variability in Arctic and Antarctic sea ice extent during January to April and July to October, 1979–2008. Data derived from the National Snow and Ice Data Center, Boulder, CO, USA (http://nsidc.org/data/seaice_index/).

foraging, reproduction, and survival. Snowmelt and sea ice loss at high latitudes create, therefore, potential for "tipping points" in their rates of change—thresholds in the rate of melting approached steadily but leading to unexpectedly large and sudden further melting. Such dramatic and rapid environmental shifts may pose the greatest and most immediate threat to the persistence of snow- and ice-dependent and ice-associated species, and for that reason we focus here on reviewing and synthesizing information on the implications of climate change for such species. We briefly summarize the condition of a few Antarctic species that may be sensitive to climate change as well. Based on previous assessments and projections posed in them, in addition to syntheses of recent observations, we conclude that the species most at risk of further population reductions due to climate change at high latitudes are polar bears, walrus, narwhal, and ivory gulls. Each of these species is dependent upon, or highly affiliated with sea ice, and its rapid loss poses the greatest challenge to their persistence. Additionally, we rate spectacled eiders worthy of serious concern because their numbers have declined rapidly and steadily in recent years. Extinctions due to climate change have not yet occurred in polar systems, but the list of species at elevated risk of extinctions is growing.

## Arctic Species

Here we review the list of arctic vertebrates most susceptible to extinction from climate change, beginning with species most at risk.

### *Polar Bear*

Polar bears (*Ursus maritimus*) are classified as ice-dependent marine mammals because reproduction, most foraging activity, and most of the developmental period occur at sea, though largely on sea ice, with parturition taking place in snow caves on offshore ice sometime during midwinter. Successful hunting of their primary prey, seals, in particular ringed seals, is largely dependent on stable sea ice, and the diminishing annual extent of sea ice throughout the Arctic is a major threat to the persistence of polar bears. Human exploitation of polar bears has declined since the International Agreement on Polar Bear Conservation came into operation in 1976, but is still practiced by indigenous groups and accounts for the loss of approximately seven hundred bears annually.

Polar bears are probably the most widely studied Arctic or Antarctic species in relation to climate change. They have been identified as one of three Arctic marine mammal species (the others being walrus and narwhal) that are most at risk of suffering substantial habitat loss due to climate change (Laidre et al., 2008). The possible role of rapid warming in recently observed declines in their numbers and productivity motivated the upgrade of their conservation status under the US Endangered Species Act to threatened (Armstrong et al., 2008) in 2008. Nonetheless, polar bears remain listed as vulnerable by the International Union for Conservation of Nature (IUCN) (table 8-1). A survey of expert polar bear researchers produced estimates of the rate and magnitude of declines from current circumarctic population sizes of 30 percent (average response) to 70 percent (greatest projected decline) by the year 2050 (O'Neill et al., 2008).

Perhaps the most extensively studied polar bear population in the world is that inhabiting Canada's Hudson Bay. Body condition, birth

TABLE 8-1. Summary of current IUCN status and likely climate change vulnerability for select Arctic and Antarctic vertebrates.

| *Species* | *IUCN Population Trend* | *IUCN Status* | *Current IUCN Population Estimate* | *Climate Change Vulnerability* |
|---|---|---|---|---|
| Polar bear | Decreasing | Vulnerable | 20,000–25,000 | Sea ice loss, including thinning, reduction of stable old ice, and prolonging of the ice-free season; population fragmentation |
| Pacific walrus | Unknown-declining | Data deficient | Unknown | Habitat loss due to sea ice retreat |
| Narwhal | Unknown | Near-threatened | 80,000 | Earlier ice melt, increased access by human hunters |
| Ivory gull | Declining | Near-threatened | 15,000–25,000 | Sea ice loss, environmental toxins |
| Spectacled eider | Unknown | Least concern | 330,000–390,000 | Pack ice changes; benthic marine biome shift |
| Ringed seal | Unknown | Least concern | 2.5 million–7 million | Reduction of stable pack ice |
| Arctic fox | Stable | Least concern; locally endangered | "Several hundred thousand" | Expansion of red foxes; indirect negative effects of warming on rodent prey abundance |

| | | | | |
|---|---|---|---|---|
| Adelie penguin | Stable-unknown | Least concern | 4 million–5.2 million | Indirect: nest-site competition with southward-expanding chinstrap penguins; possible adverse consequences of changing sea ice conditions on prey availability |
| Chinstrap penguin | Increasing | Least concern | 8 million | Southward expansion, possibly in response to shifting prey distribution |
| Emperor penguin | Stable | Least concern | 270,000–350,000 | Sea ice loss effects on reproduction |
| King penguin | Possibly increasing | Least concern | 2 million | Reductions in winter survival and offspring production associated with effects of sea ice loss on prey availability |
| Crabeater seal | Unknown | Least concern | 7 million–11 million | Sea ice loss effects on predator avoidance and pup rearing |
| Leopard seal | Unknown | Least concern | 100,000–440,000 | Indirect; possible changes in prey distribution and abundance |

rates, and the proportion of yearling polar bears in the population in western Hudson Bay have declined significantly since the 1980s (Stirling et al., 1999). These changes are associated with earlier breakup of ice in the bay that, in turn, is related to a long-term warming trend in spring temperatures (Stirling et al., 1999).

Deteriorating sea ice conditions may also influence the demography and population persistence of polar bears indirectly through population fragmentation and reduction of gene flow among populations (Crompton et al., 2008). In small, declining populations, density dependence may assume a positive role in population dynamics. In this Allee effect, positive population growth requires a minimum population size. A recent attempt to model the role of Allee effects in polar bear population dynamics—parameterized using data from Lancaster Sound, Canada—concluded that a precipitous reduction in the numbers of adult male bears in the population could trigger a reproductive collapse (Molnar et al., 2008).

The population of polar bears on the High Arctic archipelago of Svalbard, Norway, recently exhibited an increase in the average age of males and females, suggesting a decline in recruitment to that population (Derocher, 2005). From 1984 to 2004, survival of polar bears in the youngest and oldest age classes in the western Hudson Bay population of Canada also declined in concert with earlier spring ice melt in the bay, contributing to a 22 percent decline in the population between 1984 and 2004 (Regehr et al., 2007).

The persistence and quality of winter pack ice is also important to the denning ecology of female polar bears with cubs. A study tracking the distribution of den sites used by satellite collared females with cubs onshore and on pack ice documented a shift between 1985 and 2004 in the proportion of dens in both types of habitat (Fischbach et al., 2007). In the earlier period, most dens were located on pack ice, whereas in the latter period, most dens were located onshore. This shift was attributed to deterioration in ice conditions in the Beaufort Sea and an increase in the length of the pack ice melt period.

### *Walrus*

There may be three subspecies of walruses (*Odobenus rosmarus*) distributed in disconnected subpopulations across the Arctic. Of these three, all of which inhabit open water and shelf ice, the Pacific walrus, which is found offshore in Alaskan and eastern Siberian waters, may be at greatest risk of suffering adverse consequences of climate change. Estimates of population size for the Pacific walrus are scarce and spotty, but estimates dating to the mid-1980s and early 1990s placed the subspecies at approximately 200,000. The Atlantic walrus subspecies may number as few as 20,000 based on an estimate in 2006.

Walruses are ice-associated, mating and giving birth on ice flows primarily over shallow seas. They also use ice flows for haul-outs and migrate with the seasonal distribution of sea ice (Pielou, 1994). Walruses have suffered heavy exploitation by humans throughout their distribution, having undergone dramatic population declines due to hunting for meat, skin, ivory, and blubber from the eighteenth through the twentieth centuries. They were driven to extirpation in eastern Canada in the 1800s and nearly so around Svalbard. Their current conservation status has not been determined by the IUCN because of data scarcity (table 8-1).

Studies of potential effects of climate change and sea ice loss on walrus populations have been few in comparison to the other Arctic species reviewed here. However, recent observations indicate that bottom-water temperatures are increasing in the Bering Sea in association with elevated near-surface air temperatures and diminishing sea ice, resulting in a transition from Arctic to subarctic benthic conditions less favorable to walruses (Grebmeier et al., 2006). In addition, retreating ice floes due to warming necessitate longer distances between foraging and resting habitats for walruses in the Bering Sea, where walrus pups may become separated from their mothers as a consequence (Grebmeier et al., 2006). An account in the popular media reported a massive abandonment by thousands of walruses of retreating offshore ice in Alaska in 2007 (Joling, 2007). The walruses apparently began retreating to inshore haul-outs after conditions on the pack ice off of Alaska's northern coast began deteriorating in July 2007. The remaining ice had retreated to areas unsuitable for walrus foraging because it was located over water deeper than the shallow coastal water walruses prefer while foraging on bottom-dwelling benthic organisms (Joling, 2007). The National Snow and Ice Data center reported that sea ice extent in the area in September of that year was 39 percent below the long-term average (Joling, 2007). Because walruses require stable ice flows to rest upon between foraging bouts, the loss of nearshore sea ice would present them with a substantial loss of optimal habitat.

The specialized habitat requirements and dietary preferences of walruses may leave them highly vulnerable to rapid climatic warming (Bluhm and Gradinger, 2008). Walruses forage within a very specific range of water depths and distances from underwater shelves and nearshore ice edges, and are projected to be at high risk of suffering habitat loss with warming (Rausch et al., 2007; Bluhm and Gradinger, 2008). Considering the evidence collected to date, together with the highly specialized foraging and habitat ecology of this species, we regard walruses as at risk of declining with future warming in the Arctic.

### *Narwhal*

Narwhals (*Monodon monoceros*) inhabit Baffin Bay, between the west coast of Greenland and the Canadian archipelago of Baffin Island and Ellesmere Island, and the North Atlantic and Arctic Ocean eastward from the east coast of Greenland. They forage at depths of more than 1,000 meters (3,281 feet), primarily for Greenland halibut, and migrate between open, shallow-water inlets during summer and deep, ice-covered water during winter (Heide-Jørgensen and Dietz, 1995). Narwhals are currently listed as near-threatened on the IUCN Red List, with the primary pressure on this species deriving from hunting by humans.

A recent assessment of climate change sensitivity for Arctic marine mammals categorized narwhals as the most specialized and range-limited of Arctic cetaceans, and ranked them among the most at risk of suffering negative impacts of Arctic warming (Laidre et al., 2008). The greatest threats to narwhals deriving from climate change may be related to human hunting pressure. Along the coast of High Arctic Greenland, narwhal harvest has increased steadily since 2002 (Nielsen, 2009). This increased catch is not attributed by local hunters to increased effort, but rather to increased access to narwhals during summer coincident with earlier sea ice retreat in June and July (Nielsen, 2009). Although habitat specificity and dietary specialization may be the primary factors influencing vulnerability of narwhals to climate change (Laidre et al., 2008), the interaction between sea ice loss and human access to migrating narwhals is likely to assume increased importance as the Arctic continues to warm.

### *Ivory Gull*

Ivory gulls (*Pagophila eburnean*) display a nearly circumarctic distribution during the breeding season, and are found in subarctic and north-temperate pelagic zones outside of the breeding season. The species is closely associated with sea- and pack ice during winter, from which it forages mainly for fish and invertebrates, though it may also scavenge. Ivory gulls use cliffs and outcroppings for nesting colonies, though recently a breeding colony was observed on floating pack ice off the coast of northeast Greenland (Boertmann et al., 2010).

Ivory gulls are currently listed on the IUCN Red List as near-threatened, and their Arctic-wide population is believed to be in decline (table 8-1). The most notable regional population decline for this species was reported following a 2002–2003 survey in Arctic Canada of colonies that had last been surveyed in the 1980s. The recent survey estimated an 80 percent decline in numbers of nesting ivory gulls compared to the survey 20 years earlier, with complete losses of some of the formerly largest breeding colonies (Gilchrist and Mallory, 2005). This decline may have been related to loss of pack ice habitat, although mercury poisoning has also been identified in declining populations of ivory gulls in the Canadian Arctic (Braune et al., 2006). However, a more recent survey in northeast Greenland identified twenty previously unknown ivory gull colonies, resulting in an upward revision of the population estimate for this species in Greenland from 1,800 to 4,000 (Gilg et al., 2009).

Given its comparatively small population size, declining population trend, and close association with sea- and pack ice at all times of the year, we consider the ivory gull to be at risk of suffering further losses under future warming in the Arctic. Of particularly noteworthy concern is the prospect of a synergistic interaction between habitat loss due to sea ice melt and continued or even elevated exposure to environmental toxins such as mercury.

### *Spectacled Eider*

Spectacled eiders (*Somateria fischeri*) are sea ducks that overwinter in large congregations among openings in pack ice primarily in the Bering Sea, between Alaska and Siberia, but also in the Chukchi and the Beaufort Seas. During winter, they forage for mussels and other benthic invertebrates along ice edges and under the ice shelf. They breed in large coastal deltas and floodplains, including the Yukon-Kuskokwim delta in subarctic Alaska and the North Slope coastal plain in Arctic Alaska (Petersen et al., 1999; Petersen and Douglas, 2004).

This species has attracted conservation concern because its numbers in the Yukon-Kuskokwim delta have declined precipitously since the 1950s (Hodges et al., 1996). Spectacled eiders were classified as threatened under the US Endangered Species Act as of 1993 (Lovvorn et al., 2009), although the IUCN Red List classifies it as of least concern because of its large worldwide population size and range. We have included mention of it here because of its association with pack ice leads during winter.

Currently, there is disagreement as to whether the sustained population decline among spectacled eiders breeding in the Yukon-Kuskokwim delta is related to changing environmental conditions and resource availability associated with climate change. One suggestion is that this decline is associated with a major abiotic regime shift in the Bering Sea in the mid-1970s, when an onset of warmer ocean temperatures may have precipitated a reduction in food availability for spectacled eiders, perhaps in concert with increased competition with northward-expanding fish and crabs (Lovvorn et al., 2009). As well, loss of sea ice may burden spectacled eiders with greater energy demands if such loss reduces the availability of suitable resting habitat on ice because the energetic demands of resting in open water are assumed to be higher than resting on ice (Petersen and Douglas, 2004). Alternatively, declines in the Yukon-Kuskokwim delta population may be related to lead poisoning resulting from ingestion of lead pellets used by hunters (Franson et al., 1995). One analysis has suggested that declines in this species may be related to reduction of open leads in the Bering Sea due to extreme winds and dense sea ice concentrations (Petersen and Douglas, 2004).

### *Ringed Seal*

Because of their close association year-round with shore-fast ice and sea ice for mating, pup rearing, and even haul-out resting, ringed seals (*Pusa hispida*) may be considered heavily ice-associated. In particular, they are dependent upon snow dens excavated on the surface of sea ice for care and provisioning of their offspring. Their primary predator is polar bears, and an important aspect of snow denning for ringed seals is avoidance of detection by polar bears and escape under the ice through holes.

Due to their dependence on sea ice quality, distribution, and persistence, ringed seals were identified as an important indicator species for monitoring effects of climate change on Arctic marine mammals (Tynan and DeMaster, 1997). Ice-associated seals, including the ringed seal, may be most at risk from further warming in the Siberian Arctic Sea, where sea ice declines have been more extensive than in the Beaufort Sea (Tynan and DeMaster, 1997). Climate simulation runs used to project warming scenarios into the next century predict an 83 percent reduction in sea ice in the Baltic Sea, with detrimental consequences for the southernmost populations of Baltic ringed seals (Meier et al., 2004).

### *Arctic Fox*

Arctic foxes (*Alopex lagopus*) are distributed in both mainland and coastal areas of the Arctic, as well as Arctic islands. Populations of this species appear to fluctuate in concert with the dynamics of small rodents where those occur, and the collapse of rodent cycles in Fennoscandia is believed to be a major contributing factor to the decline of Arctic fox populations there. As well, Arctic foxes use sea ice for dispersal among islands and coastal areas, and the decline in sea ice extent is believed to be a contributing factor to the isolation of some Arctic fox populations and consequent degradation of genetic variability within them. In some regions, Arctic foxes may be ice-associated, traversing pack ice with polar bears and scavenging their kills. Although their thermal neutral zones differ, Arctic and red foxes overlap in parts of the distribution of Arctic foxes due to northward expansion of red foxes. In zones of overlap, red foxes appear competitively and socially dominant over Arctic foxes, and may pose a threat of partial displacement to Arctic foxes with further northward expansion.

The most imperiled populations of the Arctic fox occur in Fennoscandia, where persecution by humans at the turn of the twentieth century severely reduced their numbers, and where they are currently listed as critically endangered. In contrast, the IUCN listing for Arctic foxes worldwide is least concern, due to stable worldwide population numbers and fairly high abundance (table 8-1). Comparison of genetic variation among Arctic foxes in Scandinavia before and after the severe population decline revealed a 25 percent loss of microsatellite alleles (Nystrom et al., 2006).

The failure of Arctic foxes to recover in Scandinavia over the past several decades despite restrictions on hunting and trapping, together with worsening population declines there since the 1980s, is apparently associated with a recent and persistent absence of peaks in abundance of lemmings, upon which Arctic foxes in Scandinavia are specialist predators (Elmhagen et al., 2000). Despite the fact that Arctic foxes prey upon or scavenge a variety of species, including passerine birds, Arctic hares, and reindeer, lemming remains dominate feces of Arctic foxes in Scandinavia, even during lemming lows, and occupancy rates of Arctic fox dens display a positive correlation with abundance of lemmings (Elmhagen et al., 2000).

Arctic fox abundance is also numerically linked to lemming abundance in northeast Greenland, where predation by Arctic foxes is a primary driver of lemming cycles (Gilg et al., 2003, 2006; Schmidt et al., 2008). Sensitivity modeling of Arctic fox dynamics has revealed that the length of the interval between seasons of successful reproduction, which is linked to lemming abundance, is among the most important factors influencing extinction risk in those populations (Loison et al., 2001).

## Antarctic Species

Adelie (*Pygoscelis adeliae*) and chinstrap (*P. antarcticus*) penguins both depend on sea ice, either for primary habitat or as the substrate on which the krill food chain depends. On Penguin Island, South Shetland, Antarctica, populations of both species have declined by 75 percent and 66 percent, respectively, since 1980. In a mixed breeding colony, however, chinstrap penguin breeding pairs increased at the expense of adelie pairs by 127 percent, indicating evidence of negative consequences for adelie penguins of interspecific competition for nesting sites (Sander et al., 2007). Long-term monitoring of emperor penguin colonies revealed declines in abundance in the mid-1970s and late 1980s associated with deteriorating sea ice conditions due to episodic warming (Ainley et al., 2005). An individual-based study of marked king penguins associated oceanic warming with declining reproductive success and adult survival (Le Bohec et al., 2008). Breeding success was negatively associated with warm phases of the El Niño Southern Oscillation, which presumably reduced availability of prey near the breeding colony. This species may be at risk of extinction with further warming: a model of population growth derived from this individual-based study indicates a 9 percent decline in adult survival with each 0.26 degree Celsius additional warming (Le Bohec et al., 2008). A recent analysis indicated, however, that the use of flipper bands to mark individual king penguins appears to have contributed to declines in both survival and reproduction, rendering the association between climate change and trends in these demographic parameters more tenuous than originally thought (Saraux et al., 2011).

Crabeater (*Lobodon carcinophaga*) and leopard (*Hydrurga leptonyx*) seals are other members of the Antarctic food web sensitive to climate change, but neither is facing extinction from climate change or other factors. Crabeater seals may be the most abundant large mammal on Earth; the most current population estimates for this species range from 7 million to 12 million. Because of a singular importance of sea ice extent and physical condition, including seasonal persistence, to reproduction, offspring rearing, and escape from predation, crabeater seals may be highly vulnerable to population declines with future warming and reductions in sea ice extent (Siniff et al., 2008). The leopard seal is a true top predator of Antarctic polar and subpolar seas. Unlike crabeater seals, there is little evidence of direct negative consequences of climatic warming for leopard seals. The primary impacts of climate change, if any, on leopard seals will likely derive from indirect effects on prey species such as crabeater seals and penguins.

## Conclusions

Extinctions do not occur in vacuums; species losses are likely to precipitate changes throughout ecosystems. This may be particularly true in the Arctic, where there is low functional redundancy among species (Post et al., 2009). Of the species highlighted in this chapter, we consider polar bears to be at greatest risk of suffering extinction due to climate change, and foresee the greatest potential ecosystem consequences arising from extinction of polar bears. Because ringed seals are the primary prey of polar bears (Thiemann et al., 2008), extinction of this top predator may have positive consequences for ringed seal abundance. Adult ringed seals prey primarily upon Arctic cod and polar cod (Holst et al., 2001), and increases in numbers of ringed seals may adversely affect cod abundance and possibly even cod fisheries in the Arctic and subarctic. This, in turn, may have broader consequences for marine food webs, including zooplankton dynamics and nutrient transport. Considering the variable forces converging on Arctic foxes, including persistently low and noncyclic rodent abundance and northward expansion of red foxes, in addition to human exploitation, they too may be vulnerable to extinction despite their current IUCN status. It is likely that extinction of Arctic foxes would have wider consequences within the tundra biome as well due to their important dual roles as scavengers and generalist predators.

These examples illustrate the ecosystem consequences of extinction risk in simple systems with low functional redundancy of species. High latitude systems, especially the Arctic, have not typically garnered much attention toward biodiversity conservation in comparison to species-rich systems with high rates of endemism such as the tropics (Post et al., 2009). However, the very fact that biodiversity is comparably low in high-latitude systems ought to raise awareness of the importance of species assemblages—and of the role of individual species—in ecosystem stability in these fragile biomes. As well, the simplicity of high-latitude systems may also render them more vulnerable to trophic decoupling if climate change affects species at one trophic level more so than species dependent upon them at adjacent trophic levels (Post and Forchhammer, 2008; Post et al., 2008). Polar bears, in particular, may be vulnerable to trophic mismatch as sea ice continues to melt out earlier while emergence of parturient female bears with cubs from winter dens remains constrained by developmental period. In this case female polar bears may be unable to reach retreating sea ice and the seals it offers access to in time to meet the energetic demands of rearing their cubs. Extinction risk from climate change is a very real prospect for the near future in high-latitude systems, and we urge attention not just toward conservation of those species most likely to suffer loss of habitat with rising temperatures, but also toward conservation and management of species linked to their success or demise.

## REFERENCES

Ainley, D. G., E. D. Clarke, K. Arrigo, W. R. Fraser, A. Kato, K. J. Barton, and P. R. Wilson. 2005. "Decadal-scale changes in the climate and biota of the Pacific sector of the Southern Ocean, 1950s to the 1990s." *Antarctic Science* 17: 171–182.

Armstrong, J. S., K. C. Green, and W. Soon. 2008. "Polar bear population forecasts: A public-policy forecasting audit." *Interfaces* 38: 382–395.

Bluhm, B. A., and R. Gradinger. 2008. "Regional variability in food availability for arctic marine mammals." *Ecological Applications* 18: S77–S96.

Boertmann, D., K. Olsen, and O. Gilg. 2010. "Ivory gulls breeding on ice." *Polar Record* 46: 86–88.

Braune, B. M., M. L. Mallory, and H. G. Gilchrist. 2006. "Elevated mercury levels in a declining population of ivory gulls in the Canadian Arctic." *Marine Pollution Bulletin* 52: 978–982.

Crompton, A. E., M. E. Obbard, S. D. Petersen, and P. J. Wilson. 2008. "Population genetic structure in polar bears (*Ursus maritimus*) from Hudson Bay, Canada: Implications of future climate change." *Biological Conservation* 141: 2528–2539.

Derocher, A. E. 2005. "Population ecology of polar bears at Svalbard, Norway." *Population Ecology* 47: 267–275.

Elmhagen, B., M. Tannerfeldt, P. Verucci, and A. Angerbjorn. 2000. "The arctic fox (*Alopex lagopus*): An opportunistic specialist." *Journal of Zoology* 251: 139–149.

Fischbach, A. S., S. C. Amstrup, and D. C. Douglas. 2007. "Landward and eastward shift of Alaskan polar bear denning associated with recent sea ice changes." *Polar Biology* 30: 1395–1405.

Franson, J. C., M. R. Petersen, C. U. Meteyer, and M. R. Smith. 1995. "Lead poisoning of spectacled eiders (*Somateria fischeri*) and of a common eider (*Somateria mollissima*) in Alaska." *Journal of Wildlife Diseases* 31: 268–271.

Gilchrist, H. G., and M. L. Mallory. 2005. "Declines in abundance and distribution of the ivory gull (*Pagophila eburnea*) in Arctic Canada." *Biological Conservation* 121: 303–309.

Gilg, O., I. Hanski, and B. Sittler. 2003. "Cyclic dynamics in a simple vertebrate predator-prey community." *Science* 302: 866–868.

Gilg, O., B. Sittler, B. Sabard, A. Hurstel, R. Sane, P. Delattre, and L. Hanski. 2006. "Functional and numerical responses of four lemming predators in high arctic Greenland." *Oikos* 113: 193–216.

Gilg, O., D. Boertmann, F. Merkel, A. Aebischer, and B. Sabard. 2009. "Status of the endangered ivory gull, *Pagophila eburnea,* in Greenland." *Polar Biology* 32: 1275–1286.

Grebmeier, J. M., J. E. Overland, S. E. Moore, E. V. Farley, E. C. Carmack, L. W. Cooper, K. E. Frey, et al. 2006. "A major ecosystem shift in the northern Bering Sea." *Science* 311: 1461–1464.

Heide-Jørgensen, M. P., and R. Dietz. 1995. "Some characteristics of Narwhal, *Monodon monoceros,* diving behaviour in Baffin Bay." *Canadian Journal of Zoology—Revue Canadienne De Zoologie* 73: 2120–2132.

Hodges, J. I., J. G. King, B. Conant, and H. A. Hanson. 1996. "Aerial surveys of waterbirds in Alaska 1957–94: Population trends and observer variability." USDI National Biological Service Information and Technology Report 4.

Holst, M., I. Stirling, and K. A. Hobson. 2001. "Diet of ringed seals (*Phoca hispida*) on the east and west sides of the North Water Polynya, northern Baffin Bay." *Marine Mammal Science* 17: 888–908.

Joling, D. 2007. "Thousands of walruses abandon ice for Alaska shore." *USA Today,* Anchorage.

Laidre, K. L., I. Stirling, L. F. Lowry, O. Wiig, M. P. Heide-Jorgensen, and S. H. Ferguson. 2008. "Quantifying the sensitivity of arctic marine mammals to climate-induced habitat change." *Ecological Applications* 18: S97-S125.

Le Bohec, C., J. M. Durant, M. Gauthier-Clerc, N. C. Stenseth, Y. H. Park, R. Pradel, D. Gremillet, J. P. Gendner, and Y. Le Maho. 2008. "King penguin population threatened by Southern Ocean warming." *Proceedings of the National Academy of Sciences, USA* 105: 2493–2497.

Loison, A., O. Strand, and J. D. C. Linnell. 2001. "Effect of temporal variation in reproduction on models of population viability: A case study for remnant arctic fox (*Alopex lagopus*) populations in Scandinavia." *Biological Conservation* 97: 347–359.

Lovvorn, J. R., J. M. Grebmeier, L. W. Cooper, J. K. Bump, and S. E. Richman. 2009. "Modeling marine protected areas for threatened eiders in a climatically changing Bering Sea." *Ecological Applications* 19: 1596–1613.

Meier, H. E. M., R. Doscher, and A. Halkka. 2004. "Simulated distributions of Baltic Sea-ice in warming climate and consequences for the winter habitat of the Baltic ringed seal." *Ambio* 33: 249–256.

Molnar, P. K., A. E. Derocher, M. A. Lewis, and M. K. Taylor. 2008. "Modelling the mating system of polar bears: A mechanistic approach to the Allee effect." *Proceedings of the Royal Society B: Biological Sciences* 275: 217–226.

Nielsen, M. R. 2009. "Is climate change causing the increasing narwhal (*Monodon monoceros*) catches in Smith Sound, Greenland?" *Polar Research* 28: 238–245.

Nystrom, V., A. Angerbjorn, and L. Dalen. 2006. "Genetic consequences of a demographic bottleneck in the Scandinavian arctic fox." *Oikos* 114: 84–94.

O'Neill, S. J., T. J. Osborn, M. Hulme, I. Lorenzoni, and A. R. Watkinson. 2008. "Using expert knowledge to assess uncertainties in future polar bear populations under climate change." *Journal of Applied Ecology* 45: 1649–1659.

Petersen, M. R., and D. C. Douglas. 2004. "Winter ecology of spectacled eiders: Environmental characteristics and population change." *Condor* 106: 79–94.

Petersen, M. R., W. W. Larned, and D. C. Douglas. 1999. "At-sea distribution of spectacled eiders: A 120-year-old mystery resolved." *Auk* 116: 1009–1020.

Pielou, E. C. 1994. *A naturalist's guide to the Arctic*. Chicago: University of Chicago Press.

Post, E., and M. C. Forchhammer. 2008. "Climate change reduces reproductive success of an arctic herbivore through trophic mismatch." *Philosophical Transactions of the Royal Society of London, Series B* 363: 2369–2375.

Post, E., C. Pedersen, C. C. Wilmers, and M. C. Forchhammer. 2008. "Warming, plant phenology, and the spatial dimension of trophic mismatch for large herbivores." *Proceedings of the Royal Society B: Biological Sciences* 275: 2005–2013.

Post, E., M. C. Forchhammer, M. S. Bret-Harte, T. V. Callaghan, T. R. Christensen, B. Elberling, A. D. Fox, et al. 2009. "Ecological dynamics across the Arctic associated with recent climate change." *Science* 325: 1355.

Rausch, R. L., J. C. George, and H. K. Brower. 2007. "Effect of climatic warming on the pacific walrus, and potential modification of its helminth fauna." *Journal of Parasitology* 93: 1247–1251.

Regehr, E. V., N. J. Lunn, S. C. Amstrup, and L. Stirling. 2007. "Effects of earlier sea ice breakup on survival and population size of polar bears in western Hudson Bay." *Journal of Wildlife Management* 71: 2673–2683.

Sander, M., T. C. Balbao, E. S. Costa, C. R. dos Santos, and M. V. Petry. 2007. "Decline of the breeding population of *Pygoscelis antarctica* and *Pygoscelis adeliae* on Penguin Island, South Shetland, Antarctica." *Polar Biology* 30: 651–654.

Saraux, C., C. Le Bohec, J. M. Durant, V. A. Viblanc, M. Gauthier-Clerc, D. Beaune, Y. H. Park, N. G. Yoccoz, N. C. Stenseth, and Y. Le Maho. 2011. "Reliability of flipper-banded penguins as indicators of climate change." *Nature* 469: 203–206.

Schmidt, N. M., T. B. Berg, M. C. Forchhammer, D. K. Hendrichsen, L. A. Kyhn, H. Meltofte, and T. T. Høye. 2008. "Vertebrate predator-prey interactions in a seasonal environment." In *Advances in Ecological Research*, Vol. 40, 345–370. San Diego: Elsevier Academic Press Inc.

Siniff, D. B., R. A. Garrott, J. J. Rotella, W. R. Fraser, and D. G. Ainley. 2008. "Projecting the effects of environmental change on Antarctic seals." *Antarctic Sciences* 20: 425–435.

Stirling, I., N. J. Lunn, and J. Iacozza. 1999. "Long-term trends in the population ecology of polar bears in western Hudson Bay in relation to climatic change." *Arctic* 52: 294–306.

Thiemann, G. W., S. J. Iverson, and I. Stirling. 2008. "Polar bear diets and arctic marine food webs: Insights from fatty acid analysis." *Ecological Monographs* 78: 591–613.

Tynan, C. T., and D. P. DeMaster. 1997. "Observations and predictions of Arctic climatic change: Potential effects on marine mammals." *Arctic* 50: 308–322.

# PART IV

# Evidence from the Past

Our attention now turns to the past. What does the paleoecological record tell us about extinctions and climate change? We examine the answers to this question in time frames spanning from hundreds of millions of years to tens of thousands of years. In the first chapter of this section, Peter Mayhew looks at extinctions in deep time, from records 50 million years old and older. William Clyde and Rebecca LeCain then look at extinctions over the past 60 million years, focusing on two particularly instructive events. Barry Brook and Anthony Barnosky begin an exploration of the more recent paleoecological record, looking at Pleistocene vertebrate extinctions, and Mark Bush and Nicole Mosblech complete the section examining plant extinctions in the last few tens of thousands of years.

The picture that emerges from this paleoecological tour is of some clear associations between climate change and extinctions, but also some major climatic changes that are not associated with extinctions. The data limitations associated with events thousands or millions of years ago are evident. The deep time record comes mostly from marine fossils, leaving us to infer that events sufficient to wipe out a major proportion of marine life probably also had significant effects on land. As we move into the era of modern plants and into the Pleistocene, the record is clearer but still seldom definitive.

Most notably, recent extinctions have occurred during times of both climatic change and human impact, a possible harbinger for the future. The answers emerging from this section suggest that extinctions may be associated with major state changes in climate on the planet, but that subsequent change, even coming and going from ice ages, may have little subsequent effect. The past is a rich source of information about possible future effects, but there are no perfect past analogs to future human-caused climate change. The past offers many pieces of the puzzle, a valuable supplement to other lines of evidence.

Chapter 9

# *Extinctions in Deep Time*

PETER J. MAYHEW

Deep time is geologic time, extending to the origin of the planet. For biologists in search of an understanding of extinction, the relevant portion of deep time is that in which life has existed on the planet—about the last 4 billion years (Cowen, 2000). Extinctions are first recorded when the fossil record is robust enough to offer insights into the arrival and disappearance of groups of organisms (Benton and Harper, 2009). Extinctions in deep time can therefore be identified only over about the last 600 million years, an interval of time dominated by the Phanerozoic eon (540 million years ago to present).

This chapter will treat extinction processes, primarily in marine environments, over the length of the Phanerozoic, concentrating on events prior to 50 million years ago. The following chapter (chapter 10, this volume) picks up the story from there. The deep time record reflects multiple major extinction events. I review the evidence of their relationship to climate and discuss possible insights they provide into extinction risk from future, human-driven climate change.

## Nature of the Record

Life first flourished in the oceans, and marine fossils predate terrestrial fossils by hundreds of millions of years. Because of this, the longest data sets in the fossil record are for marine, rather than terrestrial, communities. There are other reasons why studies of biodiversity through deep time tend to be marine: preservation conditions are much more favorable in marine environments, and many marine taxa have hard shelly components that fossilize well. For these reasons, by far the most highly worked long-term data set was previously Sepkoski's genus-level synoptic compendium of marine animals and protists (Sepkoski, 2002). Terrestrial fossils are available in more recent records, and are less often included in analyses of extinctions in deep time. However, the terrestrial record has not been ignored. Terrestrial environments contain the majority of described species today, so they potentially tell us more about the totality of extinction effects, and the majority of conservation work today is on terrestrial species, hence this is where a lot of interest lies. In addition, terrestrial taxa may have undergone diversification and extinction processes that are very different from those in the marine environment (Benton, 1997; Kalmar and Currie, 2010). Benton's (1993) family-level data set is the most explored global compendium that includes terrestrial taxa.

Because only a small proportion of species have left fossilizable remains and only a small fraction of those have been discovered, the record is, obviously, more complete for higher taxonomic groups than for the individual species comprising those groups. Often it is also difficult to resolve fossil specimens to the species level, so extinctions are generally tracked at the genus or family level instead. However, this has not prevented scientists from drawing important conclusions about extinction, since genera and families go extinct as well, and it is certain that when a genus or family disappears, all species within it have become extinct.

There are other sources of bias in the record, and correcting for these may be important in interpreting the record of deep time extinctions. Possibly the most significant problem is the enormous variation across geological stages in the quantity of suitable rock records in which fossil remains can be discovered (Peters, 2005). For studies spanning hundreds of millions of years this may pose difficulties, because samples will be unstandardized, with some stages being more intensively sampled than others. Whether an observed change in taxonomic richness reflects underlying biological processes, or whether it simply reflects underlying bias in preservation or discovery rate is therefore an ever-present question. Techniques and resources to overcome these biases in large-scale studies are the subject of active research, and motivated the construction of the Paleobiology Database (PBDB) (Alroy et al., 2008). PBDB contains information on the occurrence, and sometimes abundance, of taxa in individual fossil beds, which allows researchers to implement standardized subsampling techniques to quantify richness, something that is not possible in the Sepkoski (2002) and Benton (1993) compendia, which document only first and last occurrences. Although PBDB is not taxonomically restricted in its scope, its utility is greatest for marine animals, where sample sizes are greatest. Although standardized samples are in general some improvement over unstandardized samples, the samples still contain only a small fraction of the true taxonomic richness.

Numerous extrinsic variables, including climate change, may explain variation in extinction rates, and it is helpful if variation in these variables over time can be estimated. For climate change, relevant variables include temperature, precipitation and related variables, as well as atmospheric carbon dioxide levels, which exert important direct effects on organisms in both terrestrial and marine environments. Temperature records for deep time come from proxies, such as $d^{18}O$ (Veizer et al., 2000). The uptake of $^{18}O$ is affected by the temperature of seawater. Durable remnants of marine organisms, such as coral skeletons, can be assayed for $^{18}O$. A one part per thousand change in $d^{18}O$ corresponds roughly to a temperature change of 1.5–2.0 degrees Celsius. Based on this temperature relationship, past seawater temperature and changes in seawater temperature can be inferred (e.g., Royer et al., 2004). Records for atmospheric carbon dioxide concentrations are also derived from isotopic proxies: the isotopic content of organic carbon is sensitive to atmospheric carbon dioxide concentrations, while weathering and degassing, both components of the long-term carbon cycle, are associated with extreme $^{87}Sr/^{86}Sr$ ratios (Rothman, 2002). However, explicit carbon dioxide concentration estimates are also derived from modeling changes in the long-term carbon cycle itself based on a wider variety of informative inputs (e.g., Berner and Kothavala, 2001). Changes in sea level can be inferred from documenting the extent and depth of marine sediments on continents (Hallam, 1984). Other potential environmental influences on biodiversity and their evidence are discussed below.

The evidence linking deep time events to climate is controversial, for a number of reasons: chronologies and mechanisms are obscured by an incomplete and biased geologic record; there are no experimental controls or alternative treatments for comparison; deciphering causation and mechanism from observed correlations is difficult; there are often many alternative candidate explanations. The remainder of this chapter first explores the variation in taxonomic richness and extinction across geological stages, and possible causes of this variation. Finally the evidence that climate change has played a role in generating some of that variation is examined.

## Variation in Extinction Rates

A literal reading of the compendia databases of fossil taxa over time suggests that there has been a substantial increase in taxonomic richness over the last 600 Ma on both land and sea, in which origination rates have exceeded extinction rates (Benton, 1995, 1997) (fig. 9-1).

Attempts to control for unequal sampling have somewhat mollified the increase in the marine realm in the Cenozoic (Alroy et al., 2008, but see Alroy, 2010), but less in terrestrial taxa (Kalmar and Currie, 2010, but see Davis et al., 2010). The general trends in richness are, however, punctuated by drops in which extinction rates temporarily rose to exceed origination. Per geological stage, extinction rate is approximately lognormally distributed (meaning the log of extinction rate follows a normal distribution) (Alroy, 2008). This means that extinction rates are characterized by a mean log value, but they are occasionally very high, and also occasionally very low (fig. 9-2). In addition to this variation around the mean log rate, there are apparent trends through time; extinction rates, as well as origination rates, may have decreased through time (Benton, 1995; Alroy, 2008) (fig. 9-1).

The term "mass extinction" has come to be used to describe those stages where extinction rates are unusually high given the rates in neighboring stages. In Raup and Sepkoski's (1982) original analysis, mass extinctions were those outside the 95% confidence interval of the linear regression of extinction rate through time, and hence identified the iconic "Big Five" mass extinctions in the marine realm: the end-Ordovician, late-Devonian, end-Permian, end-Triassic, and end-Cretaceous. The end-Permian and end-Cretaceous extinctions resulted in losses of 81 percent and 53 percent of marine animal genera (Alroy, 2008), and hence probably much larger proportions of species. Despite becoming popular wisdom, the special designation accorded to the rates in these geological stages is not very well justified statistically.

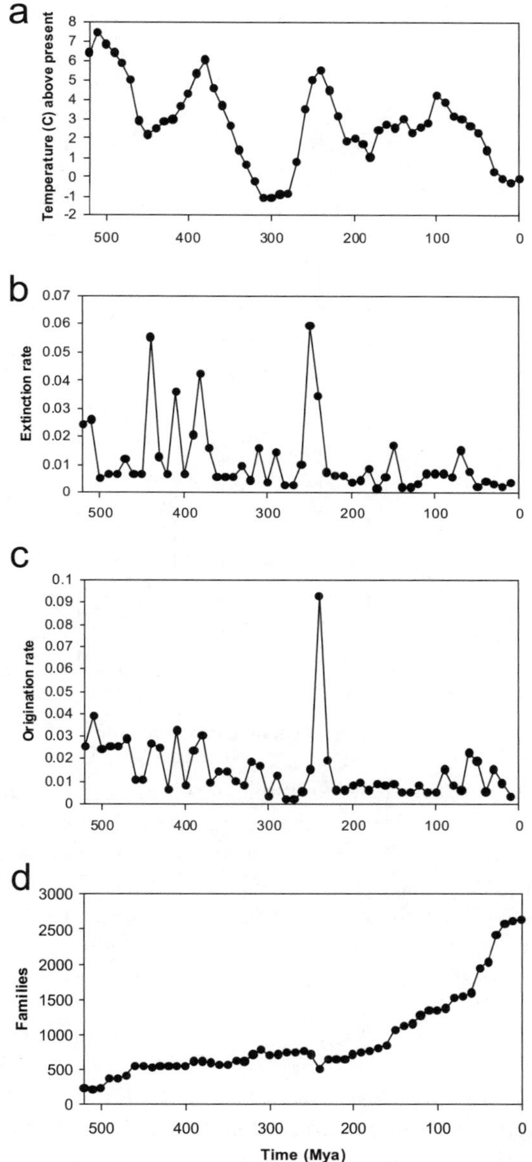


FIGURE 9-1. Time series (10 Ma intervals) of (a) estimated low-latitude sea surface temperature from Royer et al. (2004); (b) per capita extinction rate ($Ma^{-1}$) (c) per capita origination rate ($Ma^{-1}$); and (d) standing diversity of all families in Benton (1993) using the maximum dating assumption (Benton, 1995). Four mass extinctions indicated in (b) are the end-Ordovician (440Ma, O), late-Devonian (380Ma, D), end-Permian (251Ma, P), and end-Cretaceous (65Ma, C). Reprinted from Mayhew (2011) with kind permission from Cambridge University Press.

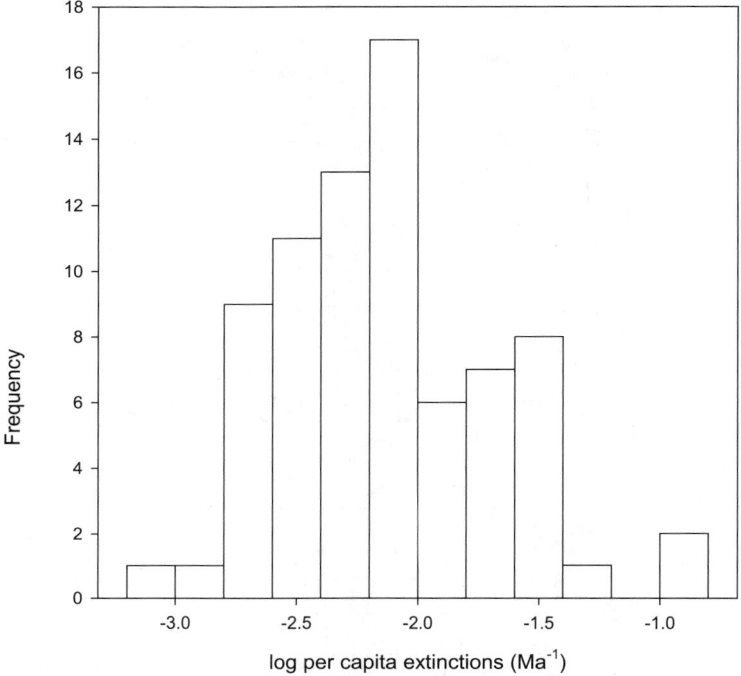


FIGURE 9-2. The approximately lognormal distribution of extinction rates in the fossil record. Data are for all families from Benton (1993) using the maximum dating assumption (Benton, 1995), with extinction quantified as Foote's (2000) per capita rate, q. Data points are geological stages. The three stages on the right-hand tail correspond to three of the iconic "Big Five" mass extinctions (from right, Tatarian (end-Permian), Rhaetian (end-Triassic), and Ashgillian (end-Ordovician). Note, however, that there are then eight stages in the next grouping to the left.

The rates are not unexpected once the data are logged, the rates do not all remain high when accounting for standing richness as well as interval duration, and they are not all repeated in both land and sea (Benton, 1995; Alroy, 2008). There is no bimodal distribution of rate intensities implied by the terms "background" and "mass" extinctions, which are now in widespread scientific, as well as popular, use, making it arbitrary as to whether to refer to a Big Five or some other number of mass extinctions. Despite this, there is evidence that the intensity of extinction may determine its selectivity (e.g., Payne and Finnegan, 2007), providing some justification for retaining the terms.

## Possible Causes

Variation in extinction rate may be caused by both intrinsic (e.g., biotic) and extrinsic (e.g., abiotic) factors—a distinction characterized by the "Red Queen" as opposed to the "Court Jester" (Benton, 2009) or "Ace of Spades" (Mayhew, 2011) paradigms. The Red Queen paradigm posits that extinction derives from biotic causes, and is ever-present. In contrast, the Court Jester and Ace of Spades posit that extinction is largely from abiotic causes, unpredictably in the case of the Court Jester (e.g., from impacts), and predictably in the case of the Ace of Spades (e.g., from climate). The fossil record suggests that both biotic and abiotic forces have affected extinction rates in deep time. There is evidence from a number of studies for density-dependent style processes. For example, in PBDB, drops in taxonomic richness tend to lead to lower extinction rates, while high extinction rates tend to lead to subsequent high origination rates (Alroy, 2008). Similar trends, though differing in detail, have been found in other data sets (Kirchner and Weil, 2000), and suggest that richness is in part a dynamic equilibrium, although one in which the equilibrium itself evolves through time, perhaps dramatically in terrestrial environments. These relationships are strongly suggestive of a role for biotic interactions such as competition.

Of the abiotic causes, both terrestrial (emanating from the earth) and extraterrestrial factors have been implicated. Terrestrial causes include sea level variation, volcanism, continental drift, atmospheric composition, and global temperature change. Extraterrestrial causes include bolide impacts and cosmic ray flux. These variables may of course be dependent on one another and result in interactive effects.

The area of marine transgressions over the continents is positively correlated with marine taxonomic diversity in the Sepkoski compendium (Purdy, 2008). More than one proximate mechanism may be responsible for this association; it may reflect variation in habitable areas, thereby affecting extinction and origination rates. Alternatively it may reflect changes in the quantity of sedimentary deposits available for study. Both are likely (Peters, 2006, 2008), the former supported by a study from PBDB (Alroy, 2010). The evidence that large igneous provinces can lead to high extinction rates is now extensive, and especially comes from the Permian-Triassic boundary, when the Siberian Traps large igneous province is known to have been active, leading to

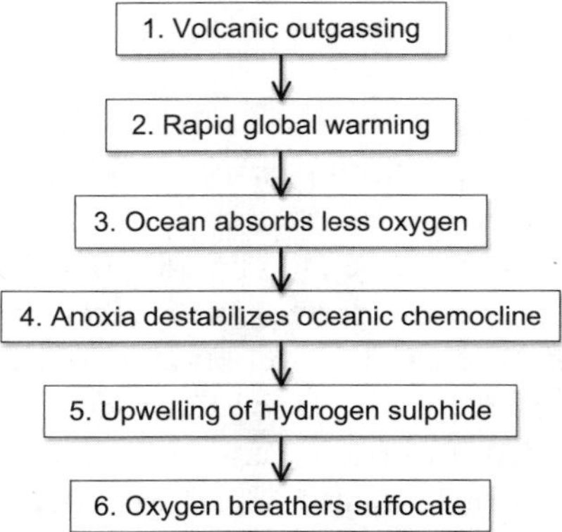


FIGURE 9-3. Scenario for the onset of the end-Permian marine crisis, based on information in Ward (2006).

rapid global warming through carbon dioxide outgassing (Wignall, 2001, 2005). Sedimentary evidence suggests strongly that this coincided with a marine crisis, or euxinium, in which the ocean surface became largely anoxic (Meyer and Kump, 2008) (fig. 9-3). A more general role for large igneous provinces, and such marine euxinia, has been suggested, but no statistical studies have been done to corroborate this (Wignall, 2005; Ward, 2006).

Volcanism is associated with plate tectonics, which is also the cause of continental drift. Over the last 500 million years, the continents first coalesced into the supercontinent of Pangea, and then subsequently split apart. The distribution of continents is important because most described species on Earth today live in shallow epicontinental seas or on land. Because of the well-supported ecological species-area relationships (Gaston and Blackburn, 2000), as well as evidence from biogeography that vicariance is an important force in adaptive radiation (McCarthy 2009), paleobiologists have long speculated that it could be one of the regulating forces of global taxonomic richness (Valentine and Moores, 1970), though its effects are difficult to quantify with certainty. Certainly continental drift has also had important indirect effects through changing sea level, changing the latitudinal distribution of continents and shallow seas—hence their local climates—and because the degree of continentality on land affects local climate.

Atmospheric composition has mainly been investigated with respect to carbon dioxide concentrations. There is a positive association between estimated carbon dioxide concentrations and extinction rates in the Sepkoski compendium and in the Benton compendium (Cornette et al., 2002; Mayhew et al., 2008). Carbon dioxide might exert an effect on extinction independent of other environmental variables (such as through oceanic acidification), or merely be a reflection of those other variables (such as global climate variation).

Though biologists, even paleobiologists, may in the past have treated extraterrestrial causes of extinction with derision, this was dramatically reversed with the discovery of worldwide iridium deposits at the Cretaceous-Paleogene boundary (Alvarez et al., 1980). It is now well established that a bolide impact at this time in Mexico is likely to have had far-reaching effects on ecosystems (Schulte et al., 2010), though whether the impact was the sole cause of extinction has been debated because of the coincident Deccan Traps large igneous province (Keller, 2001). More generally the evidence that impacts are responsible for high extinction is weak, but one study has suggested that they might interact with volcanism such that geological intervals with both large igneous provinces and impacts experience differentially high extinction (Arens and West, 2008).

Another extraterrestrial factor implicated in extinction is the variation in cosmic ray flux reaching Earth. Unlike bolide impacts, cosmic rays are sourced from outside the solar system, and show estimated cycles of 62 million years and 140 million years that correlate with fluctuations in fossil data (Rohde and Múller, 2005; Medvedev and Melott, 2007; Melott, 2008). Over hundreds of millions of years, the solar system has rotated around the galactic center, in and out of the plane of the galaxy and in and out of galactic spiral arms. Changes in cosmic neighborhood are quite capable of affecting life on a planet (Ross, 2009). One possible, though controversial, mechanism through which cosmic ray flux might affect life is through global climate (Shaviv and Veizer, 2003), to which we now turn.

## Global Climate and Its Effects on Extinction

Over the Phanerozoic, a variety of proxies suggest that global climate has fluctuated between relatively warm greenhouse modes in which evidence of high-latitude glaciation is absent, and relatively cool icehouse modes in which it was present (Frakes et al., 1992). The cycle length of greenhouse peak to greenhouse peak is approximately 140 million years. The greenhouse modes are respectively Cambrian to late-Ordovician, early Silurian to early Carboniferous, late-Permian to mid-Jurassic, and early Cretaceous to early Eocene. The icehouse phases are sandwiched in between: late-Ordovician to early Silurian, early Carboniferous to late-Permian, late-Jurassic to early Cretaceous, and early Eocene to present (fig. 9-1). It is likely that there was shorter term variation around these general trends, but these fluctuations are generally known with less confidence. For this reason, deep time studies can say much more about the effect of warm or cool temperatures on extinction than about the rate of change of climate. Recent models also suggest a general long-term cooling trend around which these modes are superimposed (Royer et al., 2004). What effect, if any, has this variation had on extinction rates?

Both the Sepkoksi and Benton compendia show associations, after detrending, with estimated long-term sea-surface temperatures, generally representing the transition from greenhouse to icehouse modes (Mayhew et al., 2008). Under greenhouse modes, or high sea-surface temperatures, extinction rates are mostly higher than average; under icehouse modes they may be low or high, leading to a general positive association between temperature and extinction rate, with moderate explanatory power (fig. 9-4). Because high atmospheric carbon dioxide concentrations are associated with high global temperature (Royer

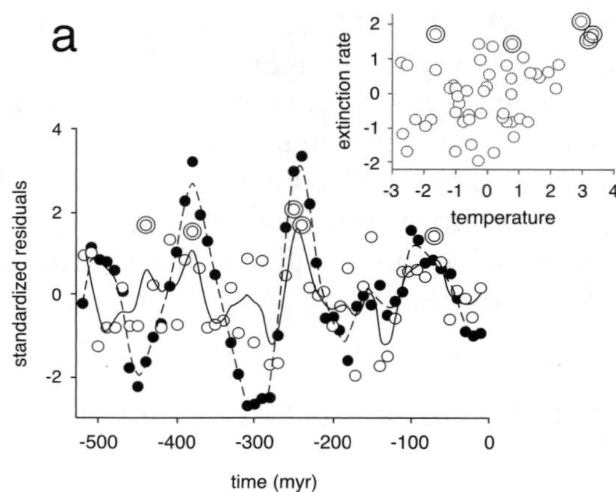


FIGURE 9-4. Associations between the time series in figure 9-1. In each case the raw data sets have been transformed to stabilize variance, then detrended, and then mean-standardized for plotting on the same scales. Inserts show the correlations between the plots. Closed circles are temperature, and open circles are extinction rate (A), origination rate (B), and standing diversity of families (C). Double

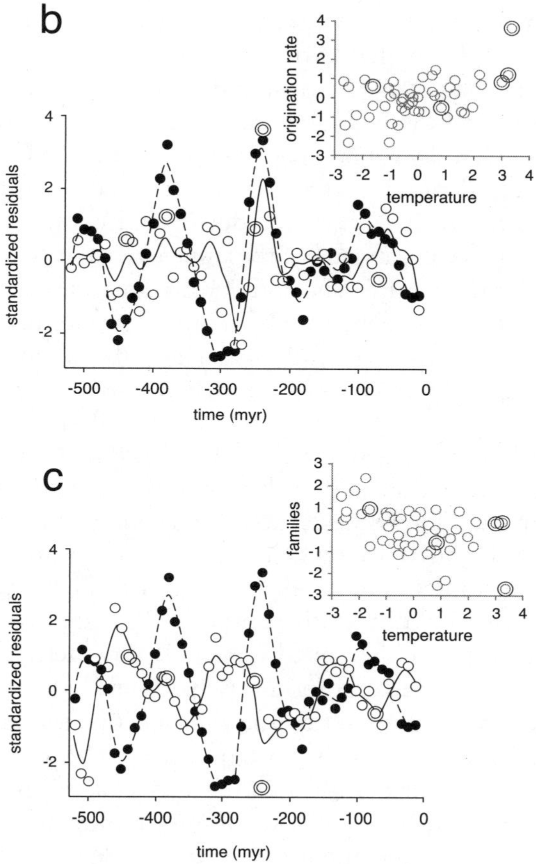


open circles indicate the intervals with the five largest extinction rate residuals, corresponding to well-known mass extinction events, which from left to right in the time series are end-Ordovician, late-Devonian, end-Permian (twice), end-Cretaceous. Lines are fitted curves using 25 degrees of freedom splines. Reprinted from Mayhew et al. (2008) with kind permission of the Royal Society of London and Blackwell Scientific.

et al., 2004), this finding is consistent with associations between taxonomic richness or extinction rates and carbon dioxide found previously (Rothman, 2001; Cornette et al., 2002). The association essentially derives from a rough and weak cycle of approximately 140 million years in fossil richness and rates, matching that of temperature, and, interestingly, that cycle appears to be preserved in PBDB (Melott, 2008). Coincident with this positive association is a negative, lagged association with richness and a positive, lagged association with origination rates, implying that extinction changes first and then taxonomic richness and origination respond later.

Extinction rates can be high with any kind of global temperature; specifically, the late-Devonian, end-Permian, and end-Cretaceous extinctions occurred close to greenhouse temperature peaks while the end-Ordovician extinction occurred during an icehouse trough. Fine-scale analyses of events at some of these stages also imply a causal relationship with temperature. The end-Permian marine crisis is likely to have been considerably enhanced by high sea-surface temperatures at this time, increasing the frequency of anoxia at the ocean surface (Benton, 2005) (fig. 9-3). In contrast, in the end-Ordovician, continental landmasses crossed the pole, triggering a rapid glaciation. Extinction has been attributed to a combination of rapid climate change, fluctuations in ocean circulation, and sea level regression (Sheehan, 2001).

Although the long-term macroevolutionary patterns outlined above cannot be used to predict future short-term ecological responses, conducting a thought experiment of this type can, in principle, be instructive for comparative purposes. Predictions can be generated by comparing the detrended (residual) short-term deviations in temperature with those in taxonomic richness and extinction rate. The slopes of these residuals describe the rise in extinction rate or loss of taxa for a 1-degree Celsius rise in temperature. I apply these slopes to the Eocene, when taxonomic richness was relatively high and background extinction rates were low, as was typical for more recent geological epochs. Using these methods, a 4-degree Celsius rise in temperature translates approximately to a 700 percent increase in family-level extinction rate in the Eocene but only to a 5 percent reduction in families globally. For marine genera a 4-degree Celsius rise in temperature translates to a 300 percent increase in extinction rate and 5 percent loss of genera. Note that uncertainty is very large because none of the relationships is very tight and that effects might be underestimated for many reasons (e.g., short-term [less than 10 million years] variation is not accounted for, origination mollifies drops in richness).

Although global analyses tend toward the opinion that high global temperatures enhance extinction rates over long time scales and reduce richness, individual taxonomic groups sometimes show the opposite relationship at least over shorter time scales. For example, Dasycladalean algal richness has fluctuated generally in line with global temperature change over the Phanerozoic (Aguirre and Ridding, 2005), and Jaramillo et al. (2006) found a similar pattern with Tertiary neotropical plant richness. Clearly, many organisms favor warm environments, and in the absence of high extinction are expected to prosper (see Woodward and Kelly, 2008).

## Conclusions

The precise scale of the current-future biodiversity crisis is uncertain but probably falls somewhere within the higher range of those recorded in the history of life (e.g., May et al., 1995; Thomas et al., 2004; Wake and Vrendenburg, 2008). Extinctions of a similar intensity have mostly occurred in deep time, hence they can inform on the likely consequences of such loss. Furthermore, over deep time there has been considerable change in the environmental variables that affect the fate of species today and in the near future, hence deep time can inform us on how these are likely to affect future extinction rates. Some of these variables, such as atmospheric carbon dioxide concentrations, are expected to change so much on a global scale that the only empirical precedents come from deep time. Deep time studies currently suggest that over the history of life, greenhouse climates are not generally favorable for the persistence of taxa, and suggest that high global temperatures can lead to ecological and geochemical processes that can cause mass extinction. Furthermore, large-scale biodiversity loss is unlikely to be replenished in the lifetime of our species (Alroy, 2008). These are sobering messages given the pace and direction of current environmental change. The richness of the environmental and biological data in the fossil record has only just begun to be examined, and many questions remain: To what extent are the fossil patterns biological, geological, or artefactual? How are the environmental variables interrelated, and what is the pattern of causation among them and the biological variables? How can we use the past to model the future? The challenge for paleobiologists is to extract the salient messages in time to make a difference to current biodiversity.

### REFERENCES

Aguirre, J., and Riding, R. 2005. "Dasycladalean algal biodiversity compared with global variations in temperature and sea level over the past 350 Myr." *Palaios* 20: 581–588.

Alroy, J. 2008. "Dynamics of origination and extinction in the marine fossil record." *Proceedings of the National Academy of Sciences, USA* 105 (Suppl. 1): 11536–11542.

Alroy, J. 2010. "Geographical, environmental and intrinsic biotic controls on Phanerozoic marine diversification." *Palaeontology* 53: 1211–1235.

Alroy, J., Aberhan, M., Bottjer, D. J., Foote, M., Fürsich, F. T., Harries, P. J., Hendy, A. J. W., et al. 2008. "Phanerozoic trends in the global diversity of marine invertebrates." *Science* 321: 97–100.

Alvarez, L. W., Alvarez, W., Asaro, F., and Michel, H. V. 1980. "Extraterrestrial cause for the Cretaceous-Tertiary extinction." *Science* 208: 1095–1108.

Arens, N. C., and West, I. D. 2008. "Press-pulse: A general theory of mass extinction?" *Paleobiology* 34: 456–471.

Benton, M. J., ed. 1993. *The Fossil Record 2*. London: Chapman & Hall.

Benton, M. J. 1995. "Diversification and extinction in the history of life." *Science* 268: 52–58.

Benton, M. J. 1997. "Models for the diversification of life." *Trends in Ecology and Evolution* 12: 490–495.

Benton, M. J. 2005. *When Life Nearly Died*. London: Thames and Hudson.

Benton, M. J. 2009. "The red queen and the court jester: Species diversity and the role of biotic and abiotic factors through time." *Science* 323: 728–732.

Benton, M. J., and Harper, D. A. T. 2009. *Introduction to Paleobiology and the Fossil Record*. Chichester: John Wiley & Sons Ltd.

Berner, R. A., and Kothavala, Z. 2001. "GEOCARB III: A revised model of atmospheric $CO_2$ over Phanerozoic time." *American Journal of Science* 301: 182–204.

Cornette, J. L., Lieberman, B. S., and Goldstein, R. H. 2002. "Documenting a significant relationship between macroevolutionary origination rates and Phanerozoic $pCO_2$ levels." *Proceedings of the National Academy of Sciences, USA* 99: 7832–7835.

Cowen, R. 2000. *The History of Life,* 4th ed. Oxford: Blackwell.

Davis, R. B., Baldauf, S. B., and Mayhew, P. J. 2010. "Many hexapod groups originated earlier and withstood extinction events better than previously realized: Inferences from supertrees." *Proceedings of the Royal Society B* 277: 1597–1606.

Foote, M. 2000. "Origination and extinction components of taxonomic diversity: General problems." *Paleobiology* 26 (4, Supplement): 74–102.

Frakes, L. A., Francis, J. E., and Syktus, J. I. 1992. *Climate Modes of the Phanerozoic*. Cambridge: Cambridge University Press.

Gaston, K. J., and Blackburn, T. M. 2000. *Pattern and Process in Macroecology*. Oxford: Blackwell.

Hallam, A. 1984. "Pre-Quaternary changes of sea-level." *Annual Review of Earth and Planetary Science* 12: 205–243.

Jaramillo, C., Rueda, M. J., and Mora, G. 2006. "Cenozoic plant diversity in the neotropics." *Science* 311: 1893–1896.

Kalmar, A., and Currie, D. J. 2010. "The completeness of the continental fossil record and its impact on patterns of diversification." *Paleobiology* 36: 51–60.

Keller, G. 2001. "The end-cretaceous mass-extinction in the marine realm: Year 2000 assessment." *Planetary and Space Science* 49: 817–830.

Kirchner, J. W., and Weil, A. 2000. "Delayed biological recovery from extinctions throughout the fossil record." *Nature* 404: 177–180.

May, R. M., Lawton, J. H., and Stork, N. E. 1995. "Assessing extinction rates." In *Extinction Rates,* edited by J. H. Lawton and R. M. May, 1–24. Oxford: Oxford University Press.

Mayhew, P. J. 2011. "Global climate and extinction: Evidence from the fossil record." In *Climate Change, Ecology and Systematics,* edited by T. Hodkinson, M. Jones, S. Waldren, and J. Parnell, 99–121. Systematics Association Special Series. Cambridge: Cambridge University Press.

Mayhew, P. J., Jenkins, G. B., and Benton, T. G. 2008. "A long-term association between global temperature and biodiversity, origination and extinction in the fossil record." *Proceedings of the Royal Society B* 275: 47–53.

McCarthy, D. 2009. *Here Be Dragons: How the Study of Animal and Plant Distributions Revolutionized Our Views of Life and Earth*. Oxford: Oxford University Press.

Medvedev, M. V., and Melott, A. L. 2007. "Do extragalactic cosmic rays induce cycles in fossil diversity?" *Astrophysical Journal* 664: 879–889.

Melott, A. L. 2008. "Long-term cycles in the history of life: Periodic biodiversity in the Paleobiology database." *PLoS ONE* 3: e4044.

Meyer, K. M., and Kump, L. R. 2008. "Oceanic euxinia in Earth history: Causes and consequences." *Annual Review of Earth and Planetary Sciences* 36: 251–288.

Payne, J. L., and Finnegan, S. 2007. "The effect of geographic range on extinction risk during background and mass extinction." *Proceedings of the National Academy of Sciences, USA* 104: 10506–10511.

Peters, S. E. 2005. "Geologic constraints on the macroevolutionary history of marine animals." *Proceedings of the National Academy of Sciences, USA* 102: 1236–1331.

Peters, S. E. 2006. "Genus extinction, origination, and the durations of sedimentary hiatuses." *Paleobiology* 32: 387–407.

Peters, S. E. 2008. "Environmental determinants of extinction selectivity in the fossil record." *Nature* 454: 626–629.

Purdy, E. G. 2008. "Comparison of taxonomic diversity, strontium isotope and sea-level patterns." *International Journal of Earth Science* 97: 651–664.

Raup, D. M., and Sepkoski, J. J., Jr. 1982. "Mass extinctions in the marine fossil record." *Science* 215: 1501–1503.

Rohde, R. A., and Muller, R. A. 2005. "Cycles in fossil diversity." *Nature* 434: 208–210.

Ross, M. 2009. *The Search for Extraterrestrials: Intercepting Alien Signals*. Berlin: Springer.

Rothman, D. H. 2001. "Global biodiversity and the ancient carbon cycle." *Proceedings of the National Academy of Sciences, USA* 98: 4305–4310.

Rothman, D. H. 2002. "Atmospheric carbon dioxide levels for the last 500 million years." *Proceedings of the National Academy of Sciences, USA* 99: 4167–4171.

Royer, D. L., Berner, R. A., Montañez, I. P., Tibor, N. J., and Beerling, D. J. 2004. "$CO_2$ as a primary driver of Phanerozoic climate." *GSA Today* 14: 4–10.

Schulte, P., Alegret, L., Arenillas, I., Arz, J. A., Barton, P. J., Bown, P. R., Bralower, T. J., et al. 2010. "The Chicxulub asteroid impact and mass extinction at the Cretaceous-Paleogene boundary." *Science* 327: 1214–1218.

Sepkoski, J. J., Jr. 2002. "A compendium of fossil marine animal genera." *Bulletins of American Paleontology* 363: 1–560.

Shaviv, N. J., and Veizer, J. 2003. "Celestial driver of Phanerozoic climate?" *GSA Today* 13: 4–10.

Sheehan, P. M. 2001. "The Late Ordovician mass extinction." *Annual Review of Earth and Planetary Sciences* 29: 331–364.

Thomas, C. D., Cameron, A., Green, R. E., Bakkenes, M., Beaumont, L. J., Collingham, Y. C., Erasmus, B. F. N., et al. 2004. "Extinction risk from climate change." *Nature* 427: 145–148.

Valentine, J. W., and Moores, E. M. 1970. "Plate-tectonic regulation of faunal diversity and sea level: A model." *Nature* 228: 657–659.

Veizer, J., Godderis, Y., and François, L. M. 2000. "Evidence for decoupling of atmospheric $CO_2$ and global climate during the Phanerozoic eon." *Nature* 408: 698–701.

Wake, D. B., and Vredenburg, V. T. 2008. "Are we in midst of the sixth mass extinction? A view from the world of amphibians." *Proceedings of the National Academy of Sciences, USA* 105: 11466–11473.

Ward, P. D. 2006. "Impact from the deep." *Scientific American* October 2006, 64–71.

Wignall, P. B. 2001. "Large igneous provinces and mass extinctions." *Earth-Science Reviews* 53: 1–33.

Wignall, P. B. 2005. "The link between large igneous province eruptions and mass extinctions." *Elements* 1: 293–297.

Woodward, F. I., and Kelly, C. K. 2008. "Responses of global plant biodiversity capacity to changes in carbon dioxide concentration and climate." *Ecology Letters* 11: 1229–1237.

## Chapter 10

# *Terrestrial Ecosystem Response to Climate Change during the Paleogene*

WILLIAM C. CLYDE AND REBECCA LECAIN

In this chapter we investigate the relationship between climatic change and extinction in continental ecosystems during the era of modern biotas. In contrast to the previous chapter, the time frame examined is shorter and therefore the number of major global extinction events is smaller. But at the same time, the more recent fossil record of the past 50 million years is more highly resolved, and it is possible to begin to examine causal linkages between climate and extinctions.

By integrating the fossil record of past biodiversity with the geological record of past climate change, it is possible to evaluate causal hypotheses that relate these dynamic parts of the earth system. Developing a baseline understanding of the relationship between climatic and biotic change in past terrestrial environments under natural conditions provides important context for the many empirical and modeling studies, like those found in later chapters of this book, that investigate the relationship of climatic and biotic change under anthropogenic conditions.

Rather than attempt to summarize the entire record of continental climatic and biotic evolution, we have chosen to focus our discussion on two specific intervals of climate change within the Paleogene period (about 65–23 million years ago)—the Paleocene-Eocene Thermal Maximum (PETM) and the Eocene-Oligocene boundary (E-O boundary, fig. 10-1). The Paleogene is a particularly good interval of

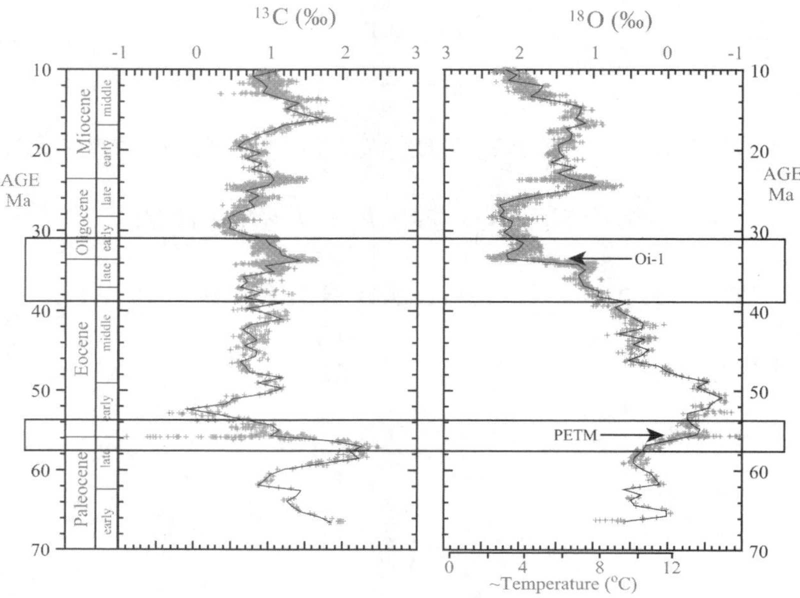


FIGURE 10-1. Global carbon and oxygen isotope record of benthic foraminifera from deep sea cores compiled by Zachos et al., 2001. Carbon isotope record represents a proxy for changes in the global carbon cycle, and oxygen isotope record represents a proxy for changes in global mean annual temperature and ice volume. Notice that both the PETM and E-O boundary interval are characterized by significant changes in both the carbon cycle and temperature, suggesting major changes to the global climate system at these times. Ma = million years ago; ‰ = parts per thousand.

geological time to investigate the role of climate change in causing biological turnover in continental ecosystems. The climate system underwent several well documented short-term and long-term global climatic changes during the Paleogene, offering geologists a choice of "natural experiments" from which to choose. In addition, well exposed and well studied stratigraphic records of this interval are numerous, particularly in the Northern Hemisphere, where continents were periodically connected by land bridges.

The PETM represents a short-term high-magnitude global warming event that coincided with a severe perturbation to the global carbon cycle about 55 million years ago (see reviews by Bowen et al., 2006; Sluijs et al., 2007). Although the PETM occurred during a greenhouse phase of Earth history and thus is not directly comparable to the current anthropogenic carbon–induced warming that is occurring against the backdrop of today's icehouse climate state, it nonetheless has many analogous qualities to the perturbations facing the modern earth system (Zachos et al., 2008). This similarity makes the PETM an especially interesting example to evaluate the biological effects of short-term climate change.

The E-O boundary, on the other hand, was a longer term climatic transition about 35 million years ago that encompassed a large-scale change from greenhouse to icehouse climate state (Berggren and Prothero, 1992). Changes in global ocean circulation and tectonism are thought to underpin the climatic changes across the E-O boundary and thus present a case study of how continental ecosystems respond to gradual, yet more permanent, climatic changes. By focusing on these two intervals of climate change, we can compare and contrast the turnover in terrestrial ecosystems in response to environmental changes of different sign (warming versus cooling) taking place over different time scales (thousands of years versus millions of years).

## The Paleocene-Eocene Thermal Maximum

About 55 million years ago, the earth system experienced a profound short-term, large-magnitude perturbation to the global carbon cycle and climate system that in some ways is analogous to current anthropogenic global warming. This event, the PETM, was characterized by an increase in global temperatures of 4–9 degrees Celsius in less than 10,000 years and coincided with the input of up to 5,000 gigatons of carbon into the mixed ocean-atmosphere carbon pool (Wing et al., 2005; Bowen et al., 2006; Sluijs et al., 2007; Zachos et al., 2008). For sake of perspective, the total amount of carbon in the earth's modern fossil fuel reservoir is about 4,000 gigatons. This disruption to the earth's carbon cycle and climate had wide-ranging effects on the rest of the earth system, including lowering latitudinal temperature gradients, fundamentally changing atmospheric moisture transport, restructuring global ocean circulation, and acidifying the ocean. Given its short time scale and large magnitude, the PETM has become the focus of considerable research into how biogeochemical cycles may respond to rapid climatic changes similar to what may occur in the near future. In the same way, it is possible to look at how the biosphere responded to this global warming event to help understand how biological communities may be affected by future anthropogenic changes to the climate system.

The PETM was first identified in marine deep-sea cores based on stable isotope analysis of foraminiferal shells (Kennett and Stott, 1991; Zachos et al., 1993). The biotic effects of the dramatic warming in the marine realm were soon identified to be far-reaching. For instance, the PETM coincides with a mass extinction of benthic foraminifera (35–50 percent) (Thomas and Shackleton, 1996; Takeda and Kaiho, 2007) and significant ecological turnover in plankton communities as well (Kelly et al., 1998; Gibbs et al., 2006; Mutterlose et al., 2007; Agnini et al., 2007). One of the most pronounced examples of marine biotic change across the PETM occurs in continental margin environments, where there is a rapid global increase in abundance of dinoflagellate cysts belonging to the subtropical genus *Apectodinium* (Crouch et al., 2001). These fundamental changes to marine ecosystems in response to the PETM demonstrate how closely integrated the marine biosphere is with the rest of the earth system. Considerable work has been carried out to understand the response of terrestrial ecosystems to the PETM as well.

For well over a century, vertebrate paleontologists have known that Holarctic mammal faunas from the Paleocene were significantly different from those in the Eocene (Lemoine, 1878; Matthew, 1914). In fact, this difference was instrumental in first recognizing these as formal epochs. Continental Eocene deposits of North America, Europe, and Asia have always been differentiated from earlier Paleocene deposits by the appearance and abundance of the so-called "modern" orders of mammals, in particular primates (sensu stricto), artiodactyls, and perissodactyls. These groups have never been unambiguously found in Paleocene rocks anywhere in the world, but they represent significant components of Eocene faunas on most continents where they have been recovered. Koch et al. (1992) recorded the PETM carbon isotope excursion in a fossiliferous continental stratigraphic section in Wyoming and discovered that the first appearance of these "modern" orders of mammals coincided precisely with the PETM. This discovery led to revitalization in research on mammalian turnover at the Paleocene-Eocene boundary and supplied an ideal scenario for testing the relationship between climatic and biotic change in continental ecosystems.

Many studies since the Koch et al. discovery have now fleshed out a widely accepted scenario for how the PETM directly and indirectly changed continental ecosystems (see Gingerich, 2006 for review). First, the basic pattern of modern mammals abruptly appearing coincident with the PETM has been replicated in many different locations across the Holarctic, indicating that these groups experienced rapid dispersal at this time. There are controversial reports of these groups from the Paleocene (e.g., Ting et al., 2007); however, these specimens are rare and do not share the derived characteristics of the taxa that form the core of the immigration event. Although the basic biogeographic pattern has been documented in Asia and Europe, the records from North America are generally more complete and have been studied in considerably more detail (Hooker, 1998; Gingerich and Clyde, 2001; Bowen et al., 2002).

The Bighorn Basin in Wyoming preserves the most complete record of the PETM from a continental environment. Here, the PETM is represented by an approximately 50-meter-thick (164 feet) sequence of fossiliferous paleosols of fluvial origin that accumulated as the basin rapidly subsided in response to the uplift of adjacent Laramide mountain ranges. This 50-meter-thick PETM interval has been studied in several different locations around the basin and records the telltale carbon isotope excursion in each case (Koch et al., 2003). Leaf margin analysis and oxygen isotope paleothermometry are consistent in estimating about 5 degrees Celsius of warming during the PETM and a maximum mean annual temperature of up to 26 degrees Celsius in the Bighorn Basin (Fricke et al., 1998; Fricke and Wing, 2004; Wing et al., 2005; Secord et al., 2010). Precipitation levels in the basin have been estimated from leaf area analysis, pedofacies analysis, and trace fossil assemblages, all of which indicate considerably drier conditions during the early part of the PETM in the Bighorn Basin with a return to moister conditions in the late PETM (Wing et al., 2005; Kraus and Riggins, 2007; Smith et al., 2008; also see Bowen et al., 2004). A transient drying has also been documented in basins farther to the south, suggesting that these hydroclimatological changes may involve changes to large-scale atmospheric circulation patterns (Bowen and Bowen, 2008; Burger, 2008). Given that these well documented climatic changes from the western interior of North America are recorded in deposits that are also very fossiliferous, it is possible to evaluate in detail the response of a continental ecosystem in this region to these dramatic climate changes.

The two best documented fossil groups from the PETM interval of the western interior of North America are plants and mammals. Both groups were clearly affected by the PETM, albeit in different ways. Early efforts to document plant communities across the PETM indicated relatively minor changes; however, these studies were hampered by a lack of preservation within the PETM interval itself (Wing, 1998; Wing and Harrington, 2001; Wing et al., 2003). Recent discovery of productive fossil plant localities within the PETM in the Bighorn Basin indicate that plants did indeed experience a rapid, transient, and profound turnover during the PETM (Wing et al., 2005). The PETM floras in the Bighorn Basin, as documented by macrofossil and pollen assemblages, are dominated by taxa that are otherwise unknown from the basin (fig. 10-2). The immigrants include several taxa from the Gulf Coast region, two taxa known from adjacent basins to the east, one taxon from Europe, and several other unknowns. The existence of short-term floral change at the PETM in the Bighorn Basin is also evidenced by a larger than normal magnitude (4–5 parts per thousand carbon isotope excursion in n-alkanes derived from leaf wax

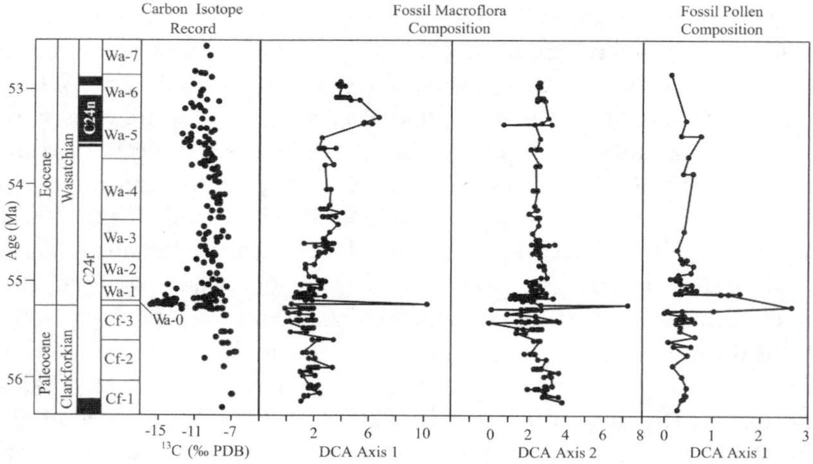


FIGURE 10-2. Summary of plant turnover in the Bighorn Basin of Wyoming across the PETM. Composite basinwide record of carbon isotope values from paleosol carbonates is shown at left (PDB = Pedee belemnite standard) with macrofossil record in center and microfossil (pollen) record at right. DCA scores are from detrended correspondence analysis of sites-by-species matrix of presence/absence data. Notice the large magnitude yet transient change in floral composition associated with the PETM carbon isotope excursion. All plant data are from Wing et al., 2005. C24r and C24n represent intervals ("Chrons") from the Geomagnetic Polarity Timescale.

lipids (Smith et al., 2007). Despite this transient turnover, the PETM in the Bighorn Basin is not marked by significantly higher rates of floral extinction compared to the rest of the Paleocene or Eocene (Wing, 1998). Although floral sampling within the PETM is still relatively coarse, these initial results indicate that plants responded individualistically over short time scales to the PETM climatic perturbation. Moreover, the response seems dominated by transient shifts in geographic range causing local immigration and emigration rather than global extinction.

The same PETM deposits in the Bighorn Basin that preserve fossil plants also preserve abundant fossil vertebrates. Mammals have received the most attention among the vertebrate groups due to excellent preservation of their readily identifiable teeth. Sampling of fossil mammals through the PETM interval is especially good here and clearly documents the Holarctic immigration of modern mammals.

Analysis of these fossil mammal assemblages has allowed the partitioning of direct from indirect effects of PETM warming on these local communities (fig. 10-3) (Clyde and Gingerich, 1998; Gingerich, 2003). First, the immigrant modern mammals immediately became a significant component of succeeding mammal faunas both in terms of species proportions but even more so in terms of relative abundance. In fact, modern mammal communities of North America still bear the strong stamp of the PETM immigration event (e.g., high proportion of artiodactyls). The addition of these PETM immigrants seems to have initially increased total diversity (species richness and evenness) in the basin until diversity re-equilibrated during the early Eocene to a level that was somewhat higher than the preceding Paleocene. The peak in first appearances during the PETM was offset by a much smaller peak in last appearances, which means that although extinction rates did increase somewhat, that increase did not lead to an overall drop in diversity because it was offset by much higher rates of origination. The introduction of these PETM immigrants also had a significant impact on community ecological structure (e.g., body-size distribution, trophic structure).

Superimposed on all of these "permanent" immigration-related effects on Bighorn Basin mammal communities was one striking direct, yet transient, effect of the PETM. Many of the native and immigrant mammal lineages during the PETM experienced profound body-size dwarfing of 40–50 percent (Gingerich, 1989; Clyde and Gingerich, 1998; Gingerich, 2003). This kind of large-magnitude

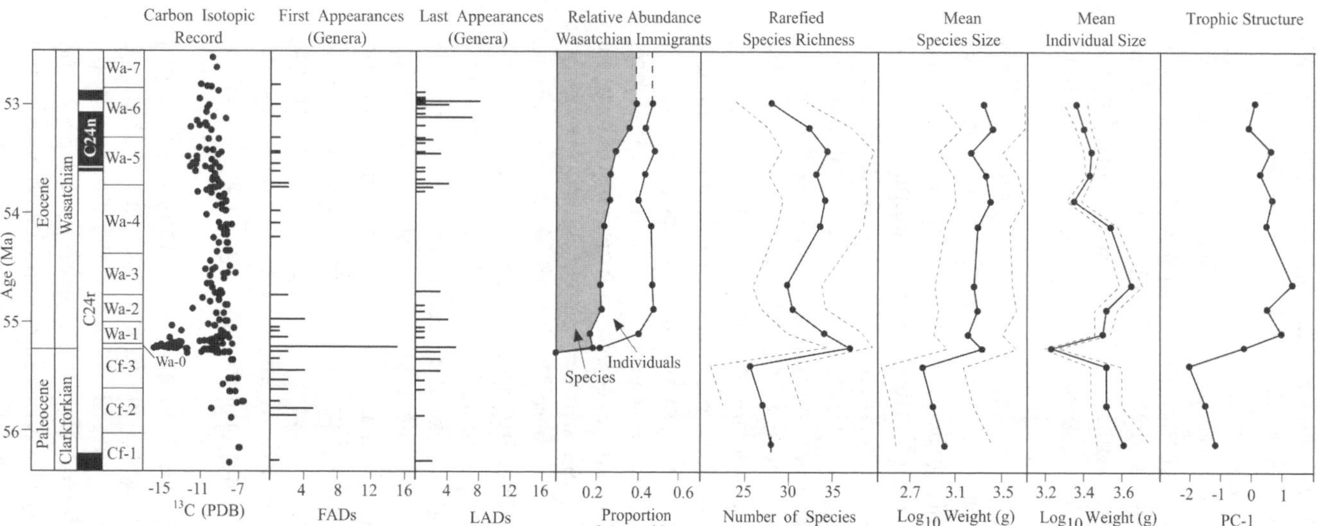


FIGURE 10-3. Summary of mammal turnover in the Bighorn Basin of Wyoming across the PETM. Composite basinwide record of carbon isotope values from paleosol carbonates (left) shows characteristic negative excursion at PETM (PDB = Pedee belemnite standard). The peak in first appearances coincident with the PETM represents the influx of Eocene immigrants (FAD = first appearance datum, LAD = last appearance datum). Immigrant species represent about 20 percent of fauna (shaded area) upon their immigration, yet come to represent about 50 percent of all individuals (white area) soon after their first appearance. Species richness rises dramatically in response to immigration event and then declines to levels in Eocene that are still higher than during preceding Paleocene time. Mean species size increases dramatically at PETM due to influx of larger bodied immigrants, whereas mean individual size decreases temporarily during brief warming of PETM due to dwarfing of many lineages. Trophic structure also undergoes significant turnover across PETM; herbivores and frugivores become more important in new Eocene community structure (PC-1 = score on first principal component). Dashed lines represent 95% confidence intervals. Figure modified from Clyde and Gingerich, 1998.

rapid morphological change occurs only during the PETM interval itself and likely exceeds the potential effects of northward "Bergman's Rule" dispersal of smaller body-size populations into the study area. Potential causes of this transient dwarfing include direct physioevolutionary responses to increases in mean annual temperature and/or poorer quality plant forage due to higher atmospheric carbon dioxide levels.

Although other terrestrial groups are not as well documented as plants and mammals during the PETM interval, they nonetheless provide important perspectives on how terrestrial ecosystems responded to this global warming event. Insect body fossils are very rare, but insect damage is readily preserved on fossil leaves, supplying a trace fossil record of insect behavior and diversity. Evidence from the Bighorn Basin indicates the PETM was characterized by a transient increase in insect herbivory and diversity (Currano et al., 2008). Analysis of fossil assemblages of nonmarine mollusks from the Bighorn Basin shows a pattern somewhat similar to the plant record in that the PETM interval is characterized by unusual taxa. However, this transient turnover did not seem to have lasting effects on community structure after the climate event is over (Hartman and Roth, 1998).

Many new detailed continental stratigraphic records across the PETM from outside of the Bighorn Basin are now being constructed. These help expand our understanding of this event and its biological consequences to regional and global scales. For instance, recent results from the Piceance Basin in Colorado, which is about 600 kilometers (373 miles) south of the Bighorn Basin, indicate that the PETM in that area was characterized by desertification and extinction of many arboreal mammal taxa (Burger, 2008). Analysis of pollen and spores from Gulf Coastal plain sediments, about 2,500 kilometers (1,553 miles) to the southeast of the Bighorn Basin, indicates high rates of floral extinction at these lower latitudes, pointing to a spatially complex pattern of turnover (Harrington, 2003; Harrington and Jaramillo, 2007). Detailed records of biotic change from other continents are also being developed that will help create a globally integrated understanding of how the biosphere reacted to this global warming event (Crouch and Visscher, 2003; Bowen et al., 2002; Bernaola et al., 2007).

In summary, the PETM represents an unusually good historical analog to projected anthropogenic changes to the earth system, albeit in the absence of *Homo sapiens* and against the backdrop of a greenhouse climate state rather than an icehouse climate state. Terrestrial ecosystems responded in dramatic fashion. Extreme arctic warming triggered intercontinental mammalian dispersal across high-latitude land bridges. These mammalian immigrants had long-lasting effects on local communities, including changes to taxonomic composition and ecological structure (e.g., diversity, trophic structure) that are still apparent today. Regional shifts in geographic ranges of plant taxa caused significant but transient turnover in some places (e.g., Bighorn Basin, Wyoming) but longer lasting loss of diversity in others (e.g., Gulf Coast). In general, it is clear that migration was an important coping mechanism for both mammal and plant taxa during the PETM. Mammals seem to have exhibited more community coherence in their biogeographic response, whereas plant taxa seem to have reacted more individualistically. This increased dispersal activity did lead to higher extinction rates in most cases but not necessarily to lower diversity, given the offsetting effects of increased origination rates. Preliminary results indicate that extinction was more profound at lower latitudes in North America; however, this will require more sampling of spatially dispersed sites.

## The Eocene-Oligocene Boundary and the Grand Coupure

The Eocene-Oligocene boundary, approximately 34 million years ago, is recognized as one of the largest global climate shifts in the last 65 million years. It was also accompanied by major regional extinctions. The late Eocene and early Oligocene were times of significant cooling (fig. 10-1), the causes of which are not yet clear. This global climate change dramatically affected local environments, which in turn led to widespread faunal extinction in Europe—the "Grand Coupure" (or "big cut")—and extensive yet heterogeneous biotic turnover elsewhere. The boundary itself is defined by an extinction of marine microfossils; however, the continental record of biotic turnover is more complex.

The Eocene climate was generally characterized by warmth; however, there were suggestions of the icehouse to come. Mountain glaciers existed intermittently in Antarctica during the mid- to late Eocene (Kump, 2005; Birkenmajer et al., 2005). These glaciers apparently did not persist, only lasting for short periods of thousands of years. E-O climate cooling has been shown to occur in several small steps rather than a single major shift (Pearson et al., 2008; Katz et al., 2008). One of the biggest climate steps was a 400,000-year cooling and glaciation in Antarctica during the early Oligocene referred to as Oi-1 (DeConto and Pollard, 2003a). This shift is recorded primarily as a 1–1.5 parts per thousand excursion in the oxygen isotope composition of foraminifera from deep-sea cores recovered from the southern ocean and corresponds to an oceanic temperature decrease of roughly 5–6 degrees Celsius (Zachos et al., 1996, 2001). The Oi-1 event has also been recorded on the northern continents, although there are contradictory reports of its impact on temperatures in Holarctic environments (Zanazzi et al., 2007; Grimes et al., 2005).

There are several proposed causes for the E-O global climatic cooling trend. Because of the long lead-up and stepwise pattern of the cooling, its cause is unlikely to be a single isolated event like that inferred for the PETM, but rather a long-lasting condition that reached a threshold, or a series of events. The leading primary causal hypothesis is declining atmospheric carbon dioxide levels (Tripati et al., 2005; Pagani et al., 2005). Decreasing atmospheric carbon dioxide would eventually lower the snow line elevation to intersect large portions of Antarctica and other mountainous regions worldwide (DeConto and Pollard, 2003b). The other possible primary cause for glaciation in Antarctica is opening of the Tasmanian Gateway between Australia and Antarctica. This led to a circumpolar current that isolated Antarctica from warmer water currents (Kennett and Exon, 2004). It is unclear how this gateway would have affected the climate across the rest of the globe, but it was likely very effective for Antarctic cooling. Other secondary causes such as orbital forcing and changing albedo could have significantly reinforced the cooling trend when coupled with these other primary climate drivers (DeConto and Pollard, 2003b).

An estimated 60 percent of endemic European taxa went extinct during the climate changes around the E-O boundary, making it one of the most severe regional extinction events in Cenozoic mammal history (Berggren and Prothero, 1992). This vertebrate response—the Grand Coupure—is characterized by significant turnover (fig. 10-4), including a vast decline in diversity of subtropical taxa such as primates and the immigration of many Asian taxa such as rhinos (Kohler and Moya-Soja, 1999; Hooker et al., 2004). Recent high-resolution results from the Hampshire Basin in the United Kingdom indicate that the Grand Coupure faunal turnover correlates closely with the Oi-1

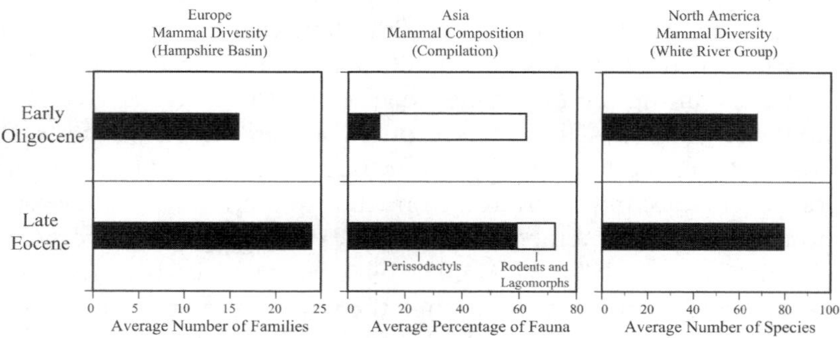


FIGURE 10-4. Bar graphs summarizing changes in mammal communities from different continents across the E-O boundary cooling episode. In Europe (left), the E-O boundary is characterized by a significant drop in diversity (the Grand Coupure), as indicated by a decline in the average number of families in the Hampshire Basin (data from Hooker, 2000). In Asia (center), the E-O boundary is marked by a significant shift from perissodactyl-dominated Eocene communities to rodent- and lagomorph-dominated Oligocene communities (the Mongolian Remodeling, data from Meng and McKenna, 1998). In North America (right), mammals experienced a relatively insignificant decrease in diversity across the E-O boundary, as indicated by the record of species richness from the White River Group (data from Prothero and Heaton, 1996).

climatic event, indicating temporal coincidence between global climate change and biotic change at this time (Hooker et al., 2004; Hooker et al., 2009). Further evidence for this linkage comes from the well studied vertebrate records of the Mongolia Plateau in Asia. Here, the perissodactyl-dominated faunas of the Eocene abruptly changed to the rodent- and lagomorph-dominated faunas of the Oligocene (Meng and McKenna, 1998; Wang et al., 2007). This so-called "Mongolian Remodeling" is closely coincident with the Grand Coupure in Europe, indicating that the Oi-1 climatic event had considerable effects on the biota across all of Eurasia. Whereas the extinction within some groups such as primates was probably the direct result of climatic cooling and drying, in many other cases the change in climate caused dispersal, which in turn led to turnover due to heightened competition. In the case of Eurasia during the E-O boundary, lower eustatic sea level due to Antarctic glaciation caused the draining of the Turgai Sea and the development of open land connections between Europe and Asia. As with the PETM, the indirect ways that climate change affected dispersal corridors were just as important as the direct effects of changes in temperature or precipitation.

Biotic changes across the E-O boundary in North America were quite different from that observed in Eurasia. In general, faunal change at this time was quite gradual, tracking the long-term decline in global temperatures from the Middle Eocene to the early Oligocene (Alroy et al., 2000). Reptiles and arboreal mammals declined in diversity during this period and were replaced by larger bodied mammalian taxa (e.g., artiodactyls) that were more suited to the newly established open grasslands (Hutchison, 1992; Prothero, 1989). No abrupt faunal changes are currently documented in North America at the time of Oi-1 (Prothero and Heaton, 1996), and it was not marked by significant intercontinental dispersal across Beringia (Prothero and Heaton, 1996; Woodburne and Swisher, 1995). Interestingly, even though lower sea levels may have caused greater exposure of Beringia, dispersal across this land bridge at this time may have been thwarted by colder high-latitude temperatures. This is in stark contrast to Europe, where the disappearance of the Turgai Sea that separated Europe from Asia at this time led to significant exchange of continental biota between these regions. The only lower latitude region with a good fossil record of terrestrial ecosystems across the E-O boundary is in Egypt. Afro-Arabian mammals do not seem to have undergone a significant extinction at the E-O boundary; however, a sharp decline in strepsirrhine primates and subsequent contraction of oligopethecid primates in Egypt may have been the result of E-O cooling (Seiffert, 2007).

Several studies have characterized the vegetation of this time period and found floral changes consistent with climatic cooling. Across the entire Holarctic, Eocene tropical floras gave way to Oligocene grasslands as plant diversity slowly declined (Retallack et al., 2004; Prothero, 1989). In Antarctica, the vegetation structure changed from largely evergreen forest to tundra as the climate became progressively cooler and drier. Vegetation-climate feedbacks may have also played an important role in facilitating the glaciation via changes in albedo and heat transfer (Thorn and DeConto, 2005).

## Conclusions

The relationship between climate change and extinction risk in continental ecosystems is not straightforward; however, the two case studies evaluated here provide some important insights. Both the PETM and E-O boundary are characterized by biotic turnover of varying magnitudes and rates on different continents. In most cases, the observed turnover is partly driven by higher rates of *per taxon* extinction. These increases in per-taxon extinction rates, however, do not always lead to net decreases in *diversity* because origination rates may simultaneously increase as well. For instance, although rates of mammalian extinction in the Bighorn Basin of Wyoming increased during the PETM, that increase was offset by a larger increase in origination rates, resulting in higher local diversity during and after the warming event (fig. 10-3). On the other hand, the record of mammals from Europe and Asia across the E-O boundary shows increased rates of per-taxon extinction during that period of cooling and these were not compensated by increased rates of origination, so net diversity decreased significantly. The lesson we can draw from these examples is that individual taxa seem to experience a higher risk of extinction during periods of unusual climate change, but that does not necessarily translate into a decrease in overall diversity because origination rates can increase at the same time. The balance between origination and extinction rates is regionally heterogeneous, making it difficult to estimate the effects of large-scale climate change on global diversity using the unevenly sampled fossil record. One of the primary paleontological goals of the twenty-first century is to increase spatial and temporal sampling density to better constrain patterns of global diversity and compare them to the well resolved proxy records of global climate change.

Reconstructing how changes in climatic variables may have forced higher rates of per-taxon extinction is difficult, but in both examples presented here, the main link seems to be associated with dispersal. As the climate changed across the PETM and E-O boundary, the dominant biotic response was manifested in shifting geographic ranges. Although it is likely that taxa initially migrated in a predictable pattern to remain within their optimal climatic window, as dispersal progressed, the existence of physical barriers or connections often led to unexpected biogeographic results. For instance, the PETM in the Holarctic is marked by the immigration of several new groups of mammals that likely diffused northward from their unknown place of origin during the PETM warming and encountered high-latitude land bridges that allowed their rapid circum-Holarctic dispersal. Similarly, mammals in Asia during the E-O boundary seem to have encountered new dispersal routes to Europe when Antarctic glaciation lowered sea level and drained the Turgai Sea. Alternatively, plants at the PETM show a pattern of northward migration but no significant dispersal across the high-latitude land bridges, resulting in an entirely different biogeographic pattern and more transient biotic turnover. In all cases, a rise in per-taxon extinction rates was likely the result of emigration and the rapid influx of climatically modulated immigrants causing an increase in local competition. In continental ecosystems, dispersal seems to be the first line of defense for taxa trying to adjust to major changes in climate. These biogeographic perturbations in turn create novel biotic interactions that lead to increased levels of turnover (e.g., extinction rates) but not necessarily lower levels of diversity.

What do these lessons from the geological past tell us about the likely effects of anthropogenic climate change on continental ecosystems? If we could isolate the climatic effects from the myriad of other anthropogenic influences on modern biotas, it would mean that current and projected climate changes would almost certainly raise per-taxon extinction rates but the effects on global diversity would be much less certain given the potentially offsetting effects of increased origination (speciation) rates. Unfortunately, given the very fast pace of anthropogenic climate change, any potential increase in origination rates would likely lag increases in extinction rates, leaving behind a period of lower overall diversity. The prevalence of human-assisted biotic dispersal and the large-scale degradation of natural habitats from rapidly expanding human populations will almost certainly amplify rates of per-taxon extinction, making it that much harder to maintain levels of global diversity.

## Acknowledgments

Aspects of this research were funded by NSF grant EAR0642291. We thank Scott Wing for sharing data and an anonymous reviewer for helpful feedback.

### REFERENCES

Agnini, C., Fornaciari, E., Rio, D., Tateo, F., Backman, J., and Giusberti, L. 2007. "Responses of calcareous nannofossil assemblages, mineralogy and geochemistry to the environmental perturbations across the Paleocene/Eocene boundary in the Venetian pre-Alps." *Marine Micropaleontology* 63: 19–38.

Alroy, J., Koch, P. L., and Zachos, J. C. 2000. "Global climate change and North American mammalian evolution." *Paleobiology* 26: 259–288.

Berggren, W. A., and Prothero, D. R. 1992. "Eocene-Oligocene climatic and biotic evolution: An overview." In *Eocene-Oligocene Climatic and Biotic Evolution,* edited by D. R. Prothero and W. A. Berggren, 1–28. Princeton, NJ: Princeton University Press.

Bernaola, G., Baceta, J. I., Orue-Etxebarria, X., Alegret, L., Martin-Rubio, M., Arostegui, J., and Dinares-Turell, J. 2007. "Evidence of an abrupt environmental disruption during the mid-Paleocene biotic event (Zumaia section, western Pyrenees)." *Geological Society of America Bulletin* 119: 785–795.

Birkenmajer, K., Gazdzicki, A., Krajewski, K. P., Przybycin, A., Solecki, A., Tatur, A., and Yoon, H. I. 2005. "First Cenozoic glaciers in West Antarctica." *Polish Polar Research* 26: 3–12.

Bowen, G. J., and Beitler Bowen, B. 2008. "Mechanisms of PETM global change constrained by a new record from central Utah." *Geology* 36: 379–382.

Bowen, G. J., Clyde, W. C., Koch, P. L., Ting, S., Alroy, J., Tsubamoto, T., Wang, Y., and Wang, Y. 2002. "Mammalian dispersal at the Paleocene/Eocene boundary." *Science* 295: 2062–2065.

Bowen, G. J., Beerling, D. J., Koch, P. L., Zachos, J. C., and Quattlebaum, T. 2004. "A humid climate state during the Palaeocene/Eocene Thermal Maximum." *Nature* 432: 495–499.

Bowen, G. J., Bralower, T. J., Delaney, M. L., Dickens, G. R., Kelly, D. C., Koch, P. L., Kump, L. R., et al. 2006. Eocene hyperthermal event offers insight into greenhouse warming. *Eos, Transactions, American Geophysical Union* 87: 165–169.

Burger, B. 2008. Extinction, migration, and the effects of global warming on fossil mammals across the Paleocene-Eocene boundary in western Colorado, U.S.A. *Journal of Vertebrate Paleontology* 28 (3 suppl.): 58a.

Clyde, W. C., and Gingerich, P. D. 1998. "Mammalian community response to the latest Paleocene Thermal Maximum; an isotaphonomic study in the northern Bighorn Basin, Wyoming." *Geology* 26: 1011–1014.

Crouch, E. M., and Visscher, H. 2003. "Terrestrial vegetation record across the initial Eocene Thermal Maximum at the Tawanui marine section, New Zealand." In *Causes and Consequences of Globally Warm Climates in the Early Paleogene,* edited by S. L. Wing, P. D. Gingerich, B. Schmitz, and E. Thomas, 351–363. Boulder, CO: Geological Society of America, Special Paper 369.

Crouch, E. M., Heilmann-Clausen, C., Brinkhuis, H., Morgans, H. E. G., Rogers, K. M., Egger, H., and Schmitz, B. 2001. "Global dinoflagellate event associated with the late Paleocene Thermal Maximum." *Geology* 29: 315–318.

Currano, E. D., Wilf, P., Wing, S. L., Labandeira, C. C., Lovelock, E. C., and Royer, D. L. 2008. "Sharply increased insect herbivory during the Paleocene–Eocene Thermal Maximum." *Proceedings of the National Academy of Sciences* 105: 1960–1964.

DeConto, R. M., and Pollard, D. 2003a. A coupled climate–ice sheet modeling approach to the Early Cenozoic history of the Antarctic ice sheet. *Palaeogeography, Palaeoclimatology, Palaeoecology* 198: 39–52.

DeConto, R. M., and Pollard, D. 2003b. "Rapid Cenozoic glaciations of Antarctica induced by declining atmospheric $CO_2$." *Nature* 421: 245–249.

Fricke, H. C., and Wing, S. L. 2004. "Oxygen isotope and paleobotanical estimates of temperature and $^{18}O$-latitude gradients over North America during the early Eocene." *American Journal of Science* 304: 612–635.

Fricke, H. C., Clyde, W. C., O'Neil, J. R., and Gingerich, P. D. 1998. "Evidence for rapid climate change in North America during the latest Paleocene Thermal Maximum, oxygen isotope compositions of biogenic phosphate from the Bighorn Basin (Wyoming)." *Earth and Planetary Science Letters* 160: 193–208.

Gibbs, S. J., Bralower, T. J., Bown, P. R., Zachos, J. C., and Bybell, L. M. 2006. "Shelf and open-ocean calcareous phytoplankton assemblages across the Paleocene–Eocene Thermal Maximum: Implications for global productivity gradients." *Geology* 34: 233–236.

Gingerich, P. D. 1989. "New earliest Wasatchian mammalian fauna from the Eocene of northwestern Wyoming: Composition and diversity in a rarely sampled high-floodplain assemblage." *University of Michigan Papers on Paleontology* 28: 1–97.

Gingerich, P. D. 2003. "Mammalian responses to climate change at the Paleocene-Eocene boundary: Polecat Bench record in the northern Bighorn Basin, Wyoming." In *Causes and Consequences of Globally Warm Climates in the Early Paleogene,* edited by S. L. Wing, P. D. Gingerich, B. Schmitz, and E. Thomas, 463–478. Boulder, CO: Geological Society of America, Special Paper 369.

Gingerich, P. D. 2006. "Environment and evolution through the Paleocene–Eocene Thermal Maximum." *Trends in Ecology and Evolution* 21: 246–253.

Gingerich, P. D., and Clyde, W. C. 2001. "Overview of mammalian biostratigraphy in the Paleocene-Eocene Fort Union and Willwood formations of the Bighorn and Clark's Fork basins." *University of Michigan Papers on Paleontology* 33: 1–14.

Grimes, S. T., Hooker, J. J., Collinson, M. E., and Mattey, D. P. 2005. "Summer temperatures of late Eocene to early Oligocene freshwaters." *Geology* 33: 189–192.

Harrington, G. J. 2003. "Geographic patterns in the floral response to Paleocene–Eocene warming." In *Causes and Consequences of Globally Warm Climates in the Early Paleogene,* edited by S. L. Wing, P. D. Gingerich, B. Schmitz, and E. Thomas, 381–393. Boulder, CO: Geological Society of America, Special Paper 369.

Harrington, G. J., and Jaramillo, C. A. 2007. "Paratropical floral extinction in the Late Palaeocene–Early Eocene." *Journal of the Geological Society, London* 164: 323–332.

Hartman, J. H., and Roth, B. 1998. "Late Paleocene and early Eocene nonmarine molluscan faunal change in the Bighorn Basin, northwestern Wyoming and south-central Montana." In *Late Paleocene–Early Eocene Climatic and Biotic Events in the Marine and Terrestrial Records,* edited by M. P. Aubry, W. A. Berggren, and S. G. Lucas, 323–379. New York: Columbia University Press.

Hooker, J. J. 1998. "Mammalian faunal change across the Paleocene–Eocene transition in Europe." In *Late Paleocene–Early Eocene Climatic and Biotic Events in the Marine and Terrestrial Records,* edited by M. P. Aubry, S. Lucas, and W. A. Berggren, 428–450. New York: Columbia University Press.

Hooker, J. J. 2000. "Paleogene mammals, crises and ecological change." In *Biotic Response to Global Change, the Last 145 Million Years,* edited by S. J. Culver and P. F. Rawson, 333–349. Cambridge: Cambridge University Press.

Hooker, J. J., Collinson, M. E., and Sille, N. P. 2004. "Eocene-Oligocene mammalian faunal turnover in the Hampshire Basin, UK: Calibration to the global time scale and the major cooling event." *Journal of the Geological Society* 161: 161–172.

Hooker, J. J., Grimes, S. T., Mattey, D. P., Collinson, M. E., and Sheldon, N. D. 2009. "Refined correlation of the UK Late Eocene–Early Oligocene Solent Group and timing of its climate history." In *The Late Eocene Earth—Hothouse, Icehouse, and Impacts,* edited by C. Koeberl and A. Montanari, 179–195. Boulder, CO: Geological Society of America, Special Paper 452.

Hutchison, J. H. 1992. "Western North American reptile and amphibian record across the Eocene/Oligocene boundary and its climatic implications." In *Eocene-Oligocene Climatic and Biotic Evolution,* edited by D. R. Prothero and W. A. Berggren, 451–463. Princeton, NJ: Princeton University Press.

Katz, M. E., Miller, K. G., Wright, J. D., Wade, B. S., Browning, J. V., Cramer, B. S., and Rosenthal, Y. 2008. "Stepwise transition from the Eocene greenhouse to the Oligocene icehouse." *Nature Geoscience* 1: 329–334.

Kelly, D. C., Bralower, T. J., and Zachos, J. C. 1998. "Evolutionary consequences of the latest Paleocene thermal maximum for tropical planktonic Foraminifera." *Palaeogeography, Palaeoclimatology, Palaeoecology* 141: 139–161.

Kennett, J. P., and Exon, N. F. 2004. "Paleoceanographic evolution of the Tasmanian Seaway and its climatic implications." *Geophysical Monograph* 151: 345–367.

Kennett, J. P, and Stott, L. D. 1991. "Abrupt deep-sea warming, palaeoceanographic changes and benthic extinctions at the end of the Palaeocene." *Nature* 353: 225–229.

Koch, P. L., Zachos, J. C., and Gingerich, P. D. 1992. "Correlation between isotope records in marine and continental carbon reservoirs near the Palaeocene/Eocene boundary." *Nature* 358: 319–322.

Koch, P. L., Clyde, W. C., Hepple, R. P., Fogel, M. L., Wing, S. L., and Zachos, J. C. 2003. "Carbon and oxygen isotope records from Paleosols spanning the Paleocene-Eocene boundary, Bighorn Basin, Wyoming." In *Causes and Conse-*

*quences of Globally Warm Climates in the Early Paleogene,* edited by S. L. Wing, P. D. Gingerich, B. Schmitz, and E. Thomas, 49–64. Boulder, CO: Geological Society of America, Special Paper 369.

Kohler, M., and Moya-Sola, S. 1999. "A finding of Oligocene primates on the European continent." *Proceedings of the National Academy of Sciences, USA* 96: 14664–14667.

Kraus, M. J., and Riggins, S. 2007. "Transient drying during the Paleocene-Eocene Thermal Maximum (PETM); analysis of paleosols in the Bighorn Basin, Wyoming." *Palaeogeography, Palaeoclimatology, Palaeoecology* 245: 444–461.

Kump, L. R. 2005. "Foreshadowing the glacial era." *Nature* 436: 333–334.

Lemoine, V. 1878. "Communication sur les ossements fossiles des terrains tertiaires inférieurs des environs de Reims." Reims: F. Keller.

Matthew, W. D. 1914. "Evidence of the Paleocene vertebrate fauna on the Cretaceous-Tertiary problem." *Geological Society of America Bulletin* 25: 381–402.

Meng, J., and McKenna, M. C. 1998. "Faunal turnovers of Palaeogene mammals from the Mongolian Plateau." *Nature* 394: 364–367.

Mutterlose, J., Linnert, C., and Norris, R. D. 2007. "Calcareous nannofossils from the Paleocene-Eocene Thermal Maximum of the equatorial Atlantic (ODP site 1260B); evidence for tropical warming." *Marine Micropaleontology* 65: 13–31.

Pagani, M., Zachos, J., Freeman, K. H., Bohaty, S., and Tipple, B. 2005. "Marked decline in atmospheric carbon dioxide concentrations during the Paleogene." *Science* 309: 600–603.

Pearson, P. N., McMillan, I., Wade, B. S., Dunkley Jones, T., Coxall, H. K., Bown, P. R., and Lear, C. H. 2008. "Extinction and environmental change across the Eocene-Oligocene boundary in Tanzania." *Geology* 36: 179–182.

Prothero, D. R. 1989. "Stepwise extinctions and climatic decline during the later Eocene and Oligocene." In *Mass Extinctions: Processes and Evidence,* edited by S. K. Donovan, 211–234. New York: Columbia University Press.

Prothero, D. R., and Heaton, T. H. 1996. "Faunal stability during the Early Oligocene climatic crash." *Palaeogeography, Palaeoclimatology, Palaeoecology* 127: 257–283.

Retallack, G. J., Orr, W. N., Prothero, D. R., Duncan, R. A., Kester, P. R., and Ambers, C. P. 2004. "Eocene-Oligocene extinction and paleoclimatic change near Eugene, Oregon." *Geological Society of America Bulletin* 116: 817–839.

Secord, R., Gingerich, P. D., Luhmann, K. C., and MacLeod, K. G. 2010, "Continental warming preceding the Palaeocene-Eocene Thermal Maximum." *Nature* 467: 955–958.

Seiffert, E. R. 2007. "Evolution and extinction of Afro-Arabian primates near the Eocene-Oligocene boundary." *Folia Primatologica* 78: 314–327.

Sluijs, A., Bowen, G. J., Brinkhuis, H., Lourens, L. J., and Thomas, E. 2007. "The Palaeocene–Eocene Thermal Maximum super greenhouse: Biotic and geochemical signatures, age models and mechanisms of global change." In *Deep-Time Perspectives on Climate Change: Marrying the Signal from Computer Models and Biological Proxies,* edited by M. Williams, A. M. Haywood, F. J. Gregory, and D. N. P. Schmidt, 323–349. London: The Geological Society.

Smith, F. A., Wing, S. L., and K. H. Freeman. 2007. "Magnitude of the carbon isotope excursion at the Paleocene-Eocene Thermal Maximum: The role of plant community change." *Earth and Planetary Science Letters* 262: 50–65.

Smith, J. J., Hasiotis, S. T., Kraus, M. J., and Woody, D. T. 2008. "Relationship of floodplain ichnocoenoses to paleopedology, paleohydrology, and paleoclimate in the Willwood Formation, Wyoming, during the Paleocene-Eocene Thermal Maximum." *Palaios* 23: 683–699.

Takeda, K., and Kaiho, K. 2007. "Faunal turnovers in Central Pacific benthic Foraminifera during the Paleocene-Eocene Thermal Maximum." *Palaeogeography, Palaeoclimatology, Palaeoecology* 251: 175–197.

Thomas, E., and Shackleton, N. J. 1996. "The Paleocene-Eocene benthic foraminiferal extinction and stable isotope anomalies." In *Correlation of the Early Paleogene in Northwest Europe,* edited by R. W. O. Knox, R. M. Corfield, and R. E. P. Dunay, 401–441. London: Geological Society [London] Special Publication 101.

Thorn, V. C., and DeConto, R. 2005. "Antarctic climate at the Eocene/Oligocene boundary—climate model sensitivity to high latitude vegetation type and comparisons with the palaeobotanical record." *Palaeogeography, Palaeoclimatolology, Palaeoecology* 231: 134–157.

Ting, S., Meng, J., Qian, L., Wang, Y., Tong, Y., Shiebout, J. A., Koch, P. L., Clyde, W. C., and Bowen, G. J. 2007. "*Ganungulatum xincunliense,* an artiodactyl-like mammal (Ungulata, Mammalia) from the Paleocene, Chijiang Basin, Jiangxi, China." *Vertebrata PalAsiatica* 45: 279–286.

Tripati, A., Backman, J., Elderfield, H., and Ferretti, P. 2005. "Eocene bipolar glaciations associated with global carbon cycle changes." *Nature* 436: 341–346.

Wang, Y. Q., Meng, J., and Ni, X. J. 2007. "Major events of Paleogene mammal radiation in China." *Geological Journal* 42: 415–430.

Wing, S. L. 1998. "Paleocene/Eocene floral change in the Bighorn Basin, Wyoming." In *Late Paleocene–Early Eocene Climatic and Biotic Events in the Marine and Terrestrial Records,* edited by M. P. Aubry, S. Lucas, and W. A. Berggren, 380–400. New York: Columbia University Press.

Wing, S. L., and Harrington, G. J. 2001. "Floral response to rapid warming in the earliest Eocene and implications for concurrent faunal change." *Paleobiology* 27: 539–563.

Wing, S. L., Harrington, G. J., Bowen, G. J., and Koch, P. L. 2003. "Floral change during the Paleocene/Eocene Thermal Maximum in the Powder River Basin, Wyoming." In *Causes and Consequences of Globally Warm Climates in the Early*

*Paleogene,* edited by S. L. Wing, P. D. Gingerich, B. Schmitz, and E. Thomas, 425–440. Boulder, CO: Geological Society of America, Special Paper 369.

Wing, S. L., Harrington, G. J., Smith, F. A., Bloch, J. I., Boyer, D. M., and Freeman, K. H. 2005. "Transient floral change and rapid global warming at the Paleocene-Eocene boundary." *Science* 310: 993–996.

Woodburne, M. O., and Swisher, C. C. 1995. "Land mammal high-resolution geochronology, intercontinental overland dispersals, sea level, climate, and vicariance." In *Geochronology Time Scales and Global Stratigraphic Correlation,* 335–364. SEPM (Society for Sedimentary Geology) Special Publication No. 54.

Zachos, J. C., Lohmann, K. C., Walker, J. C. G., and Wise, S. W. 1993. "Abrupt climate change and transient climates during the Paleogene: A marine perspective." *Journal of Geology* 101: 191–213.

Zachos, J. C., Quinn, T. M., and Salamy, K. A. 1996. "High-resolution (104 years) deep-sea foraminiferal stable isotope records of the Eocene-Oligocene climate transition." *Paleoceanography* 11: 251–266.

Zachos, J. C., Pagani, M., Sloan, L., Thomas, E., and Billups, K. 2001. "Trends, rhythms, and aberrations in global climate 65 Ma to Present." *Science* 292: 686–693.

Zachos, J. C., Dickens, G. R., and Zeebe, R. E. 2008. "An early Cenozoic perspective on greenhouse warming and carbon-cycle dynamics." *Nature* 451: 279–283.

Zanazzi, A., Kohn, M. J., MacFadden, B. J., and Terry, D. O. 2007. "Large temperature drop across the Eocene-Oligocene transition in central North America." *Nature* 445: 639–642.

Chapter 11

# *Quaternary Extinctions and Their Link to Climate Change*

BARRY W. BROOK AND ANTHONY D. BARNOSKY

Millennia before the modern biodiversity crisis—a worldwide event being driven by the multiple impacts of anthropogenic global change—a mass extinction of large-bodied fauna occurred. After a million years of severe climatic fluctuations, during which the earth waxed and waned between frigid ice ages and warm interglacials, with apparently few extinctions, hundreds of species of mammals, flightless birds, and reptiles suddenly went extinct over the course of the last 50,000 years (Barnosky, 2009). Due both to our intrinsic fascination with huge prehistoric beasts and to the possible insights these widespread species losses might lend to the modern extinction problem, the mystery of the "megafaunal" (large animal) extinctions have led to much theorizing, modeling, and digging (for their fossils or environmental proxies) over the last 150 years (Martin, 2005). The topic continues to invoke strong scientific interest (Koch and Barnosky, 2006; Grayson, 2007; Gillespie, 2008; Barnosky and Lindsey, 2010; Nogues-Bravo et al., 2010; Price et al., 2011).

In this chapter, we focus on recent work that explicitly considers the relative role of natural climate change compared to nonclimate human-caused threatening processes (such as habitat loss and hunting) in driving the megafaunal extinctions. We begin with a short review of the global pattern of Quaternary extinctions and summarize some general reasons why large animals might be particularly vulnerable to direct human impacts and climate change. We then explore how the pattern of megafaunal extinction corresponds to both the chronology of human expansion and climate change and their projected impacts. Taken together, this body of information leads us to conclude that climate change alone did not drive the mass extinction of late Quaternary megafauna, but overlain on direct and indirect human actions, it exacerbated overall extinction risk tremendously. The take-home message is that the synergy of fast climate change with more direct human impacts can have particularly fatal consequences for many nonhuman species—and this is particularly true today, when human influences, including climate disruption, are so dramatically greater than they have ever been in the past.

## Extinction and Vulnerability of Megafauna

The end-Quaternary (late Pleistocene and Holocene) die-offs comprised a significant global mass extinction event, which led to the elimination of half of all mammal species heavier than 44 kilograms (100 pounds) and other large-bodied fauna across most continents (Australia, Eurasia, North and South America) and large islands (West Indies, Madagascar, and New Zealand), between 50,000 and 600 years before present (Koch and Barnosky, 2006). The losses included large mammals (e.g., mammoth, genus *Mammuthus*), reptiles (e.g., giant lizards such as *Megalania*), and huge flightless birds (e.g., New Zealand moa and Australian *Genyornis*). In Australia, around fifty species, including rhinoceros-size wombats, short-faced kangaroos, and predatory possums disappeared (MacPhee, 1999). In North America, the death toll was some sixty species of large mammals plus the largest birds and tortoises, and South America saw the disappearance of at least sixty-six large-mammal species. Eurasia and Africa were less hard hit, but nevertheless saw major losses in their large-mammal fauna, fifteen and seventeen species, respectively. Region by region, these extinction events followed within a few centuries to a few millennia the first dispersal of *Homo sapiens* to new lands, and were particularly severe when they were also entwined with changes in the regional or global climate system (fig. 11-1).

So what was the causal mechanism behind these extinctions—climate, humans, or both? The drivers of biotic extinctions, past and present, are often surprisingly difficult to pin down (McKinney,

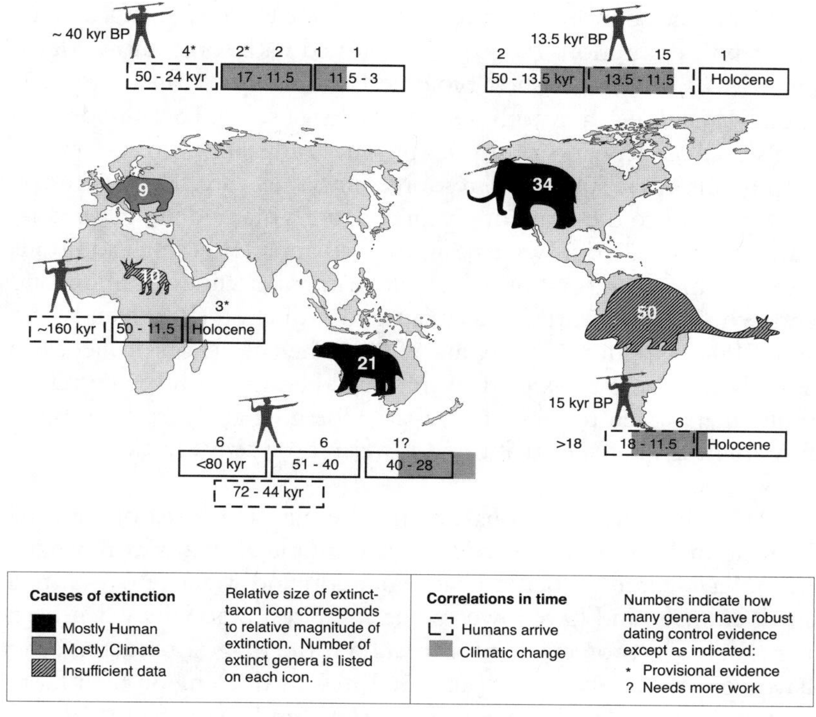


FIGURE 11-1. Late Quaternary megafaunal extinctions, human hunting, and climate change on the continents. The dashed box indicates the credible bounds of the first arrival of modern people, *Homo sapiens sapiens,* with the best estimate adjacent to the human figure (the latest estimate for Australia is about 48,000 years ago [kyr BP]). Substantial climate change events, predominantly the last glacial maximum and Holocene warming, are indicated by gray shading inside boxes. Source: Koch and Barnosky, 2006.

1997), but plausibly include: (i) being outcompeted by newly evolved or invasive species; (ii) failing to adapt to long-term environmental change (e.g., climatic shifts); and (iii) reduction in abundance caused by random disturbance events (e.g., epidemics, severe storms) with a subsequent failure to recover to a viable population (Blois and Hadly, 2009). A commonly cited generalization is that larger-bodied vertebrates (with the extreme recent form being the Quaternary megafauna) are more extinction-prone than smaller bodied ones (Bodmer et al., 1997; McKinney, 1997). Because body size is inversely correlated with population size, large-bodied animals tend to be less abundant and so more intrinsically vulnerable to rapid change and demographic disruption. Indeed, when armed with some knowledge of empirically well established biological scaling rules (allometry; Damuth, 1981), such a hypothesis makes a lot of sense. Large-bodied animals such as elephants or whales produce only a few, precocious offspring, but invest substantial resources into their care. This life-history strategy leads to the death of juveniles being a major demographic setback. On a population-wide basis, even an apparently small additional level of chronic mortality can result in rapid declines in abundance and, within a few centuries, a collapse to extinction (Brook and Johnson, 2006; Nogues-Bravo et al., 2008). The extinction proneness of large-bodied animals is further enhanced because of other correlated traits such as their requirement of large foraging area, greater food intake, high habitat specificity, and lower reproductive rates (West and Brown, 2005).

Why then (in evolutionary terms) be big? Three reasons are that large animals are long-lived (so have multiple attempts at reproduction), have relatively better heat regulation and water retention than small animals, and have lower predation rates, especially when herding. Their size protects them from all but the biggest predators, they have a great capacity to ride out hard times by drawing on their fat reserves, they can migrate long distances to find water or forage, and they can opt not to reproduce in times when environmental conditions are unfavorable, such as during a drought (Brook et al., 2007). Thus in the majority of circumstances, being big is good, because it acts as a demographic buffer. Indeed, such ecological specialization tends to evolve repeatedly because, in relatively stable environments, specialist species tend to be better than generalists at particular narrow tasks. However, when an environment is altered abruptly at a rate above normal background change, specialist species with narrow ecological preferences bear the brunt of progressively unfavorable conditions such as habitat loss, degradation, and invasive competitors or predators (Balmford, 1996; Harcourt et al., 2002). An extreme event, such as a bolide strike from space (Haynes, 2008) or an intelligent, weapon-wielding bipedal ape (Martin, 2005), that also widely alters landscapes by practices such as burning and farming, can be the lever that unhinges the optimality of this regularly evolved strategy of large body size.

The environmental context and type of threat also helps dictate an organism's response to change or novel stressors. For instance, when hunted by invading prehistoric people in Pleistocene Australia, arboreal (tree-dwelling) species occupying closed forests suffered far fewer extinctions than savanna (grassland) species, and of the latter group, those with high per capita population replacement rates (e.g., grey kangaroos; *Macropus giganteus*) or the ability to escape to refuges such as burrows (e.g., wombats; *Vombatus ursinus*) were best able to persist (Johnson, 2005).

## The Role of Human Arrivals

During the last 100,000 years, modern humans have spread across the world from their center of origin in Africa, reaching the Middle East by 90,000 years ago, Australia by 48,000 years ago (based on the most secure evidence presently known, Gillespie et al., 2006), Europe by 40,000–50,000 years ago, South America by 14,600 years ago, North America by 13,000 years ago, most of the Pacific Islands by 2,000 years ago, and New Zealand by 800 years ago. (For dates estimated by radiocarbon dating, the radiocarbon age is calibrated to calendar years.) This wave of human dispersal was likely to have been mediated by climate change: a wet penultimate interglacial probably encouraged the spread of early *Homo sapiens* out of Africa, and in the Northern Hemisphere, end-Pleistocene immigration into the Americas was facilitated by glacial ice sequestering water and lowering sea levels, which in turn exposed a land bridge between Eurasia and North America and opened coastal migration routes. At the very end of the Pleistocene, it was global warming that melted ice and opened an ice-free corridor through central Canada for a wave of Clovis hunters.

A striking feature of the megafaunal extinctions is that, in every major instance where adequate data exist, the extinction follows the first arrival of people on a "virgin" continent or large island within a few hundred to a few thousand years (fig. 11-1). This point is further underscored in figure 11-2, which shows the short overlap period for well dated megafaunal remains and archeological artifacts, in New Zealand, North America, and Australia, based on the latest dating and sample selection protocols (Gillespie, 2008). (Note the different time scales on panels A, B, and C—these three events were not synchronous in time.) Coincidence alone is not sufficient evidence for causation, but this consistency at the very least provides strong circumstantial support for the idea that a human presence was a necessary precondition for accelerated

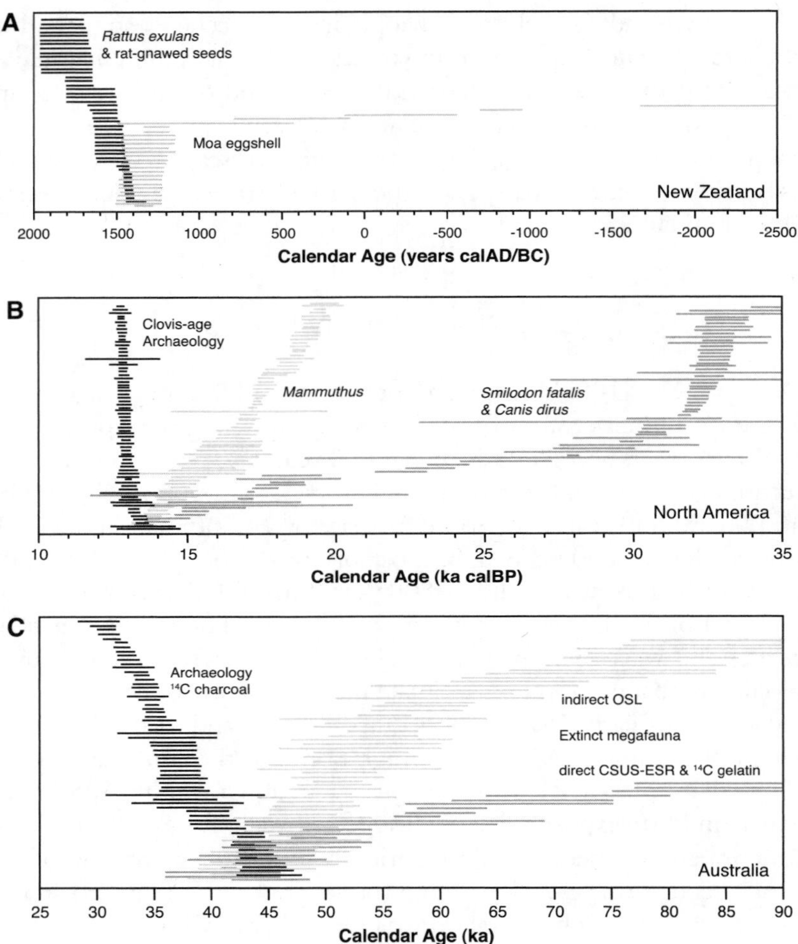


FIGURE 11-2. Dating data on human-megafauna overlap in New Zealand (A), North America (B), and Australia (C). The dates are stacked from youngest (top) to oldest (bottom) for the archeology (dark shading) and oldest (top) to youngest (bottom) for the animal remains or proxies (light gray shading). Bars represent dating uncertainties. Source: Gillespie, 2008 (includes detailed legend).

megafaunal extinction, especially given the evidence that most of the extinct taxa survived through previous, equally pronounced environmental perturbations before humans arrived.

A further line of indirect evidence comes from assessing jointly the global rise in human abundance and the precipitous loss of megafauna. We are a species that broke a fundamental ecological rule: large predators and omnivores are typically rare (Tudge, 1989). A recent analysis by one of us (Barnosky, 2008) has shown that in achieving ecological dominance, a rising biomass of people ultimately and permanently displaced the once-abundant biomass of megafauna. The point, well illustrated in figure 11-3, is that when the species richness of megafauna crashed to today's low levels, their equivalent total biomass was replaced by one species (*Homo sapiens*). Indeed, we surpassed the normal prehistoric levels of megafaunal biomass when the Industrial Revolution commenced, and now, when combined with our livestock, vastly outweigh the biomass of mammal faunas of the deep past—an explosion of living tissue supported primarily by the use of fossil energy (which, for example, makes it possible to produce and distribute inorganic fertilizers). The energetic trade-off between a large human biomass (lots of people) and a large nonhuman biomass (lots of other species) demonstrated by this Pleistocene history has a clear conservation implication: to avoid losing many more species as the human population grows in the very near future, it will be necessary to formulate policies that recognize and guard against an inevitable energetic trade-off at the global scale. The pressing need is to consciously channel some measure of natural resources toward supporting other species, rather than

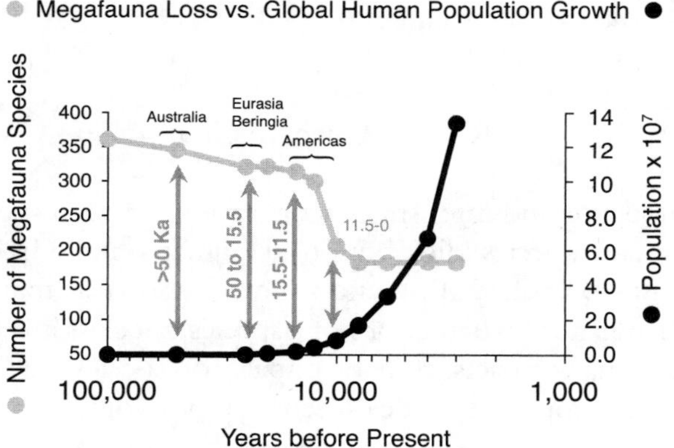


FIGURE 11-3. Decline in global megafauna biodiversity (number of species; light gray) over the last glacial-interglacial cycle, plotted against the increase in world population size of *Homo sapiens*. Major extinction events by continent are indicated by dark gray arrows. Ka = thousands of years before present) Source: Barnosky, 2008.

solely toward humans, for example, in the form of enhanced sustainable farming practices and stepped-up efforts to protect and expand existing nature reserves. Also critical will be developing alternatives to fossil fuels for the energy that currently sustains the global ecosystem, especially humans, so far above its pre-anthropogenic level of megafauna biomass.

Human impacts on late Quaternary environments were many and varied (Barnosky et al., 2004; Lyons et al., 2004a). The role of prehistoric people as hunters of big and small game has been reviewed extensively (Martin, 2005; Surovell et al., 2005; Grayson, 2007); meat was clearly a component of the hunter-gatherer lifestyle (Bulte et al., 2006), but killing may have also occurred for reasons beyond subsistence (e.g., hunter prestige). Beyond direct predation, however, humans seem to have stressed megafauna by burning vegetation on a landscape scale (and in doing so, perhaps radically altering local climate: Miller et al., 2005) and by introducing commensal species such as dogs (Fiedel, 2005), rats (Duncan et al., 2002), and disease (Lyons et al., 2004b). Overkill, the hunting of a species at a level sufficient to drive it to extinction, with or without an additional pressure from factors such as habitat modification and climate change, has been shown to be a viable killing mechanism for megafaunal species (fig. 11-4) if the hunters also could use other species when they deplete the original target species below viable abundances (Bodmer et al., 1997; Alroy, 2001, Brook and Johnson, 2006).

## Role of Climate Change

Niche modeling indicates strong correlation between specific climate variables and species distributions (Hijmans and Graham, 2006; Nogues-Bravo et al., 2008), and it now seems clear that climate is a key determinant of whether or not a species can exist in a given locale. Just like human impacts, climatic impacts on species are direct and indirect. Direct impacts include exceeding physiologically imposed temperature and precipitation limits on a species, such as critical temperature thresholds for musk oxen or pikas, which have limited heat-loss abilities. Indirect impacts include mismatch of life history strategy with timing of seasons or other climatic parameters (phenology), for example, emerging from hibernation too early in the spring, before

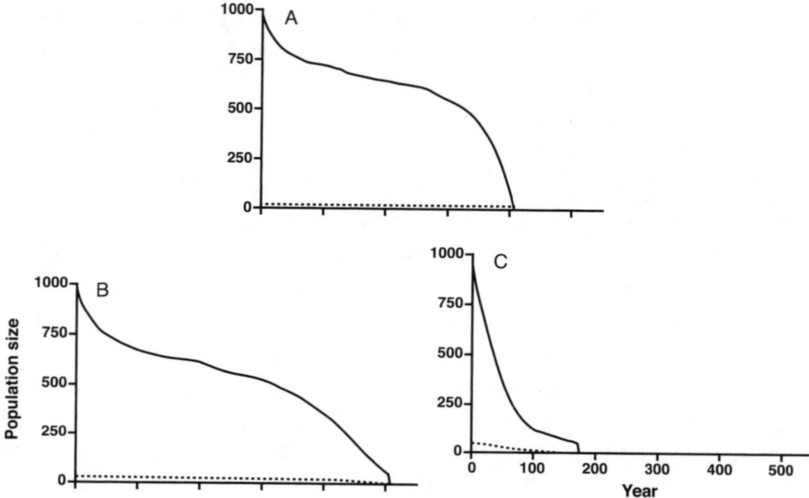


FIGURE 11-4. Overkill by the selective harvest of juveniles (less than 6 years old) of a simulated population of the extinct giant marsupial *Diprotodon optatum*. Solid line is the total regional population (carrying capacity = 1,000) and the (barely visible toward the bottom of each graph) dotted line is the annual number of juveniles killed by hunting (human population size = 150). (A) Constant hunting offtake. (B) Type II functional response (assumes prey are naïve). (C) Type III functional response (assumes adaptive prey and higher hunting pressure). Source: Brook and Johnson, 2006.

snowmelt has exposed critical food resources (Parmesan, 2006; Barnosky, 2009).

Although numerous examples of climatic change stimulating changes in local abundance or geographic range changes exist, there are few examples of climate change causing worldwide extinction in the absence of any other biotic stressor. Examples such as the golden toad (*Bufo periglenes*) and harlequin frogs (genus *Atelopus*) may qualify (Parmesan, 2006) for recent times, and in deeper time, the demise of Irish elk (*Megaloceras*) in Ireland, and horses (*Equus*) and short-faced bears (*Arctodus*) in Beringia seems attributable mainly to late Pleistocene climate changes (Barnosky, 1986; Guthrie, 2003; Barnosky et al., 2004; Koch and Barnosky, 2006). Although available models fail to adequately simulate megafaunal extinctions based on climate change alone (Brook and Bowman, 2004; Lyons et al., 2004a), modeling and empirical evidence has shown climate change alone to cause extinctions *if* species ranges are restricted by barriers that prevent them from moving to track their needed climate space (Barnosky, 1986, 2009; Thomas et al., 2004). It is precisely this latter situation in which the world's fauna (and flora) today find themselves.

The late Quaternary was a period of major natural climate change (fig. 11-5). The most prominent events were the glacial-interglacial cycles, which have repeated thirty-nine times over the last 1.8 million years; the last nine cycles show about a 100,000-year periodicity. During these shifts in climate, the globally averaged temperature changed by 4–6 degrees Celsius—comparable in magnitude to but at a much slower rate than that predicted for the coming century due to anthropogenic global warming under the fossil fuel–intensive, business-as-usual scenario (A1FI; http://www.ipcc.ch: IPCC, 2007). Triggered by orbital forcing and reinforced by albedo changes (ice-sheet retreat or growth) as well as the feedback of terrestrial and oceanic greenhouse

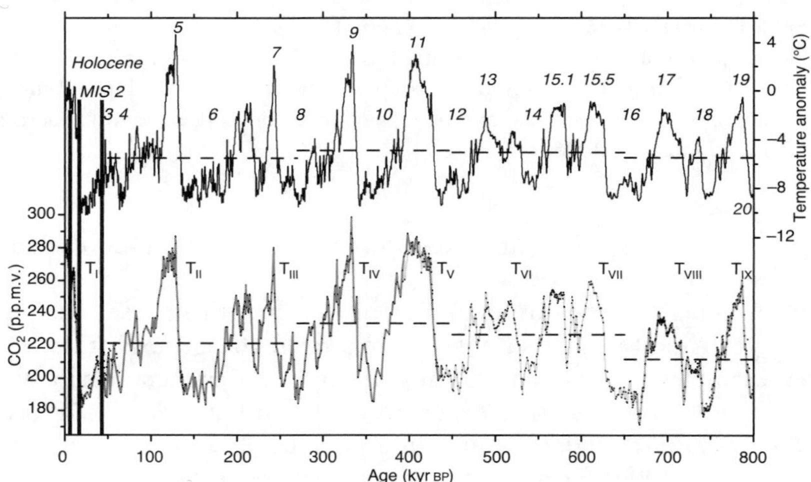


FIGURE 11-5. Antarctic ice core record of polar temperature (top; deuterium data) and carbon dioxide concentration (bottom) for the past 800,000 years. Horizontal lines show mean temperature and carbon dioxide values over different intervals. Marine isotope stages are in italics and glacial terminations by Tx (e.g., $T_I$). The vertical black lines show the timing of megafaunal extinctions in New Zealand, North America, and Australia (left to right). Source: Modified from Luthi et al., 2008.

gas release, the longer-term glacial cycles also were punctuated by numerous short-lived (and likely regional-scale) abrupt climatic changes, such as the Younger Dryas, Dansgaard-Oeschger, and Heinrich climate events (Overpeck et al., 2003). These short-term, high-magnitude climatic changes probably exacerbated any stresses that the larger-scale glacial-interglacial shifts were placing on species, although all of these kinds of cyclical changes seem within bounds of what species have evolved to withstand in the absence of impermeable geographic barriers (Barnosky, 2001; Barnosky et al., 2003; Benton, 2009).

Mechanistically, climate change over the last 100,000 years changed vegetation substantially in many parts of the world, although the nature and magnitude of the changes were different in different places (Barnosky et al., 2004). In central North America, for example, the end-Pleistocene witnessed a relatively rapid transition of vegetational structure and composition from a heterogeneous mosaic to a more zonal pattern that was relatively less suitable to large herbivores (Graham and Lundelius, 1984; Guthrie, 1984). Abrupt events such as the Younger Dryas probably superimposed even more rapid vegetation shifts (Stuart et al., 2004). In Australia, the climate became more arid as the depth of an ice age was approached, and the surface water available to large animals would have become scarcer and more patchily distributed (Wroe and Field, 2006). Yet, most megafauna species appear to have persisted across multiple glacial-interglacial transitions, only to become extinct within a few thousand years of, and in some cases, coincident with, the most recent one (fig. 11-5; extinctions marked with black vertical bars).

The resilience of species can be inferred from the fossil record and molecular markers (Lovejoy and Hannah, 2005). In the Northern Hemisphere, populations shifted ranges southward as the Fennoscandian and Laurentide ice sheets advanced (or persisted in locally equable refugia; Hewitt, 1999), and then reinvaded northern realms during interglacials. Some species may have also persisted in locally favorable refugia that were otherwise isolated within the tundra and ice-strewn landscapes (Hewitt, 1999). In Australia, large-bodied mammals were able to persist throughout the Quaternary (Prideaux et al., 2007b), even in remarkably arid landscapes such as the Nullarbor Plain (Prideaux et al., 2007a).

There were many times during the last 100,000 years when the climate apparently shifted from cool-dry to warm-wet conditions, and back again (fig. 11-6, based on the Greenland ice core data), a point

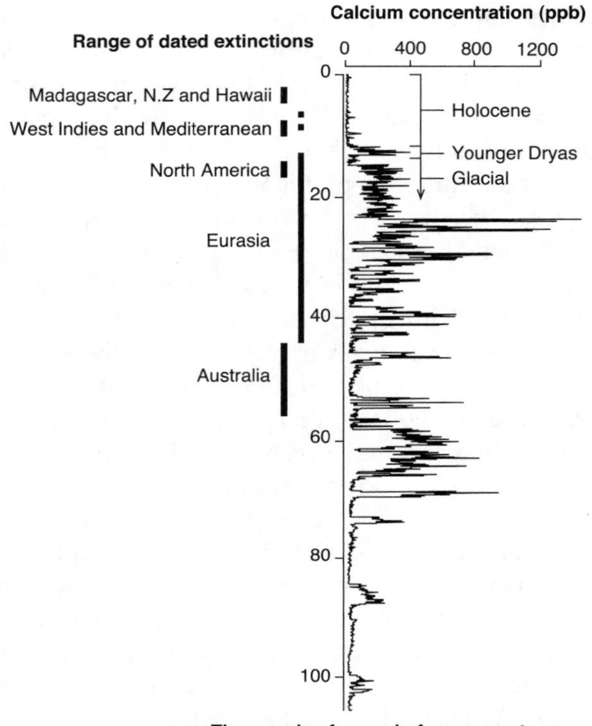


FIGURE 11-6. Greenland ice core calcium concentrations (parts per billion) over the last 100,000 years. Low values indicate wet-warm conditions with relatively denser vegetative cover, and high values point to a cool-dry climate with sparser global vegetation. Also marked are the last glacial maximum, Younger Dryas abrupt cooling event, and the Holocene warm period. The timing of extinctions on islands and continents is indicated; also shown are the earliest and latest extinctions in Beringia with Eurasia. Source: Burney and Flannery, 2005.

reinforced by new stable isotope data from Australia, as described in Brook et al. (2007) and summaries presented in recent reviews (Barnosky et al., 2004; Koch and Barnosky, 2006). Although such changes undoubtedly led to the disappearance of various species in local areas and altered their abundance where they remained on the landscape, nevertheless they persisted regionally or globally until the die-offs clustered in the last few tens of millennia of the Pleistocene and into the Holocene. If climate change were a driver of those extinctions, what was so different as to make the seemingly normal global warming (in comparison to previous glacial-interglacial transitions) at that time negatively affect such a wide range of species and habitats (Burney and Flannery, 2005; Johnson, 2005) to the extent that once-abundant, ecologically dominant animals simply disappeared? The answer to this question probably lies in threat synergies.

## Threat Synergies, Past and Present

The Pleistocene megafaunal die-offs provide a salutary lesson about the future of biodiversity under projected global warming scenarios. Over most of the last 2 million years, there was a lack of widespread extinctions, particularly of plants (Willis et al., 2004), despite regular bouts of extreme climatic fluctuations (fig. 11-5). So what made the last glacial cycle different? We believe it was the synergy of mutually reinforcing events brought by the double blow of anthropogenic threats and natural climate change. Together, these produced a demographic-ecological pressure of sufficient force and persistence to eliminate a sizeable proportion of the world's megafauna species (Barnosky et al., 2004; Brook, 2008; Barnosky, 2009; Blois and Hadly, 2009)—a group whose evolved life-history strategy left them particularly vulnerable to chronic mortality stress from a novel predator and modifier of habitats (Brook and Bowman, 2005). Without humans on the scene, climate change would not have been enough.

A good example of this interaction, using a method of coupling bioclimate envelopes and demographic modeling in woolly mammoth (Nogues-Bravo et al., 2008), shows how the human-climate synergy probably operated in the High Arctic. The model indicates that mammoths survived multiple Pleistocene climatic shifts by condensing their geographic range to suitable climate space during climatically unfavorable times. Finally, however, the new presence of modern humans during the late-Pleistocene and Holocene, at the same time as a climatically triggered retraction of steppe-tundra reduced maximally suitable habitat by some 90 percent (fig. 11-7), resulted in extinction. The important message is that mammoth populations' resilience was weakened by habitat loss and fragmentation, as it may well have been in previous interglacials, but during that last range reduction the mammoths were unable to cope because of the addition of predatory pressure (and possibly other landscape modifications) by human hunters.

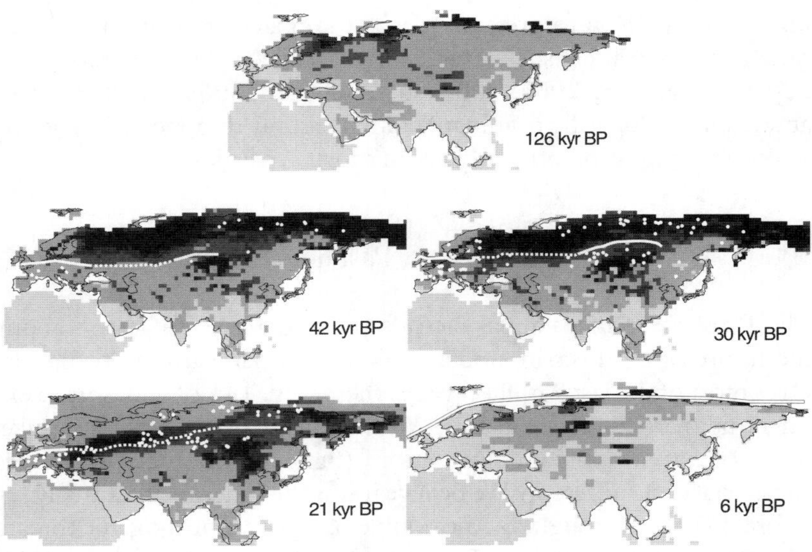


FIGURE 11-7. Climate envelope model of habitat suitability in Eurasia for woolly mammoth (*Mammuthus primigenius*) at five times over the last interglacial-glacial-interglacial cycle. Darker shading indicates higher suitability. Full glacial conditions occurred at 21,000 years before present (kyr BP), warm conditions (as warm or warmer than today) at 126 and 6 kyr BP. The white lines indicate likely northern limit of people. Line is dotted where there is uncertainty about the limit of modern humans. Source: Nogues-Bravo et al., 2008.

In principle, the same sort of fatal synergy is now attacking many species, but in a much magnified way. Modern climate change is occurring at a much faster rate than past events (Barnosky et al., 2003) and began in a world that was already relatively hot because warming started in an interglacial rather than in a glacial. By 2050, the planet is projected to be hotter than it has been at any time since humans evolved as a species. And the backdrop of human pressures on which this extreme climate change is taking place is more pronounced than ever before; in the twenty-first century the human enterprise reaches into all corners of the planet (Brook et al., 2008). Not only are we causing the climate itself to change (Miller et al., 2005), but thanks to our already high population density and ongoing population growth (fig. 11-3), extensive appropriation of natural capital, and technological expansion (Steffen et al., 2007), we are limiting more than ever before other species' ability to track their needed habitats as climate zones rapidly shift across the earth's surface. In short, we are witnessing a similar collision of human impacts and climatic changes that caused so many large animal extinctions toward the end of the Pleistocene. But today, given the greater magnitude of both climate change and other human pressures, the show promises to be a wide-screen, technicolor version of the (by comparison) black-and-white letterbox drama that played out the first time around.

## Conclusions

The important message from the late Quaternary megafaunal extinctions is not so much that humans caused extinctions in many (maybe most) places and climate caused them in others. Rather, the key point is that where direct human impacts and rapid climate change coincide, fatalities are higher and faster than where either factor operates alone. It is the synergy that presents the biggest problem, and that synergy is exactly what we find ourselves in the middle of today. Indeed, synergies between seemingly different causal mechanisms seem to characterize mass extinctions in general (Barnosky et al., 2011).

Today, that intelligent predatory ape, the human species, is driving a planetwide loss and fragmentation of habitats, overexploitation of populations, deliberate and accidental introduction of alien species beyond their native ranges, release of chemical pollution, and the global disruption of the climate system. Most damaging of all is the interactions among these different threats, which mutually reinforce each individual impact. Are the modern extinctions resulting from these processes a much magnified version of what already happened once to cause the late Quaternary megafauna extinctions, and can this perspective illuminate how to chart the future to avoid an even more severe biotic collapse? The emerging consensus quite clearly says yes, and that conclusion, in turn, implies that only a systems-based approach to threat abatement will be effective in staving off future extinctions.

Conversely, coming at the problem from trying to figure out what caused Quaternary extinctions, the question "Was it humans or natural climate change that forever ended the evolutionary journey of hundreds of megafaunal species?" is the wrong one to ask. That question anticipates a unicausal mechanism, which might be appealing on parsimonious grounds, but cannot be supported by fossil, archeological, climatological, and modeling evidence. Just as for our modern global biodiversity crisis, one factor (e.g., overhunting) may dominate in one place, and a second factor somewhere else (e.g., a species disappearing off a mountaintop that heats up too much). But at the global scale, synergy among the distinct proximate causes adds up to more than the sum of each individual cause. If one insists on a minimalistic answer for what caused the late Quaternary extinctions, it seems to be this: the actions of colonizing and expanding prehistoric humans (primarily hunting and habitat modification) seems omnipresent in the past global extinction (Brook et al., 2007; Gillespie, 2008), but in many cases, species were left much more vulnerable because of climate-induced range contractions and changes in habitat quality (Guthrie, 2006; Nogues-Bravo et al., 2008).

The degree to which climate change was the "straw that broke the camel's back" probably differed to some extent for each species of extinct Quaternary megafauna, and will only be really understood after detailed study of each extinct species (Koch and Barnosky, 2006). But the fact that even "natural" climate change synergistically exacerbated extinctions when human pressures first increased is worrisome in the modern context. The climate change is now far outside the bounds of what is normal for ecosystems (Barnosky, 2009), and the other kinds of human pressures on species are so much greater than Earth has ever seen. In the end, it will not only be the extent to which we can minimize each individual cause of extinction—increasing human population and attendant resource use, habitat fragmentation, invasive species, and now, global warming—but also the degree to which we can minimize the synergy *between* each separate cause that will determine just how many species we lose.

## Acknowledgments

We thank Marc Carrasco, Kaitlin Maguire, Lee Hannah, and two anonymous reviewers for constructive comments. BWB's research on this topic was supported by Australian Research Council grant DP0881764, and ADB's by grant DEB-0543641 from the US National Science Foundation.

## REFERENCES

Alroy, J. 2001. "A multispecies overkill simulation of the end-Pleistocene megafaunal mass extinction." *Science* 292: 1893–96.

Balmford, A. 1996. "Extinction filters and current resilience: The significance of past selection pressures for conservation biology." *Trends in Ecology and Evolution* 11: 193–96.

Barnosky, A. D. 1986. "'Big game' extinction caused by Late Pleistocene climatic change: Irish elk (*Megaloceros giganteus*) in Ireland." *Quaternary Research* 25: 128–35.

Barnosky, A. D. 2001. "Distinguishing the effects of the Red Queen and Court Jester on Miocene mammal evolution in the northern Rocky Mountains." *Journal of Vertebrate Paleontology* 21: 172–185.

Barnosky, A. D. 2008. "Megafauna biomass tradeoff as a driver of Quaternary and future extinctions." *Proceedings of the National Academy of Sciences, USA* 105: 11543–48.

Barnosky, A. D. 2009. *Heatstroke: Nature in an Age of Global Warming*. Washington, D.C.: Island Press.

Barnosky, A. D., and E. L. Lindsey. 2010. "Timing of Quaternary megafaunal extinction in South America in relation to human arrival and climate change." *Quaternary International* 217: 10–29.

Barnosky, A. D., E. A. Hadly, and C. J. Bell. 2003. "Mammalian response to global warming on varied temporal scales." *Journal of Mammalogy* 84: 354–68.

Barnosky, A. D., P. L. Koch, R. S. Feranec, S. L. Wing, and A. B. Shabel. 2004. "Assessing the causes of Late Pleistocene extinctions on the continents." *Science* 306: 70–75.

Barnosky, A. D., N. Matzke, S. Tomiya, G. O. U. Wogan, B. Swartz, T. B. Quental, C. Marshall, et al. 2011. "Has the Earth's sixth mass extinction already arrived?" *Nature* 471: 51–57.

Benton, M. J. 2009. "The Red Queen and the Court Jester: Species diversity and the role of biotic and abiotic factors through time." *Science* 323: 728–732.

Blois, J. L., and E. A. Hadly. 2009. "Mammalian response to Cenozoic climate change." *Annual Review of Earth and Planetary Sciences* 37. doi:10.1146/annurev.earth.031208.100055.

Bodmer, R. E., J. F. Eisenberg, and K. H. Redford. 1997. "Hunting and the likelihood of extinction of Amazonian mammals." *Conservation Biology* 11: 460–66.

Brook, B. W. 2008. "Synergies between climate change, extinctions and invasive vertebrates." *Wildlife Research* 35. doi:10.1071/wr07116.

Brook, B. W., and D. M. J. S. Bowman. 2004. "The uncertain blitzkrieg of Pleistocene megafauna." *Journal of Biogeography* 31: 517–23.

Brook, B. W., and D. M. J. S. Bowman. 2005. "One equation fits overkill: Why allometry underpins both prehistoric and modern body size–biased extinctions." *Population Ecology* 42: 147–51.

Brook, B. W., and C. N. Johnson. 2006. "Selective hunting of juveniles as a cause of the imperceptible overkill of the Australian Pleistocene 'megafauna.'" *Alcheringa Special Issue* 1: 39–48.

Brook, B. W., D. M. J. S. Bowman, D. A. Burney, T. F. Flannery, M. K. Gagan, R. Gillespie, C. N. Johnson, et al. 2007. "Would the Australian megafauna have become extinct if humans had never colonised the continent?" *Quaternary Science Reviews* 26: 560–64.

Brook, B. W., N. S. Sodhi, and C. J. A. Bradshaw. 2008. "Synergies among extinction drivers under global change." *Trends in Ecology and Evolution* 23: 453–60.

Bulte, E., R. D. Horan, and J. F. Shogren. 2006. "Megafauna extinction: A paleoeconomic theory of human overkill in the Pleistocene." *Journal of Economic Behavior & Organization* 59: 297–323.

Burney, D. A., and T. F. Flannery. 2005. "Fifty millennia of catastrophic extinctions after human contact." *Trends in Ecology & Evolution* 20: 395–401.

Damuth, J. 1981. "Population density and body size in mammals." *Nature* 290: 699–700.

Duncan, R. P., T. M. Blackburn, and T. H. Worthy. 2002. "Prehistoric bird extinctions and human hunting." *Proceedings of the Royal Society of London B—Biological Sciences* 269: 517–21.

Fiedel, S. J. 2005. "Man's best friend—mammoth's worst enemy? A speculative essay on the role of dogs in Paleoindian colonization and megafaunal extinction." *World Archaeology* 37: 11–25.

Gillespie, R. 2008. "Updating Martin's global extinction model." *Quaternary Science Reviews* 27: 2522–29.

Gillespie, R., B. W. Brook, and A. Baynes. 2006. "Short overlap of humans and megafauna in Pleistocene Australia." *Alcheringa Special Issue* 1: 163–85.

Graham, R. W., and E. L. Lundelius, Jr. 1984. "Coevolutionary disequilibrium and Pleistocene extinction." In *Quaternary Extinctions: A Prehistoric Revolution*, edited by Paul S. Martin and Richard G. Klein, 223–49. Tucson: University of Arizona Press.

Grayson, D. K. 2007. "Deciphering North American Pleistocene extinctions." *Journal of Anthropological Research* 63: 185–213.

Guthrie, R. D. 1984. "Alaskan megabucks, megabulls, and megarams: The issue of Pleistocene gigantism." *Contributions in Quaternary Vertebrate Paleontology: A Volume in Memorial to John E. Guilday*. Carnegie Museum of Natural History Special Publication 8: 482–510.

Guthrie, R. D. 2003. "Rapid body size decline in Alaskan Pleistocene horses before extinction." *Nature* 426: 169–71.

Guthrie, R. D. 2006. "New carbon dates link climatic change with human colonization and Pleistocene extinctions." *Nature* 441: 207–09.

Harcourt, A. H., S. A. Coppeto, and S. A. Parks. 2002. "Rarity, specialization and extinction in primates." *Journal of Biogeography* 29: 445–56.

Haynes, C. V. 2008. "Younger Dryas 'black mats' and the Rancholabrean termination in North America." *Proceedings of the National Academy of Sciences, USA* 105: 6520–25.

Hewitt, G. M. 1999. "Post-glacial re-colonization of European biota." *Biological Journal of the Linnean Society* 68: 87–112.

Hijmans, R. J., and C. H. Graham. 2006. "The ability of climate envelope models to predict the effect of climate change on species distributions." *Global Change Biology* 12: 2272–81.

IPCC. 2007. *Intergovernmental Panel on Climate Change: Fourth Assessment Report (AR4)*. Available at http://www.ipcc.ch.

Johnson, C. N. 2005. "What can the data on late survival of Australian megafauna tell us about the cause of their extinction?" *Quaternary Science Reviews* 24: 2167–72.

Koch, P. L., and A. D. Barnosky. 2006. "Late Quaternary extinctions: State of the debate." *Annual Review of Ecology, Evolution and Systematics* 37: 215–50.

Lovejoy, T. E., and L. Hannah, eds. 2005. *Climate Change and Biodiversity*. New Haven: Yale University Press.

Luthi, D., M. Le Floch, B. Bereiter, T. Blunier, J.-M. Barnola, U. Siegehnthaler, D. Raynaud, et al. 2008. "High-resolution carbon dioxide concentration record 650,000–800,000 years before present." *Nature* 453: 379–82.

Lyons, S. K., F. A. Smith, and J. H. Brown. 2004a. "Of mice, mastodons and men: Human-mediated extinctions on four continents." *Evolutionary Ecology Research* 6: 339–58.

Lyons, S. K., F. A. Smith, P. J. Wagner, E. P. White, and J. H. Brown. 2004b. "Was a 'hyperdisease' responsible for the late Pleistocene megafaunal extinction?" *Ecology Letters* 7: 859–68.

MacPhee, R. D. E. 1999. *Extinctions in Near Time: Causes, Contexts, and Consequences*. New York: Kluwer Academic/Plenum Publishers.

Martin, P. S. 2005. *Twilight of the Mammoths: Ice Age Extinctions and the Rewilding of America*. Berkeley: University of California Press.

McKinney, M. L. 1997. "Extinction vulnerability and selectivity: Combining ecological and paleontological views." *Annual Review of Ecology and Systematics* 28: 495–516.

Miller, G. H., M. L. Fogel, J. W. Magee, M. K. Gagan, S. J. Clarke, and B. J. Johnson. 2005. "Ecosystem collapse in Pleistocene Australia and a human role in megafaunal extinction." *Science* 309: 287–90.

Nogues-Bravo, D., J. Rodiguez, J. Hortal, P. Batra, and M. B. Araujo. 2008. "Climate change, humans, and the extinction of the woolly mammoth." *PLoS Biology* 6: 685–92.

Nogues-Bravo, D., Ohlemuller, R., Batra P., and Araujo, M. B. 2010. "Climate predictors of late Quaternary extinctions." *Evolution* 64: 2442–49.

Overpeck, J. T., C. Whitlock, and B. Huntley. 2003. "Terrestrial biosphere dynamics in the climate system: Past and future." In *Paleoclimate, Global Change and the Future,* edited by R. S. Bradley, T. F. Pedersen, K. D. Alverson, and K. F. Bergmann, 81–103. Berlin: Springer-Verlag.

Parmesan, C. 2006. "Ecological and evolutionary response to recent climate change." *Annual Review of Ecology Evolution and Systematics* 37: 637–69.

Price, G. J., G. E. Webb, J. X. Zhao, Y. X. Feng, A. S. Murray, B. N. Cooke, S. A. Hocknull, and I. H. Sobbe. 2011. "Dating megafaunal extinction on the

Pleistocene Darling Downs, eastern Australia: The promise and pitfalls of dating as a test of extinction hypotheses." *Quaternary Science Reviews* 30: 899–914.

Prideaux, G. J., J. A. Long, L. K. Ayliffe, J. C. Hellstrom, B. Pillans, W. E. Boles, M. N. Hutchinson, et al. 2007a. "An arid-adapted middle Pleistocene vertebrate fauna from south-central Australia." *Nature* 445: 422–25.

Prideaux, G. J., R. G. Roberts, D. Megirian, K. E. Westaway, J. C. Hellstrom, and J. I. Olley. 2007b. "Mammalian responses to Pleistocene climate change in southeastern Australia." *Geology* 35: 33–36.

Steffen, W., P. J. Crutzen, and J. R. McNeill. 2007. "The Anthropocene: Are humans now overwhelming the great forces of nature?" *Ambio* 36: 614–21.

Stuart, A. J., P. A. Kosintsev, T. F. G. Higham, and A. M. Lister. 2004. "Pleistocene to Holocene extinction dynamics in giant deer and woolly mammoth." *Nature* 431: 684–89.

Surovell, T., N. Waguespack, and P. J. Brantingham. 2005. "Global archaeological evidence for proboscidean overkill." *Proceedings of the National Academy of Sciences, USA* 102: 6231–36.

Thomas, C. D., A. Cameron, R. E. Green, M. Bakkenes, L. J. Beaumont, Y. C. Collingham, B. F. N. Erasmus, et al. 2004. "Extinction risk from climate change." *Nature* 427: 145–48.

Tudge, C. 1989. "The rise and fall of *Homo sapiens sapiens*." *Philosophical Transactions of the Royal Society of London B* 325: 479–88.

West, G. B., and J. H. Brown. 2005. "The origin of allometric scaling laws in biology from genomes to ecosystems: Towards a quantitative unifying theory of biological structure and organization." *Journal of Experimental Biology* 208: 1575–92.

Willis, K. J., K. D. Bennett, and D. Walker. 2004. "The evolutionary legacy of the Ice Ages." *Philosophical Transactions of the Royal Society of London B—Biological Sciences* 359: 157–58.

Wroe, S., and J. Field. 2006. "A review of the evidence for a human role in the extinction of Australian megafauna and an alternative interpretation." *Quaternary Science Reviews* 25: 2692–703.

# Chapter 12

## *Quaternary Tropical Plant Extinction: A Paleoecological Perspective from the Neotropics*

Mark B. Bush and Nicole A. S. Mosblech

We have found no examples of global plant extinctions from the tropics within the Quaternary. Examples of extinctions over longer periods of time are readily documented within the fossil record, with the loss of whole families evident between Eocene and modern times (Morley, 2000, 2007). Herein lies a clue to the problem of detecting extinction of tropical plants—the taxonomic resolution of the fossil record.

Most of the paleobotanical records that we have from the tropics are based on fossil pollen, plus a few on wood, and even less on seeds and other macrofossils. With a few exceptions, fossil pollen identifications are at the genus or family level, and so an extinction sufficient to remove an entire genus would be the minimum detectable level of loss. Because many tropical genera contain congeners that occupy very different habitats, losing all of them requires a huge change in the ecosystem, or a lot of bad luck. Over long enough periods of time, evolution, luck, and continental-scale modifications of climate are possible, and extinction does become evident. Because of this taxonomic bias, we actually have a clearer vision of extinction that took place between the Eocene and the Miocene than we do across the much shorter timescale of the Quaternary. We can see at that scale that major climatic events and spread of fire initiated cycles of species loss and speciation. It is not unreasonable to suppose that the spread of fire within Amazonia as a result of interglacial droughts and, later, human activity, has induced extinctions, but there is simply no datum to support this contention.

In this chapter we will offer some thoughts on the prognosis for tropical plant extinction based on our developing understanding of past climatic volatility.

## The Paleoecological Record and Assessing Extinction Risk

Any paleoecological record will contain a blend of taxa with both broad and narrow geographic ranges. Increasing the number of study sites provides more replicates of broad-ranging taxa and more complete coverage of rare species. Because the science of paleoecology frequently infers patterns from a very limited number of sites, narrowly distributed taxa are strongly underrepresented relative to broad-ranging ones. This pattern provides two biases promoting underestimations of extinction. First, there are many more rare, narrowly distributed taxa than common, broad-ranging ones (Preston, 1948). Second, basic biogeographic theory predicts that narrowly distributed taxa are inherently more vulnerable to extinction than broad-ranging ones (MacArthur and Wilson, 1967). For example, a species that has evolved to occupy particular edaphic conditions may find that as climate changes, all suitable soil types lie outside its bioclimatic envelope. Under the new conditions the species will undergo rapid directional selection in favor of new climate conditions, adaptation to a new soil type, or extinction.

In the context of extinction under climate change, paleoecological records can also identify taxa that may be extinction-prone. Truly rare species (i.e., consistently low population and very restricted distribution) are probably most at risk, but they form a small fraction of any given species pool, and are rather unlikely to be included in paleoecological records. More information is likely to be found regarding moderately abundant species. Here, paleoecologists can identify taxa that go through seemingly wild, stochastic population changes versus those that are relatively stable. Again basic predictions from evolutionary biology lead us to postulate that populations with erratic numbers have the highest probability of going extinct (the concept of gambler's ruin; Raup, 1992).

## The Importance of Landscape

High biodiversity has long been associated with regions of environmental heterogeneity, especially within the tropics (Hewitt, 2000; Rosenzweig, 1995). Debate over the latitudinal diversity gradient often centered on the role of the tropics as "museums" with low extinction rates, or "cradles" with high speciation rates (Mittelbach et al., 2007; Stebbins, 1974). For example, Fjeldsa (1994) observed that avian phylogenies in the humid lowlands of South America and Africa were rich in basal taxa, with many of the derived forms living in the adjacent highlands. This disparity led to the proposal that the lowlands functioned as museums, whereas the highlands or ecotonal areas of foothills were cradles of Plio-Pleistocene speciation. The suggestion that topographical complexity promoted isolation and allopatry was supported by the ratio of recently derived to basal species being about 1.5:1 in the lowlands and greater than 3:1 in the high Andes. However, it should be noted that Fjeldsa assumed that the pattern was the product of geographically different rates of speciation rather than extinction. This choice was probably robust, but given that 99.99 percent of all species ever to live on the planet have gone extinct (Raup, 1992), we should not ignore the phylogenetic importance of extinction just because it is essentially invisible. Colwell and Rangel (2010) suggested that the midelevations have the greatest capacity to avoid climatically induced extinction through migration. These species could move up- and downslope to maintain near equilibrium. In contrast to this, glacial cooling may have eliminated some lowland taxa because they could not go further downslope to escape cooling. Interglacial warming may have caused extinction of some montane species that could not migrate higher to escape the warmth.

Another way to look at climate change is that it could induce a migratory response to the coming and going of glaciers, effectively priming a speciation pump. Although discredited in its original form, the refugial hypothesis (Haffer, 1969) was popular because it provided an elegant means for the tropics to produce many species in response to subtle changes in climate. Haffer (1969, 2001) suggested that glacial-age aridity caused the contraction of rain forest species ranges into isolated fragments surrounded by savanna. In these refugia, the rain forest species underwent allopatric speciation. Empirical data show that rain forests were the dominant glacial vegetation, and there was no such contraction of taxa into refugia (Bush and de Oliveira, 2006; Colinvaux et al., 2001). Rather, the data show that during ice ages glaciers descended as much as 1,500 meters (4,921 feet) in the Andes, as temperatures dropped 7–8 degrees Celsius (Clapperton, 1987; Smith et al., 2005; Groot et al., 2011) and the lowlands cooled by as much as 5–10 degrees Celsius (Liu and Colinvaux, 1985; Bush et al., 2004). With each swing in temperatures, better-adapted species competed for space at lower elevations during warming or at higher elevations during cooling. However, in the middle of the Amazon lowlands, distance would have prevented the migration of cold-tolerant competitors in the time available; hence even poorly adapted lowland taxa could survive. As interglacial conditions warmed beyond the range of modern temperatures, there were no species adapted to warmer-than-modern niches to offer superior competition. In the absence of intense competition, species that were suboptimally adapted for the niche could survive (sensu Scheffer and van Nes, 2006). These species in the lowlands were dubbed "diehards" because they must have weathered all the Quaternary climate change with minimal ability to accommodate to altered growing conditions through migration (Bush and Colinvaux, 1990).

We now know that these climate changes, while broadly paced by glacial-interglacial activity, were not constant periods of cooling or warming. Indeed, precessional forcing (19,000–23,000-year cycles) set up long-term cycles of wetter or drier conditions in most tropical areas (e.g., Clement et al., 2004). Superimposed on these events were millennial-scale cycles of precipitation change in response to changes in the circulation of the Atlantic Ocean. Recent data from the Yucatan Peninsula reveal that even the short-lived Heinrich events (decadal scale) induced cold, dry conditions that influenced tropical vegetation (Correa-Metrio, 2010). Thus, the old view of the tropics as pulsing dry (glacial) and wet (interglacial) is wrong, and so too is the view of the tropics as being uniformly 5 degrees Celsius cooler than today throughout the glacial periods and as warm as modern times in the interglacials. Systems responded to forcings at a variety of superimposed temporal scales to produce extremely complex climate histories. Given this kind of heterogeneity of climate, the Quaternary is not a promising time to look for sustained evolutionary pressure toward a given climatic extreme that would last tens of tree generations.

Under these short-term stresses populations may have become small and stressed, a few may have gone extinct, or they could have undergone rapid adaptation and speciated. A probable example for this last response is evident in the apparent burgeoning Quaternary diversity of the speciose genus *Inga* (Richardson et al., 2001).

On mountainsides, there are abundant opportunities for isolation that under one circumstance could lead to extinction, while subtly different conditions might lead to speciation. Thus, one hypothesis would be that at the ecological level, speciation and extinction may be closely related phenomena. Certainly some evidence for such rapid evolution is apparent in animals, especially where natural selection is reinforced by sexual selection (e.g., Hoskin et al., 2005; Mendelson and Shaw, 2005). Logically, as diversity shows a strong decline with elevation in most bird groups, and given the exuberant speciation rates in mountainous areas, rates of extinction in the mountains must be correspondingly high (e.g., Colwell and Rangel, 2010).

Alternatively, the Andes could be generating new species but have low extinction rates. The key factor here is that the Andes is a young landscape. Even in the mid-Miocene the elevation of the mountains allowed forest cover from east to west (Hoorn et al., 1995). Consequently, all the high elevation specialists are by definition "young" species. Modern low diversity at the highest elevations, but high rates of endemicity could simply indicate low rates of evolution. Thus, extinction rates need not be high, because the age difference of the landscapes could account for the difference in the ratios of young to old species. It is also plausible that both of these scenarios are at work, according to the family or the subhabitats being considered—for example, the species of hummingbirds in *Polylepis* woodlands above tree line are being driven by different factors than the antbirds in foothill forests.

Although mountain systems are arguably more dynamic in terms of community turnover and speciation, the topographic complexity lends itself to niche continuity. Over millions of years, species might avoid extinction in montane settings because niches were continuously available. Niche continuity does not refer to a lack of change in geographic location, but rather, over time, microclimates form a shifting mosaic of habitats, within which suitable habitat for a species may change location and extent, but nonetheless is constantly present within the landscape. Small isolated populations might persist in such sheltered localities, or microrefugia (McGlone and Clark, 2005; Mosblech et al., 2011; Rull, 2009) until conditions change and their populations can expand once more. In the Andes, for example, a wide variety of soil types exists in fine geographic mosaics (Young et al., 2007). Not only does this fine-grained landscape offer chances for speciation (Gentry, 1988), but as bioclimatic envelopes move, there is still the possibility that the new area of tolerable climate will encompass a suitable soil type. In this way the dynamic environments of montane settings can serve as evolutionary "museums."

## *Diehards, Coming Climate Change, and Nonlinear Responses*

Loarie et al. (2009) demonstrated that although mountains may experience the most radical climate changes, it is lowland species that must migrate most quickly to maintain climatic equilibrium. Mountains provide steep environmental gradients. Therefore, to stay within its bioclimatic envelope, a montane species needs only migrate at a moderate pace. Even on the steep slopes of the Andes, plants are failing to keep pace with current climate change. Feeley et al. (2010) documented an average required rate of migration of 5.5–7.5 vertical meters (18–25 feet) per year to keep pace with measured warming of the last 30 years. However, preliminary data show that 62 percent of observed genera of montane trees are increasing in their mean elevation, but migrating at an average rate of only 2.5 vertical meters (8 feet) per year. The observed rates are very similar to those inferred for red maple and beech during deglaciation in the American Midwest (McLachlan and Clark, 2005) and may represent typical high rates of migration for moderate- to large-seeded species. Among the more abundant montane forest genera, upslope migration appears to be occurring more rapidly and is perhaps indicative of how tropical taxa will be redistributed in a warmer than modern landscape (Feeley et al., 2010). Rates of migration required for lowland species to leave Amazonia and gain refuge in cooler climes would be so huge as to be wholly unrealistic.

Colwell et al. (2008) suggested that as conditions warm, Amazonian species will have nowhere to which they can migrate. Furthermore, there may be no cue for seed dispersers to migrate the shortest distance toward more favorable conditions. It can be argued that because Amazonian plants—the diehards—survived the ice-age cooling and the following 5-degree Celsius warming at the onset of the Holocene, they have managed to keep pace with rapid change before. However, Colwell et al. observed that because the warming of the Holocene may be close to the maxima (say, within 1–3 degrees Celsius) of other interglacials, plants may already be near their upper thermal limit. Consequently, the next 5 degrees Celsius of warming is quite different from the 5 degrees Celsius of warming that plants experienced at the onset of the Holocene. Whether or not the physiological argument is true (this probably varies considerably from genus to genus), a 3–5-degree Celsius warming in Amazonia and the increased probability of strong drought years, such as that of 2005, greatly increases the risk of fire, for which most Amazonian plants have no tolerance.

The dieback hypothesis of Amazonian vegetation, which predicts that as much as 80 percent of Amazonia will be bare ground or savanna by 2100, is based on the simulations of the HadCM3LC outputs (Cox et al., 2000, 2004; Harris et al., 2008). This fully coupled ocean-atmosphere model includes carbon feedbacks from vegetation and soil, and it is the inclusion of this biotic component that really accelerates the "savannization" of Amazonia (Cox et al., 2000, 2004). This model certainly has some critics, who point out problems in the initial assessment of precipitation and an apparent bias in the model toward inducing exaggeratedly dry conditions (e.g., Cochrane and Laurance, 2008; Malhi et al., 2009). The HadCM3LC output can also be criticized for producing "permanent" El Niño conditions in the eastern equatorial Pacific. For every time period when high resolution paleoclimate data are available, El Niño appears to exist as a pseudocyclic phenomenon. In other words, contrary to the HadCM3LC output, ENSO does not lock into a permanent state (e.g., Moy et al., 2002; Tudhope et al., 2001). Perhaps the greatest danger of the dieback hypothesis is inducing a sense of inevitable doom among policy makers. The output is just a model, and its proponents would not defend it more strongly. The extinction of Amazonian species is far more likely to come about from the deadly cocktail of land clearance, accidental fire, and climate change than from climate change alone. Because policy can influence both land clearance and fire, it is critically important to leave policy makers with the sense of being empowered rather than just waiting for the climate shoe to drop.

An important realization for those engaged in conservation is that species' migrations and likelihoods of extinction are not linear. Despite the nice, smooth curves, of the Intergovernmental Panel on Climate Change's (IPCC) (2007) projections, the reality is a lot messier. Plants do not simply traipse upslope at a steady pace. Indeed, unforeseen feedbacks can completely alter migrational and local climatic patterns. For example, during two of the last four interglacial periods (about 120,000 and 330,000 years before present, respectively), the onset of warming in the Andes caused species to start an upslope migration. At Lake Titicaca in the Peru/Bolivian Altiplano, the Puna grasslands of the preceding glacial period gave way to invasion by *Polylepis* and, during one of these episodes at about 330,000 years ago, *Juglans* (walnut) (Hanselman et al., 2011). However, in each case, before a full woodland developed around the lake, the upslope migration halted amid aridity. Fire eliminated both the *Polylepis* and the *Juglans,* and the warm, dry conditions caused the area of the great lake to contract by 85 percent. Under what is thought to have been a steady warming trend, the system had flipped from being a mosaic parkland landscape to a salt pan. Because we know modern tree line lies between 3,500 and 3,700 meters (11,483–12,139 feet) in this section of the Andes, and the lake elevation is at 3,810 meters (12,500 feet), it is possible to parameterize this tipping point as being within about 1–1.5 degrees Celsius of modern temperatures (Bush et al., 2010). Interestingly, in the context of this chapter, the *Juglans* observed in the pollen record of Lake Titicaca might represent an extinction. We are unaware of a *Juglans* species that currently grows above the regular Andean tree line. The pollen is distinctive, so this could be a case of a species evolving rapidly to change its niche, or perhaps more likely this may be the first whiff of a climate-based extinction in the Quaternary of the Andes.

## *No-Analog Climates and Extinction*

A no-analog climate is defined as a combination of temperature, seasonality, or precipitation regime that differs from any found today. No-analog climates were first inferred from unfamiliar groupings of pollen found in the climate changes associated with the last deglaciation (Overpeck, 1992). The occurrence of these odd assemblages of plants in North America seemed to center on the Younger Dryas event of about 12,000–11,500 calendar years before present, suggesting that rapid climate change might induce them (e.g., Davis, 1986; Huntley and Birks, 1983). In the tropics, similar no-analog mixtures of species were reported, but their occurrence was not limited to the last portion of the deglacial (e.g., Bush and Colinvaux, 1990; Bush et al., 1990; Colinvaux, 1986; Piperno et al., 1990). Indeed for much of the glacial period the lowlands contained a mixture of true "lowland" species with a strong component of more cold-tolerant species characteristic of modern uplands, such as *Hedyosmum, Podocarpus, Myrsine, Symplocos,* and, in Panama, *Quercus*. Thus, no-analog floras and no-analog climates are the norm. Indeed from a late Pleistocene biologist's perspective, the vast majority of Earth would today be a no-analog system.

Analysis of global climate model predictions under either the A2 or B1 scenarios (IPCC, 2007 report) suggests that almost all tropical climates are going to change significantly by 2100. Williams et al. (2007) analyzed the extent to which the new climates would lie outside the range of modern ones, determining a category of no-analog climates. The A2 scenario induces novel climates across the center of the Amazon Basin, the Congo, eastern Tanzanian arc, and most of southeast Asia. Under the B1 scenario, the extent of novelty is much less. However, if a filter is imposed to find a modern pixel with the postulated future climate within 500 kilometers (311 miles)—a first-order approximation of how far plants might be expected to migrate—then once again the tropical lowlands are seen to support no-analog climates regardless of scenario. Williams et al. (2007) are rightly cautious about suggesting that extinction would be associated with no-analog settings, rather they suggest that the disappearing climates of high latitudes and mountaintops are the most problematic for species survival.

Because the modern climates of wide flat expanses of landscape—for example, the Amazon Plain—generally do not have very steep environmental gradients, they appear, at the scale of global climate model simulations, to be uniform. If these systems then change their climate, vast areas shift from a familiar one to a novel one, and because of the assumed uniformity and large scale, the 500-kilometer migrational barrier is not likely to be breached. These data reaffirm that species living on wide, flat plains will have vast migration distances to cover to keep pace with climate change.

Trying to predict whether the postulated climate change will lead to extinction is fraught with uncertainty. First, these are equilibrium models that expect plants to migrate and stay in equilibrium with climate. As Colwell et al. (2008) note, plants may already be close to their upper thermal limit, which would force them to migrate accordingly. However, if enough species are not in that state, or are capable of rapid speciation or adaptation, forest cover may not be breached as the model suggests. If the forest cover is not breached, then the biotic feedbacks of the model—for example, soil shading, evapotranspirative rates, and carbon stocks in biomass and soils—would remain "forest-like" rather than "savannalike," even under the new forcing.

Second, the reduction of a landscape into pixels brings into question the extent to which models truly capture the climatic variation of an area. It is fairly obvious that in mountains, strong topography, aspect, and springs seeping from hillsides all offer microclimates at spatial scales that cannot be modeled. Overall, therefore, the image of macroclimatic change greatly overstates the necessary migratory response of species, which, to a large extent, rely on microclimates for survival as seedlings. Plant populations will respond by migration away from areas where stress is high, but others will be located in these favorable microclimates, and their forced response will be much slower, perhaps lagging by millennia in the more exposed locations. However, such climatic diversity is less apparent in the Amazon lowlands.

Many models predicting migratory requirements rely primarily upon temperature as the driving force behind species responses. There are wide temperature differentials across Amazonia, without obvious correlations to diversity (Silman, 2007). However, low water availability does broadly correlate with low tree diversity (Clinebell et al., 1995; Gentry, 1988; Silman, 2007; Steege et al., 2000). Although warming a forest by several degrees Celsius is not of itself hugely problematic, changes in seasonality and fire regime may be much more influential in determining species' survival. Within the Amazon, the availability of groundwater may be a critical component in forest stability (e.g., Malhi et al., 2009). Plants thermoregulate by pumping water and evapotranspiring that moisture into the atmosphere. The scale of this response is evident in the estimate that about 50 percent of rainfall received over western Amazonia has its source in evapotranspired moisture.

The driest sections of Amazonia lie in an approximate north-south corridor between Boa Vista and Santarém, and the lowest arboreal diversities in Amazonia are associated with this corridor (Silman, 2007). This section has the highest probability of transitioning to savanna, but has also probably wavered between forest and drier ecosystems in the past. Hence the species found there today have a greater probability of being drought- and fire-tolerant than those of wetter forest areas. An insight from this observation is that the history of sites may have led to preadaptations that will lessen impacts of postulated climate change.

Classic adaptations to dry forest systems include corky bark to withstand fire and deeper rooting to access water (Cochrane, 2009; Cochrane et al., 1999; Nepstad et al., 1994), but such adaptations are rare in true rain forest. Plants have withstood considerable changes in precipitation and temperature at the peak of previous interglacials, and even during the Holocene. Between about 8,000 and 4,500 calendar years before present, a major drought event caused forest contraction at the savanna ecotone and widespread fire within southern Amazonia and the southern tropical Andes (Mayle and Power, 2008). This period would undoubtedly have produced no-analog climates in the direction of those proposed for the future. During such times of thermal stress, the lack of homogeneity in Amazonia may become apparent. Swamps, watercourses, and floodplains, and even relatively subtle landscape depressions, may have provided access to water and acted as firebreaks, ensuring species survival. Indeed, it is apparent that modern diversity survived the droughts, suggesting that Amazon forests may be more resilient to change in precipitation and fire regimes than equilibrial models suppose.

## Conclusions

Overall, there is a distinct lack of data relating to extinctions in Amazonia. We know that climate change has forced the system through all combinations of warm, wet, cold, and dry episodes, but the taxonomic resolution and incompleteness of the paleoecological record does not allow us to identify global extinctions. What are clearly evident are non-analog climates and assemblages of plants and animals (see chapter 11, megafaunal extinctions) as the neotropical ecosystems responded to climatic forcing. Because climatic change has a track record of inducing novel combinations and proportions of plant species, more no-analog systems can be expected in the future. A major issue here is whether some species are already close to their maximum physiological limit for temperature. Again, much will depend on water availability. A general assumption (though not rigorously established yet) is that prior interglacials were 1–3 degrees Celsius warmer than modern conditions. If such a warming occurred, and given the temperature-mitigating effects of increasing carbon dioxide fertilization, we will probably not leave the Quaternary bioclimatic envelope until midcentury. If the only forcing influencing populations is climatic, we probably have at least 40 years before seeing major extinction pressures in Amazonia.

Neotropical systems are water-cooled, and warming alone is relatively unlikely to induce broad-scale extinctions. However, if warming is combined with reduction in precipitation, lengthening of the dry season, and increased fire frequency, extinctions are virtually ensured. Parameterizing the threatened extinction, because we have been asked to do so, exposes the enormity of our uncertainty. At best, climate-induced extinctions by 2100 might be close to zero. If species are preadapted to somewhat warmer conditions and rainfall increases (about half the models show this), the extinction might be minimal. At worst, if precipitation decreases and there is 85 percent reduction in Amazonian forest cover, species-area curves would suggest about a 70 percent loss of species. However, that number is certainly too high. Riparian corridors and floodplains would still exist, providing habitat for thousands of species. Some species from the drier Amazonian forests are also found in wetter savanna habitats, and are likely to persist. Furthermore, the highest diversity of species is in the western Amazon, with relatively few endemics in the drier regions of eastern Amazonia. Hence, the true loss of diversity under Amazon dieback might be closer to 10 percent than 70 percent, but such speculation is really moot.

A blithe optimist could suppose that climate change will pressure systems independently of the ugly handmaidens of deforestation and human-induced wildfire. Realistically, however, the synergy of the three pressures will produce a much greater and more imminent extinction threat than climate change alone.

## REFERENCES

Bush, M. B., and P. A. Colinvaux. 1990. "A long record of climatic and vegetation change in lowland Panama." *Journal of Vegetation Science* 1: 105–119.

Bush, M. B., P. A. Colinvaux, M. Wiemann, D. Piperno, and K.-B. Liu. 1990. "Late Pleistocene temperature depression and vegetation change in Ecuadorian Amazonia." *Quaternary Research* 34: 330–345.

Bush, M. B., and P. E. de Oliveira. 2006. "The rise and fall of the Refugial Hypothesis of Amazonian speciation: A paleoecological perspective." *Biota Neotropica* DOI: 10.1590/S1676-06032006000100002.

Bush, M. B., J. A. Hanselman, and W. D. Gosling. 2010. "Non-linear climate change and Andean feedbacks: An imminent turning point?" *Global Change Biology* DOI: 10.1111/j.1365-2486.2010.02203.x.

Bush, M. B., M. R. Silman, and D. H. Urrego. 2004. "48,000 years of climate and forest change in a biodiversity hot spot." *Science* 303: 827–829.

Clapperton, C. M. 1987. "Maximal extent of the late Wisconsin glaciation in the Ecuadorian Andes." In *Quaternary of South America and Antarctic Peninsula 5,* edited by J. Rabassa, 165–180. Rotterdam, Netherlands: A.A. Balkema.

Clement, A. C., A. Hall, and A. J. Broccoli. 2004. "The importance of precessional signals in the tropical cimate." *Climate Dynamics* 22: 327–341.

Clinebell, R. R., O. L. Phillips, A. H. Gentry, N. Stark, and H. Zuuring. 1995. "Prediction of neotropical tree and liana species richness from soil and climatic data." *Biodiversity and Conservation* 4: 56–90.

Cochrane, M. A. 2009. "Fire, land use, land cover dynamics, and climate change in the Brazilian Amazon." In *Tropical Fire Ecology: Climate Change, Land Use, and Ecosystem Dynamics,* edited by M. A. Cochrane, 389–426. Berlin: Springer.

Cochrane, M. A., A. Alencar, M. D. Schulze, C. M. Souza Jr., D. C. Nepstad, P. Lefebvre, and E. A. Davidson. 1999. "Positive feedbacks in the fire dynamic of closed canopy tropical forests." *Science* 284: 1832–1835.

Cochrane, M. A., and W. F. Laurance. 2008. "Synergisms among fire, land use, and climate change in the Amazon." *Ambio* 37: 522–527.

Colinvaux, P. 1986. "Amazon diversity in light of the paleoecological record." *Quaternary Science Reviews* 6: 93–114.

Colinvaux, P. A., G. Irion, M. E. Räsänen, M. B. Bush, and J. A. S. Nunes de Mello. 2001. "A paradigm to be discarded: Geological and paleoecological data falsify the Haffer and Prance refuge hypothesis of Amazonian speciation." *Amazonia* 16: 609–646.

Colwell, R. K., and T. F. Rangel. 2010. "A stochastic, evolutionary model for range shifts and richness on tropical elevational gradients under Quaternary glacial cycles." *Philosophical Transactions of the Royal Society B: Biological Sciences* 365: 3695–3707.

Colwell, R. K., G. Brehm, C. L. Cardelus, A. C. Gilman, and J. T. Longino. 2008. "Global warming, elevational range shifts, and lowland biotic attrition in the wet tropics." *Science* 322: 258–261.

Correa-Metrio, A. 2010. *An 86,000-Year History of Paleoecological and Paleoclimatic Change from the Yucatan Peninsula, Guatemala*. Melbourne, FL: Florida Institute of Technology.

Cox, P. M., R. A. Betts, C. D. Jones, S. A. Spall, and I. J. Totterdell. 2000. "Acceleration of global warming due to carbon-cycle feedbacks in a coupled climate model." *Nature* 408: 184–187.

Cox, P. M., R. A. Betts, M. Collins, P. P. Harris, C. Huntingford, and C. D. Jones. 2004. "Amazonian forest dieback under climate-carbon cycle projections for the 21st century." *Theoretical and Applied Climatology* 78: 137–156.

Davis, M. B. 1986. "Climatic instability, time lags, and community disequilibrium." In *Community Ecology,* edited by J. Diamond and T. J. Case, 269–284. New York: Harper & Row.

Feeley, K. J., M. R. Silman, M. B. Bush, W. Farfan, K. G. Cabrera, Y. Malhi, P. Meir, M. S. Revilla, M. N. R. Quisiyupanqui, and S. Saatchi. 2010. "Upslope migration of Andean trees." *Journal of Biogeography* DOI: 10.1111/j.1365-2699.2010.02444.x.

Fjeldsa, J. 1994. "Geographical patterns for relict and young species of birds in Africa and South America and implications for conservation priorities." *Biodiversity and Conservation* 3: 207–226.

Gentry, A. H. 1988. "Changes in plant community diversity and floristic composition on environmental and geographical gradients." *Annals of the Missouri Botanical Garden* 75: 1–34.

Groot, M. H., R. G. Bogotá, L. J. Lourens, H. Hooghiemstra, M. Vriend, J. C. Berrio, E. Tuenter, et al. 2011. "Ultra-high resolution pollen record from the northern Andes reveals rapid shifts in montane climates within the last two glacial cycles." *Climate of the Past* 7: 299–316.

Haffer, J. 1969. "Speciation in Amazonian forest birds." *Science* 165: 131–137.

Haffer, J., and G. T. Prance. 2001. "Climatic forcing of evolution in Amazonia during the Cenozoic: On the refuge theory of biotic differentiation." *Amazonia* 16: 579–608.

Hanselman, J. A., M. B. Bush, W. D. Gosling, A. Collins, C. Knox, P. A. Baker, and S. C. Fritz. 2011. "A 370,000-year record of vegetation and fire history around Lake Titicaca (Bolivia/Peru)." *Palaeogeography, Palaeoclimatology, Palaeoecology* 305: 201–214.

Harris, P. P., C. Huntingford, and P. M. Cox. 2008. "Amazon basin climate under global warming: The role of the sea surface temperature." *Philosophical Transactions of the Royal Society B* 363: 1753–1759.

Hewitt, G. M. 2000. "The genetic legacy of the Quaternary ice ages." *Nature* 405: 907–913.

Hoorn, C., J. Guerrero, G. A. Sarmiento, and M. A. Lorente. 1995. "Andean tectonics as a cause for changing drainage patterns in Miocene northern South America." *Geology* 23: 237–240.

Hoskin, C. J., M. Higgie, K. R. McDonald, and C. Moritz. 2005. "Reinforcement drives rapid allopatric speciation." *Nature* 437: 1353–1356.

Huntley, B., and H. J. B. Birks. 1983. *An Atlas of Past and Present Pollen Maps of Europe: 0–13,000 Years Ago*. Cambridge: Cambridge University Press.

IPCC. 2007. *Climate Change 2007: Climate Change Impacts, Adaptation and Vulnerability*. Geneva: IPCC.

Liu, K.-B., and P. A. Colinvaux. 1985. "Forest changes in the Amazon basin during the last glacial maximum." *Nature* 318: 556–557.

Loarie, S. R., P. B. Duffy, H. Hamilton, G. P. Asner, C. B. Field, and D. D. Ackerly. 2009. "The velocity of climate change." *Nature* 462: 1052–1055.

MacArthur, R. H., and E. O. Wilson. 1967. *The Theory of Island Biogeography*, Monographs in Population Biology, v. 1. Princeton, NJ: Princeton Press.

Malhi, Y., L. E. O. C. Aragão, D. Galbraith, C. Huntingford, R. Fisher, P. Zelazowski, S. Sitch, C. McSweeney, and P. Meir. 2009. "Exploring the likelihood and mechanism of a climate-change-induced dieback of the Amazon rainforest." *Proceedings of the National Academy of Sciences, USA* 106: 20610–20615.

Mayle, F. E., and M. J. Power. 2008. "Impact of a drier Early-Mid-Holocene climate upon Amazonian forests." *Philosophical Transactions of the Royal Society B* 363: 1829–1838.

McGlone, M. S., and J. S. Clark. 2005. "Microrefugia and macroecology." In *Climate Change and Biodiversity,* edited by T. Lovejoy and L. Hannah, 157–159. New Haven: Yale University Press.

McLachlan, J. S., and J. S. Clark. 2005. "Molecular indicators of tree migration capacity under rapid climate change." *Ecology* 86: 2088–2098.

Mendelson, T. C., and K. L. Shaw. 2005. "Sexual behaviour: Rapid speciation in an arthropod." *Nature* 433: 375–376.

Mittelbach, G. G., D. W. Schemske, H. V. Cornell, A. P. Allen, J. M. Brown, M. B. Bush, S. P. Harrison, et al. 2007. "Evolution and the latitudinal diversity gradient: Speciation, extinction, and biogeography." *Ecology Letters* 10: 315–331.

Morley, R. J. 2000. *Origin and Evolution of Tropical Rain Forests*. Chichester, UK: Wiley and Sons.

Morley, R. J. 2007. "Cretaceous and Tertiary climate change and the past distribution of megathermal rainforests." In *Tropical Rainforest Responses to Climatic Change*, edited by M. B. Bush and J. R. Flenley, 1–31. Chichester, UK: Praxis.

Mosblech, N. S., M. B. Bush, and R. van Woesik. 2011. "On metapopulations and microrefugia: Palaeoecological insights." *Journal of Biogeography* 38: 419–429.

Moy, C. M., G. O. Seltzer, D. T. Rodbell, and D. M. Anderson. 2002. "Variability of El Niño/Southern Oscillation activity at millennial timescales during the Holocene epoch." *Nature* 420: 162–164.

Nepstad, D. C., C. R. de Carvalho, E. Davidson, P. Jipp, P. Lefebvre, G. H. Negreiros, E. D. da Silva, et al. 1994. "The role of deep roots in water and carbon cycles of Amazonian forests and pastures." *Nature* 372: 666–669.

Overpeck, J. T., R. S. Webb, et al. (1992). "Mapping eastern North American vegetation change of the past 18 Ka: No-analogs and the future." *Geology* 20 (12): 1071–1074.

Piperno, D. R., M. B. Bush, and P. A. Colinvaux. 1990. "Paleoenvironments and human occupation in late-glacial Panama." *Quaternary Research* 33: 108–116.

Preston, F. W. 1948. "The commonness and rarity of species." *Ecology* 29: 254–283.

Raup, D. 1992. *Extinction: Bad Genes or Bad Luck*. Chicago: W.W. Norton & Company.

Richardson, J. E., R. T. Pennington, T. D. Pennington, and P. M. Hollingsworth. 2001. "Rapid diversification of a species-rich genus of neotropical rain forest trees." *Science* 293: 2242–2245.

Rosenzweig, M. L. 1995. *Species Diversity in Space and Time*. Cambridge: Cambridge University Press.

Rull, V. 2009. "Microrefugia." *Journal of Biogeography* 36: 481–484.

Scheffer, M., and E. H. van Nes. 2006. "Self-organized similarity, the evolutionary emergence of groups of similar species." *Proceedings of the National Academy of Sciences* 103: 6230–6235.

Silman, M. 2007. "Plant species diversity in Amazonian forests." In *Tropical Rainforest Responses to Climatic Change,* edited by M. B. Bush and J. R. Flenley, 269–294. Berlin: Springer.

Smith, J. A., G. O. Seltzer, D. L. Farber, D. T. Rodbell, and R. C. Finkel. 2005. "Early local last glacial maximum in the tropical Andes." *Science* 308: 678–681.

Stebbins, G. L. 1974. *Flowering Plants: Evolution above the Species Level*. Cambridge, MA: The Belknap Press of Harvard University Press.

Steege, H. T., D. Sabatier, H. Castellanos, T. V. Andel, J. Duivenvoorden, A. A. de Oliveira, R. Ek, et al. 2000. "An analysis of the floristic composition and diversity of Amazonian forests including those of the Guiana Shield." *Journal of Tropical Ecology* 16: 801–828.

Tudhope, A. W., C. P. Chilcott, M. T. McCulloch, E. R. Cook, J. Chappell, R. M. Ellam, D. W. Lea, et al. 2001. "Variability in the El Niño-Southern Oscillation through a glacial-interglacial cycle." *Science* 291: 1511–1517.

Williams, J. W., S. T. Jackson, and J. E. Kutzbach. 2007. "Projected distributions of novel and disappearing climates by 2100 AD." *Proceedings of the National Academy of Sciences, USA* 104: 5738–5742.

Young, K. R., B. Leon, P. M. Jorgenson, and C. Ulloa-Ulloa. 2007. "Tropical and subtropical landscapes of the Andes." In *The Physical Geography of South America*, edited by T. T. Veblen, K. R. Young, and A. R. Orme. Oxford: Oxford University Press.

# PART V

## Predicting Future Extinctions

We now come to the future. The first two parts of this book examined modeling views of the future. But as we've seen from contemporary extinctions and the record of the past, there is no perfect model of future impacts. So we turn to a more qualitative analysis by groups of experts on a variety of natural systems. We look at freshwater and marine systems, tropical forests and invertebrates and species interactions. Robert Dunn and Matt Fitzpatrick look at invertebrate extinctions in a chapter devoted to the taxa that are likely to suffer the most numerous extinctions under any estimates. Yadvinder Malhi complements the insect extinctions chapter with a look at extinction of tropical forests, where most of the invertebrates reside. The marine chapters address a full spectrum of impacts. Ove Hoegh-Guldberg focuses on possible future impacts on coral reefs, the system most heavily affected by climate change over the past half century. Benjamin Halpern and Carrie Kappel take on the rest of the oceans. Lesley Hughes concludes this part by looking at the complexities that species interactions may introduce.

Although none of these views of the future provide the precise quantitative estimates of extinction risk possible in the modeling studies, they all offer unique perspectives, often troubling, on the fate of biodiversity as climate changes. The threat to coral reefs in the near future is perhaps the clearest, but many other taxa are entering a similarly perilous future.

# Chapter 13

## *Every Species Is an Insect (or Nearly So): On Insects, Climate Change, Extinction, and the Biological Unknown*

ROBERT R. DUNN
AND MATTHEW C. FITZPATRICK

Any estimate of the number of species on Earth at risk from climate change must begin with the question of how many species can be found on Earth, and because most species are insects, how many insect species in particular. The question of how many species of insects live on Earth and where they can be found is an old one. Linnaeus was aware of variation from place to place in the diversity of insects but believed that most insect species could be named in his lifetime. One of Linnaeus's students (he called them apostles), Daniel Rolander, traveled to Surinam, however, and encountered there a diversity of insect life that he found overwhelming. Rolander began to wonder in confronting such diversity whether the species he saw would ever all be collected (the task he had been given) and named (the task Linnaeus would take for himself when Rolander returned) (Dunn, 2009c). Rolander's experience was a hint of what was to come. Nearly two hundred years later, in the late 1970s, Terry Erwin was studying beetles in Panama when Peter Raven asked for his best guess as to the number of species of insects in a hectare of tropical forests. Erwin had been fogging tropical forest canopy trees to collect carabid beetles (also known, somewhat ironically, as ground beetles). Erwin had a very physical sense that the answer, whatever it might be, was a big number. That sense came from the observation that as he fogged one tree and then another, the overlap in species from tree to tree was often very low. Erwin's great insight in thinking about Raven's question was that if one knew the number of beetle or other insect species specialized on a given tree species, then the algebra necessary to figure out the number of tropical canopy insect species on Earth, assuming host specificity doesn't vary in space, and so forth, was simple.

Erwin estimated that on the basis of his knowledge at the time of host specificity, that there might be a total of 30 million tropical insect species on Earth (Erwin, 1982). Prior to Erwin's paper, estimates of the number of all species on Earth, not just tropical insect species, had consistently been conservative (Erwin, 1982; Stork, 1988; Dunn, 2009c). As humans we tend to consistently imagine that we know most of what remains to be known and that what is left around the corner is just details, a species here or there, and that most species are like us. In the first few years after Erwin's paper, there were diverse critiques of his estimate (Stork, 1988; Gaston, 1991; Hodkinson and Hodkinson, 1993; Basset et al., 1996). With time though, studies have come to focus on Erwin's parameter estimates, rather than critiquing or improving his approach, with a particular emphasis on measuring host specificity (Novotny et al., 2002). Studies subsequent to Erwin's have tended to suggest values of host specificity lower than those he measured and hence also lower estimates of the global diversity of tropical arthropod species, with a range in recent studies from 2.8 million (which would essentially mean most species are already named) to 10 million (meaning that 80 percent of species are unnamed) (Odegaard, 2000; Odegaard et al., 2000; Novotny et al., 2002). Such uncertainty magnifies as one steps back to consider the terrestrial world more generally. If there are 2.8 million to 10 million tropical arthropod species, how many total arthropod species might there be? Twice that? Several-fold more? As Odegaard concluded in his 2000 review, "Uncertainty is too high, and data sets are still too few. We certainly have a long way to go."

As a consequence of our remaining ignorance, the difference between the known number of species and the total number of species out there in the world—the biological unknown—walking, flying, and crawling around is somewhere between 2.8 million species and, at the opposite extreme, 10 million or even tens of millions of species. We do not know enough yet to distinguish between these possibilities, nor have we shown ourselves in the years since Erwin's original paper to be very dedicated at distinguishing these possibilities, excellent work at a few study sites notwithstanding (Longino et al., 2002, Novotny et al., 2006, 2007). No single hectare of tropical forest anywhere in the world has yet been exhaustively surveyed for arthropods (a prerequisite, Sir Robert May once argued, for having any confidence in any global estimates [May, 1992]). We are not even aware of any temperate forest habitats that one might conclude have been exhaustively sampled. Similarly, none of the existing methods of estimating global diversity have been compared to any sort of simulated world to test their accuracy or power.

If most species are insects and total insect diversity at a global scale is unknown and perhaps unknowable, what are we to do when it comes to the question of estimating the number of species likely at risk to extinction due to anthropogenic climate change? In the context of climate change, as with other questions, our approach to date has been to deal with this problem of the biological unknown by focusing on well known species and lineages in well known places. In most cases, this approach has led us to focus on vertebrates or more rarely plants. Here, we use a well studied insect taxon (but see Current Temporal and Geographic Trends in Ant Diversity section below), ants, to explore patterns of distribution and the potential impacts of climate change.

We will argue here—using data from North American ants—that the species and regions where we know most about insects (cold and otherwise marginal climates) are precisely those that are least important in the context of estimating the number of extinctions due to climate change. Instead, it is the unknown species and regions, which tend to be hot, tropical, and far away from the homes of most biologists, that are simultaneously most important to estimates of climate-induced insect extinctions and furthest from being well understood. Before moving to our discussion of the future of insect diversity, we first ask: What do we know about present patterns?

## Current Temporal and Geographic Trends in Ant Diversity

Most insect species remain unnamed, even ants, which are, compared to other taxa—even taxa as economically important as mosquitoes—well studied. The number of species described by year with the proportion of those species described in the tropics (fig. 13-1A) and the number of species described by biogeographic region (fig. 13-1B), with predominantly tropical biogeographic regions in black and other regions in gray. Data are binned into 10-year intervals. The extent to which ants remain incompletely known is indicated by the continuing

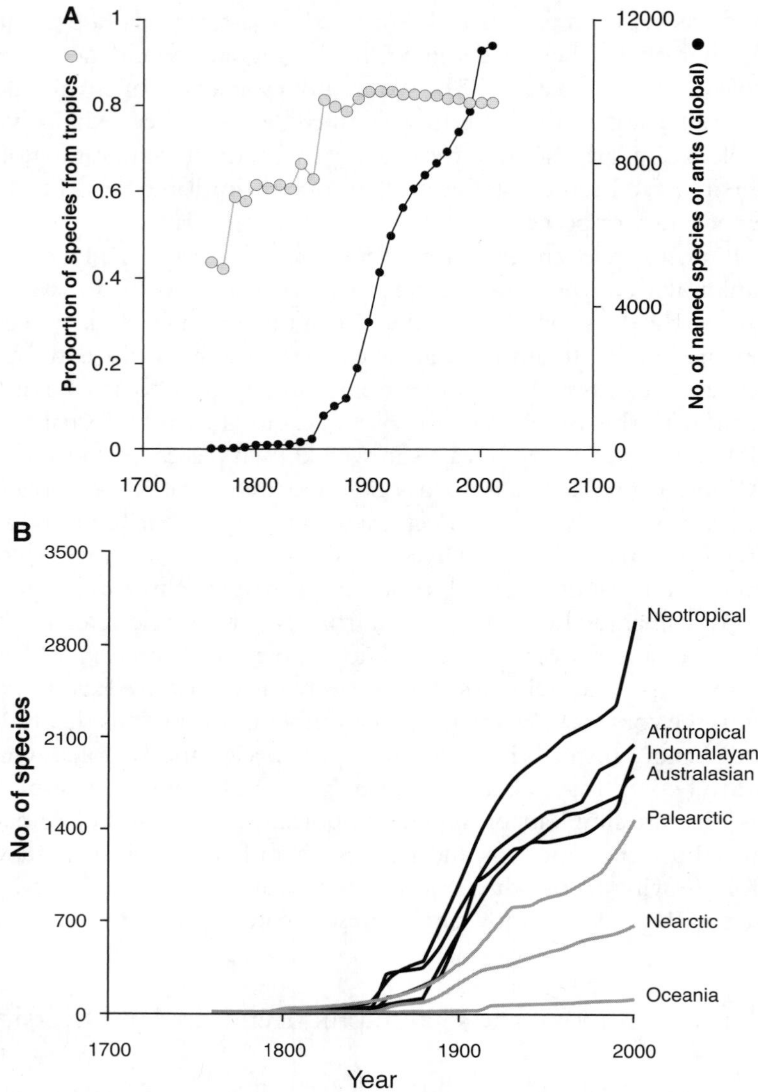


FIGURE 13-1. The number of species described by year with the proportion of those species described in the tropics (A) and the number of species described by biogeographic region (B). Tropics are represented by gray in (A) but black in (B).

increase in the number of named ant species, in nearly every biogeographic region. In well studied places, this increase in the number of species comes both from the discovery of cryptic species and from the discovery and naming of totally new (yet obvious once encountered) finds. In contrast, discoveries in other regions tend to come from the work of individual systematists who finally find the time to name species in their collection drawers or even backyards. In recent years, several new genera of ants have been found in the tropics and even an entirely new subfamily. Interestingly, although the total number of named species continues to rise, with no sign of plateau, the proportion of species from tropical biogeographic regions seems to have reached at least a temporary asymptote. Whether this asymptote is real or not, time will tell, but at the very least it conveys the extent to which most named ant species are, and will likely remain, tropical.

## Four Responses to Climate Change

Arguably, there are just four things that can happen to species in response to climate change. Their ranges can shrink (with possible extinctions), their ranges can expand (with the possibility of speciation), they can move (with extinction as a possibility if they are unable to move), or they can evolve. In different biomes, different processes are likely to predominate. We consider, in turn, regions in which the fate of species historically present is likely to be dominated by range contraction, range expansion, and range shifts, respectively. In addition, we consider those regions where novel conditions and hence novel biomes are likely to emerge, so-called no-analog biomes and climates.

### *Range Contraction*

Most of the work to date on species ranges and climate change has focused on biomes and regions in which most of the species historically present have shrinking geographic ranges. Such studies include excellent work on high latitude biomes (tundra, taiga, and so on) (Henry and Molau, 1997), high elevation biomes (Brown, 1971; Hill et al., 2002; Walther et al., 2005; Sekercioglu et al., 2008), and biomes that are regionally marginal, such as both temperate and tropical forests in Australia (Williams et al., 2003; Fitzpatrick et al., 2008). In these regions, species have few choices but to adapt to new conditions. Although such shrinking biomes are home to a number of bird and mammal species, including charismatic species such as polar bears, because they tend to be in cooler climates or at high elevations (with important exceptions), they include a very small proportion of all ant species and probably insect species more generally (Lessard et al., 2007; Colwell et al., 2008). For example, figure 13-2A shows the number of ant species in North America as a function of the maximum mean annual temperature at which they occur. Very few ant species are found only in those coolest high elevation and high latitude conditions expected to contract as a consequence of climate change (fig. 13-2A).

Like ants, all the other insect taxa studied to date at global scales also show the concentration of the vast majority of species in tropical and to a lesser extent arid biomes (e.g., Stork, 1988). Consequently, the climatically marginal regions where biomes will shrink may have large proportions of their species go extinct, but those extinctions will never represent a large proportion of the global fauna. In fact, marginal climatic conditions may generally tend to have few species (relative to global totals), as suggested by simulation models (Rangel et al., 2007) and by observations that marginal climates that have experienced climate change in the past tend to already have fewer species today than they might have otherwise had, both for insects such as ants (Dunn et al., 2009a) and more generally (Jansson and Dynesius, 2002; Jansson, 2003). One exception to this general pattern may be the Mediterranean and other relatively dry regions where less arid middle elevations sometimes have more species than do lower elevations (Sanders et al., 2003; Wilson et al., 2005; Botes et al., 2006).

To summarize, although interesting for many reasons, we suspect shrinking biomes are irrelevant to global estimates of insect (and other) extinctions from climate change. A formal analysis of the proportion of species of different insect taxa (or even genera of different insect taxa) confined to expanding versus shrinking biomes would be telling, but is for now conceivable for only a small subset of groups, which also tend to be those, like ants, that are less diverse to start with.

### *Range Expansion*

Perhaps the least studied set of biomes and species with respect to climate change is those that will, under new climates and global change,

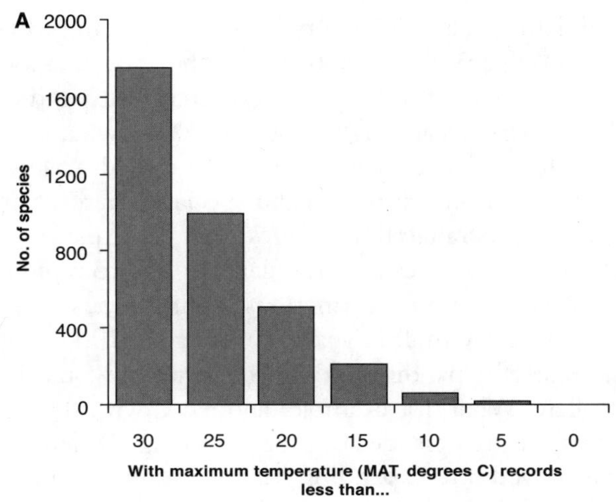


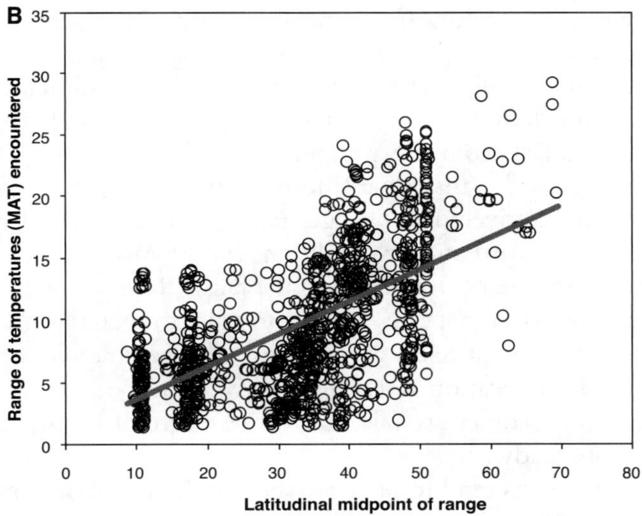


FIGURE 13-2. The number of species with no records at sites above 0 degrees Celsius, 5 degrees, 10 degrees, and so on in terms of mean annual temperature (A) and the relationship between the range of mean annual temperatures a species occurs at and the latitudinal midpoint of its geographic range (B). In (A), nearly all species are recorded at least somewhere within their range from sites with mean annual temperatures above 25 degrees Celsius. In (B), species from more southern latitudes tend to have ranges that encompass a much narrower range of conditions.

expand (Williams et al., 2007). Dry habitats, for example, are expanding globally (Williams et al., 2007), and the species associated with such habitats and climates should be expected to expand their ranges both with and perhaps among regions. To know what effect such expansions will have on net extinction rates, one needs to know how expansion will affect both extinction and speciation rates. Studies of the radiation of the Australian flora suggest that the expansion of biomes in general, but arid biomes in particular, can lead to rapid net diversification (speciation minus extinctions), with thousands of species evolving in just a few million years (Crisp et al., 2004). Other examples of the fate of expanding biomes historically would be useful to study or collate. What, for example, happened with the expansion of cool climates during episodes of global cooling? Where did the new species that moved into expanding climates come from and how rapidly did they diversify? Although the dominant effect of climate change will undoubtedly be extinctions, speciation events will also occur, and if we are simply interested in tallying totals, such novel lineages are also part of the story. The rates of diversification seen in association with drying in Australia are slow on a timescale of human planning, but fast from the perspective of Earth history. The Australian case suggests that diversification in expanding biomes may proceed from relatively few lineages particularly able to tolerate and disperse across the expanding conditions (in the Australian plant case, for example, species of *Acacia* and of the family Myrtacaeae). The rate of origin of new insect species is likely to pale compared to the rate of extinction of insect species. Nonetheless, the geographic areas over which net diversification rates are positive may be relatively large, such that these biomes are relevant to the future of biodiversity and deserve more study.

Because insects tend to be more thermophilic than are other animals that receive study, they may be more likely to be favored over the long term in biomes that are expanding due to warming. For example, almost half of the ant species in North America can be found at sites with mean annual temperatures above 25 degrees Celsius, such that depending on the levels of precipitation in those warm climates that expand, the number of species with expanding ranges may be great. Nonetheless, from the strict perspective of extinction rates, expanding biomes and the species associated with them are unlikely to be a significant component of the global story.

### *Range Shifts*

Many biomes and the species that compose them will move (or at least need to, if they are to survive) in light of global change (Malcolm et al., 2002). Some may both move and shrink. Others may move and expand. We will concentrate on the movement per se here. Arguably the first question in understanding the migration of species and shifts in biomes in response to climate change is where species would or should go given perfect dispersal (Colwell et al., 2008).

Key to understanding the extent to which species in general and insect species in particular track changes in climate has been observations of shifts of temperate insect species with modern climate change (Parmesan, 2006). Observations of shifts in insect species ranges have to date shown that temperate butterfly species shift up in latitude and elevation in response to warming (Wilson et al., 2005). On elevational gradients, evidence from at least one study suggests that the colonization of newly favorable sites may be slower than the loss from newly unfavorable sites (Wilson et al., 2005), whether because of dispersal limitation, the fragmentation of landscapes, or the interaction of the two. In contrast to studies of recent climate change, studies of prehistoric shifts in the distribution of insect species in response to climate change tend to suggest that insect species track climate very well at both the cool and warm edges of their geographic ranges and tend to persist even at climatic margins rather than go extinct (Coope and Wilkins, 1994; Coope, 2004). A key difference between contemporary and historic changes in climate has to do with the projected rate of change and the relative availability of different habitat types. Today, more so than in the past, many natural habitat types cover relatively small geographic areas, such that corridors for migration and migration rates may be slower than simple distribution models assume and historical examples suggest. (One interesting exception in this regard may be for species that do well in urban and disturbed environments, for which cities may provide excellent conduits for movement; e.g., Menke et al., 2011).

Clearly, there remains much to be learned in terms of how insect species move (or don't move) with climate change. Perhaps the bigger problem is the geographic distribution of studies of insect movement and climate change. Both distribution models of insects and studies of responses of insects to paleoclimatic changes are nearly all from temperate regions (Coope and Wilkins, 1994; Coope, 2004; but see Fitzpatrick et al., 2011), but insect species are nearly all from tropical and subtropical regions—at least that is our understanding to date, an understanding that holds up, for example, when considering the patterns of diversity in ant genera, nearly 80 percent of which are tropical.

Studies of the migration of tropical species in response to climate change, whether prehistorical or modern, would be disproportionately valuable. We need to understand local changes in species ranges much better than we do now. One potential barrier to movement that is unique (or magnified anyway) in the tropics relative to other biomes is the geographic distance that species need to move to track their climates. Because climatic gradients are shallower in the tropics than in temperate realms (i.e., less change in temperature occurs over similar distances; Colwell et al., 2008; Loarie et al., 2009), the distances species need to move in the tropics may be much larger than in other biomes. Although species can (over much shorter geographic distances) move up in elevation, such moves entail, almost inevitably, contractions in area (Colwell et al., 2008). In addition, our work with ants suggests that the closer one gets to the tropics, the narrower the range of climatic conditions a species is likely to experience. For example, high latitude species tend to occur at a much larger range of temperatures than do more tropical species (fig. 13-2B). If true and general, this result combined with the shallow tropical climate gradients may mean that tropical species (which is to say, most species) may have to move much farther to avoid extinction than do better studied temperate species.

In addition, we need to understand the frequency of large-scale dispersal events—whether most insect species are capable of moving the distances necessary to persist under changing climate. Historically, even when insects have moved in response to climate change, some barriers have proven difficult to cross. As a simple measure of such barriers, only 15–25 percent of ant genera are shared (depending on how one defines the two realms) between the tropical Americas and tropical Africa, despite the similarity in their climates (based on: http://www.antmacroecology.org/ant_genera/index.html). By a similar token, when we have modeled the distribution of ant species endemic to temperate forests of eastern North America using software such as MaxEnt (Phillips et al., 2006), such species are often predicted to occur in eastern Oregon and Washington, which have similar climates. Yet, the overlap in composition between these regions is very low, undoubtedly due to the barriers between them (e.g., the Rockies). More recently, we modeled the distribution of ant genera in North America using a suite of approaches (Fitzpatrick et al., 2011). In all of these approaches, the ant genus *Atta* (leaf-cutter ants) was predicted to occur in Florida on the basis of the places it is known to occur and on the climate of Florida (fig. 13-3). It is not currently found in Florida, perhaps again simply due to dispersal limitation. Future models predict that *Atta* will move even farther north (fig. 13-3C), but just as for the current distribution, dispersal limitation may be significant.

At the same time that barriers have historically hindered the movement of insects among biogeographic regions and may continue to do so as species respond to climate change, human influences have removed such barriers for some species. Many thousands of species have been introduced from one region to another around the world. The question that emerges in the context of climate change is what proportion of global species can and will move with humans. Intuition suggests that even with humans moving species from place to place the proportion of species that are moved may remain small. Intuition may be wrong. Tens of thousands of plant species, perhaps as many as one in ten species globally, are moved around the world as ornamentals. Those species may drag their herbivores with them. But movement does not require the shipping of plants. For ants, two studies—one in the United States, the other in New Zealand—have now looked at the number of ant species arriving at ports of entry. In both cases, most of the individual ants captured at the ports of entry represented unique species (in other words, each new ant that arrived tended to be a unique species) (Lester, 2005; Suarez et al., 2005). Similarly, it was recently noticed that there is a new invasive ant species spreading rapidly across Texas. This ant species, now being called in newspapers "the strawberry ant," has yet to be given a scientific name (it is a species of the genus *Paratrechina*). It is a new species and, despite its growing consequences, no one has had a chance to name it.

These studies suggest that even though a large number of the ant species are being moved from place to place, most are either poorly known or entirely unknown. As much of the earth warms, the number will increase as more of the earth comes within the potential range of tropical lineages—the lineages that now include the vast majority of species. Right now, we can do little more than guess at what the consequences of these introductions might pose for biodiversity.

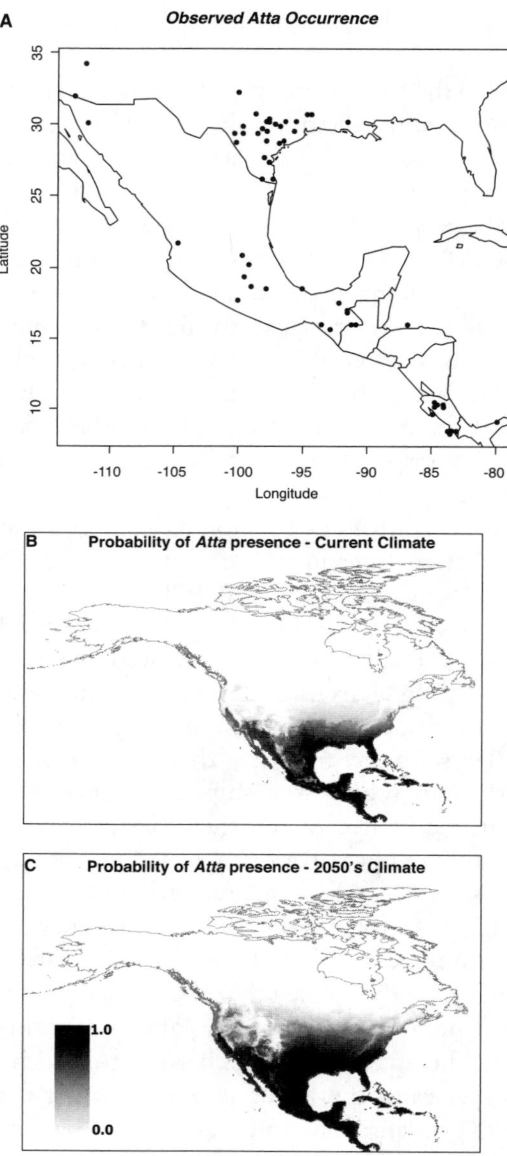


FIGURE 13-3. Recorded occurrences of species of *Atta* (leaf-cutter ants) in North and Central America (A), and modeled current (B) and projected future (2050) (C) probability of *Atta* presence based on a weighted consensus of nine different statistical algorithms within the BIOMOD framework (Thuiller et al., 2009). Darker shades indicate higher probability of *Atta* presence. Future climate was derived from the HadCM3 global circulation model forced using the A2a emissions scenario (IPCC, 2007).

Ants may not be representative. It may be that ants are better at getting around than are other taxa of insects. It is certainly true that within the taxon of ants, some lineages are poor at dispersing to new sites, such as army ants (Kumar and O'Donnell, 2007). If, as climates change, insects differ in their relative abilities to disperse, extinctions of insects could be nonrandom with regard to their functions (Fitzpatrick et al. 2011), as would be the case were army ants slow to colonize newly suitable climates. Figs (*Ficus* spp.), to take another example, are nearly as dense in the city of Hong Kong as in surrounding forests, but tend to lack pollinators (Corlett, 2006). That pollinators have not yet colonized trees so near to the forest suggests they may also be relatively unable to migrate effectively with climate change, even if their host trees do. Another key disparity among insect groups with regard to migration may be between species that favor disturbed habitats and those that favor intact habitats, the two of which differ greatly in the availability of corridors and the availability, however unintentional, of human-assisted dispersal.

## *Novel Biomes*

In large parts of the world, new sets of climatic and other conditions and hence novel biomes are emerging (Williams and Jackson, 2007; Williams et al., 2007). In some cases, these biomes are due to our own active management—for example, urban habitats. In other cases, they are the result of climate changes. In the Amazon, climatic conditions are predicted to be different by 2100 or even 2050 than in any existing biome (Williams and Jackson, 2007; Williams et al., 2007). Notably, just what constitutes a novel biome with no-analog climate depends on the dispersal ability of the taxon being studied. For example, in some models the future climate of Florida has analogs globally, but not regionally, so whether those global analogs can serve as source pools is entirely contingent on the ability of species to move on their own or aided by us. An enormously important question, perhaps the most important with regard to understanding global estimates of extinction numbers, is just what these new climatic conditions will be like and whether many or any species are preadapted to them.

Data like those we show for ants (fig. 13-2B) suggest that tropical species may tend to have narrower niches than temperate species, but observations of the conditions where species now occur do not necessarily correlate well with where they could occur. To date, more direct evidence of the ability of insect species to respond to new biomes comes from two sources: studies of physiological tolerances and studies of urban environments. The only study that has compared the physiological tolerances of species across latitudes has suggested that of species that have been studied to date (forty-eight relatively widespread pest species), low latitude species tend to have relatively narrow thermal tolerances compared to temperate species (in line with Rapoport's rule) (Deutsch et al., 2008). The data analyzed in Deutsch et al. (2008) are very preliminary and may have systematic biases (for example, if the environmental tolerance of pest species differs in some consistent way from that of nonpest species). However, if they are right, they suggest that relatively few tropical species will be preadapted to warmer conditions than already exist. Whether the conclusions of Deutsch et al. (2008) are justified based on their data, the authors are clearly asking an important question and one that begs further analyses and more data, particularly on the physiological tolerances of species. In addition, much depends on whether "few" species being preadapted means only tens of species or, as seems more likely, "just" hundreds of thousands. The difference between these scenarios is verbally subtle, but ecologically consequential. If there are 500,000 species in the Amazon (there are at least that many) and none are preadapted to the novel conditions, 500,000 species now have to move to survive. If, as an opposite extreme, those species are all preadapted, the new biome of the Amazon arises with 500,000 preexisting species. Obviously the truth is between these two extremes, but just where is unknown and perhaps unknowable.

Urban environments offer another example of no-analog environments, albeit in this case no-analog in terms of their general habitat structure rather than simply climate. Nonetheless, urban habitats provide a context for understanding how and how quickly species colonize novel biomes. Perhaps surprisingly, given their nearness to us, the origins of urban species remain relatively poorly studied (Shochat et al., 2006). Where it has been considered, however, it has been suggested that urban plants (Lundholm and Marlin, 1996) and also insects (Larson et al., 2004) come disproportionately from cliffside, cave, and more generally rocky environments. If supported, this hypothesis suggests that new biomes tend to be populated by species with preadaption for those biomes. Alternatively, species in urban and other no-analog biomes may evolve novel traits to tolerate novel conditions or simply be generalist species that are flexible in their diets, habitat preferences, and behaviors (Shochat et al., 2006; Hegglin et al., 2007). Although it tends to be assumed that species in urban environments are generalists, this is often not the case (Shochat et al., 2006). For example, the ant *Tetramorium caespitum* is very common in North American cities and often thought of as a generalist introduced ant species, but where studied it appears to be common only under cement or pavers (hence its common name, the pavement ant). In a recent study on medians on Broadway in Manhattan, the best predictor of its abundance was the amount of cement, a habitat that, although expanding, is actually very specific in its characteristics.

Knowing how many species will tolerate the novel biomes created by climate change and global change more generally is key to estimating the extinction rates of insects. Put in another way, what we need to know is the number of species that currently live in biomes that will move and lack both the ability to track changes in climate (move) and the ability to persist in the new conditions where they once lived. That the number of such species is in the hundreds of thousands seems conceivable, but until we have better knowledge of the range of physiological tolerances of species in such regions *and* a better handle on the conditions that will exist in such regions in the future (how hot, how wet, how seasonal), it is hard to put great faith in any estimate more specific than "lots."

## Insect Specificity and Insect Tolerance

Most research on the effects of climate change assumes that climate and dispersal limitation exclusively determine where species can live. For some species this may be true, but for many species, and in particular many insect species, it may be that other characteristics of the environment are required in addition to appropriate climates for survival. For example, ground-nesting insects might require both suitable climates and suitable soil types, and host-dependent taxa may require both the appropriate climate and the presence of the appropriate host. To the extent that is generally true, insects might tend to have narrower requirements than do other animals and than is implied simply based on their geographic distributions (Dunn, 2005; New, 2008a, b). It is common for habitats with small geographic areas to have endemic insects, but no endemic vertebrates (New, 2008b), which could be the result of insects having requirements in addition to simply the appropriate climates. To the extent that insects do sometimes or generally have additional habitat requirements, there may be more insect extinctions than predicted by broad-scale analyses. At the same time, such small ranges would also be an indication of the greater ability of insects than of other taxa to persist in small habitat patches (small is, of course, relative, such that a larger number of individual insects than elephants fit in a 10-hectare [25-acre] reserve). The relative risk of insects due to their potentially narrow distributions (Dunn, 2005; New, 2008b) versus their tolerance (Samways, 2006) has been argued and will probably continue to be. What are needed are better data on the geographic distributions of insects, particularly relative to their hosts, when they are host-dependent. Even for well studied groups such as ants, data remain incomplete.

Host specialization poses a similar problem to that of habitat specificity. More so than other taxa, insects often depend on the presence not only of favorable climate or habitat characteristics, but also on the presence of a host or mutualist partner (Dunn, 2009a; Dunn et al., 2009b). Even if host-dependent species, such as many beetles, are able to effectively track climate, they may still be at risk if their hosts do not. Such host and parasite or host and mutualist or guest relationships can be elaborate. Ants, the taxon we have focused most on here, are only rarely dependent on particular host plant species. However, within ants a relatively large number of species are social parasites on other ant species and in turn dependent on those species. Such social parasites inevitably have smaller geographic ranges than their hosts (though in most cases the geographic ranges of social parasites are poorly known, because they are not easy to encounter using standard methods). In addition, they have smaller population sizes. All of these factors may predispose a subset of host-dependent social parasite ants and dependent insect species more generally to extinction due to cascading effects of climate change. It is perhaps telling that one of the best studied examples of extinction, albeit local (from the United Kingdom), in insects is that of the large blue (*Maculinea arion*), a butterfly that in its larval stages depends on a particular ant species (Thomas et al., 1998; Mouquet et al., 2005). That ant species became rare in the United Kingdom due to changes in its habitat, and although the ant itself did not go extinct, the butterfly did. (It has now been reintroduced from other populations). Analogous to the situation of narrow habitat specialists, what happens to narrow host specialists depends on their host specificity, their ability to evolve, and their ability to disperse.

## Conclusions

In the end, the bad news is that if there are a million or even millions of species at risk of extinction, they are mostly insects and so mostly events that will, even when noticed, attract little attention. The Antioch Dunes, for example, are a patch of ancient sand, long isolated from other similar habitat, outside the town of Antioch, California. The dunes were once home to perhaps as many as nineteen and no fewer than nine endemic insect species as recently as 1940 (Rentz, 1977; Dunn, 2005, 2009b). The species were known to be there, known to be endemic, and known to be at risk, and yet largely ignored. In the 1970s, the dunes were turned into a national wildlife refuge, ostensibly because of the presence of an endangered subspecies of plant and an endangered subspecies of butterfly. Yet although those two species have been monitored, in many years no one has even looked, according to the reserve manager, for the other species once endemic to the dunes (pers. comm.). The good news is that these species are near at hand and can be studied. Anyone could do it, with some permits and some time. To understand the fate of mammals with climate change, one may need to go to faraway places. To understand the fate of insects, much can be learned in our literal backyards. These are the species that a young boy or girl could go out and see and, in watching, make observations new to the world. One of the species endemic to the Antioch Dunes, but now extinct, the Antioch Dunes katydid, for example, was discovered by Dave Rentz when he was still in high school (Rentz, 1977).

Because of our ignorance about insect diversity, distribution, and physiological tolerances, estimates of the number of extinctions of insects may continue to be difficult. One productive way forward may be to try to know some groups, be they butterflies, dung beetles, or ants, as well as we can. This is the approach that medical research has taken in knowing well fruit flies, rats, and mice, for example. From a few well known cases, one can begin to better understand details and on-the-ground realities. For these focal groups, what we ideally want are data not just on distribution, but also on the particular histories of diversification and adaptation. Such stories are interesting, but also speak clearly to what has happened, which we need to know if we are to understand what will happen. With a few well understood lineages we might finally be able to say how many species are on Earth, at least for those groups, and even, just maybe, how many of those species are at risk from climate change.

## REFERENCES

Basset, Y., G. A. Samuelson, A. Allison, and S. E. Miller. 1996. "How many species of host-specific insects feed on a species of tropical tree?" *Biological Journal of the Linnean Society* 59: 201–216.

Botes, A., M. A. McGeoch, H. G. Robertson, A. van Niekerk, H. P. Davids, and S. L. Chown. 2006. "Ants, altitude, and change in the northern Cape Floristic Region." *Journal of Biogeography* 33: 71–90.

Brown, J. H. 1971. "Mammals on mountaintops—nonequilibrium insular biogeography." *American Naturalist* 105: 467.

Colwell, R. K., G. Brehm, C. L. Cardelus, A. C. Gilman, and J. T. Longino. 2008. "Global warming, elevational range shifts, and lowland biotic attrition in the wet tropics." *Science* 322: 258–261.

Coope, G. R. 2004. "Several million years of stability among insect species because of, or in spite of, Ice Age climatic instability?" *Philosophical Transactions of the Royal Society of London B* 209–214.

Coope, G. R., and A. S. Wilkins. 1994. "The response of insect faunas to glacial-interglacial climatic fluctuations." *Philosophical Transactions of the Royal Society of London Series B-Biological Sciences* 344: 19–26.

Corlett, R. T. 2006. "Figs (*Ficus,* Moraceae) in urban Hong Kong, south China." *Biotropica* 38: 116–121.

Crisp, M., L. Cook, and D. Steane. 2004. "Radiation of the Australian flora: What can comparisons of molecular phylogenies across multiple taxa tell us about the evolution of diversity in present-day communities?" *Philosophical Transactions of the Royal Society of London Series B-Biological Sciences* 359: 1551–1571.

Deutsch, C. A., J. J. Tewksbury, R. B. Huey, K. S. Sheldon, C. K. Ghalambor, D. C. Haak, and P. R. Martin. 2008. "Impacts of climate warming on terrestrial ectotherms across latitude." *Proceedings of the National Academy of Sciences, USA* 105: 6668–6672.

Dunn, R. R. 2005. "Modern insect extinctions, the neglected majority." *Conservation Biology* 19: 1030–1036.

Dunn, R. R. 2009a. "Coextinction: Anecdotes, models, and speculation." In *Holocene Extinctions,* edited by S. Turvey. Oxford, UK: Oxford University Press.

Dunn, R. R. 2009b. "Dune buggies. What is the sound of one katydid stridulating?" *Natural History Magazine* September.

Dunn, R. R. 2009c. *Every Living Thing*. New York: Harper Collins/Smithsonian.

Dunn, R. R., D. Agosti, A. N., Andersen, X. Arnan, C. A. Bruhl, X. Cerda, A. M. Ellison, et al. 2009a. "Climatic drivers of hemispheric asymmetry in global patterns of ant species richness." *Ecology Letters* 12: 324–333.

Dunn, R. R., N. C. Harris, R. K. Colwell, L. P. Koh, and N. S. Sodhi. 2009b. "The sixth mass coextinction: Are most endangered species parasites and mutualists?" *Proceedings of the Royal Society of London Series B-Biological Sciences* 276: 3037–3045.

Erwin, T. L. 1982. "Tropical forests: Their richness in Coleoptera and other Arthropod species." *The Coleopterists Bulletin* 36: 47–75.

Fitzpatrick, M. C., A. D. Gove, N. J. Sanders, and R. R. Dunn. 2008. "Climate change, plant migration, and range collapse in a global biodiversity hotspot: The Banksia (Proteaceae) of Western Australia." *Global Change Biology* 14: 1337–1352.

Fitzpatrick, M. C., N. J. Sanders, S. Ferrier, J. T. Longino, M. D. Weiser, and R. R. Dunn. 2011. "Forecasting the future of biodiversity: A test of single- and multi-species models for ants in North America." *Ecography* 34: 1–12.

Gaston, K. J. 1991. "The magnitude of global insect species richness." *Conservation Biology* 5: 283–296.

Hegglin, D., F. Bontadina, P. Contesse, S. Gloor, and P. Deplazes. 2007. "Plasticity of predation behaviour as a putative driving force for parasite life-cycle dynamics: The case of urban foxes and *Echinococcus multilocularis* tapeworm." *Functional Ecology* 21: 552–560.

Henry, G. H. R., and U. Molau. 1997. "Tundra plants and climate change: The International Tundra Experiment (ITEX)." *Global Change Biology* 3: 1–9.

Hill, J. K., C. D. Thomas, R. Fox, M. G. Telfer, S. G. Willis, J. Asher, and B. Huntley. 2002. "Responses of butterflies to twentieth century climate warming: Implications for future ranges." *Proceedings of the Royal Society of London* 269: 2163–2171.

Hodkinson, I. D., and E. Hodkinson. 1993. "Pondering the imponderable—a probability-based approach to estimating insect diversity from repeat faunal samples." *Ecological Entomology* 18: 91–92.

Jansson, R. 2003. "Global patterns in endemism explained by past climatic change." *Proceedings of the Royal Society of London Series B-Biological Sciences* 270: 583–590.

Jansson, R., and M. Dynesius. 2002. "The fate of clades in a world of recurrent climatic change: Milankovitch oscillations and evolution." *Annual Review of Ecology and Systematics* 33: 741–777.

Kumar, A., and S. O'Donnell. 2007. "Fragmentation and elevation effects on bird–army ant interactions in neotropical montane forest of Costa Rica." *Journal of Tropical Ecology* 23: 581–590.

Larson, D. W., U. Matthes, P. E. Kelly, J. T. Lundholm, and J. A. Gerrath. 2004. *The Urban Cliff Revolution: New Findings on the Origins and Evolution of Human Habitats*. Markham, Canada: Fitzhenry and Whiteside.

Lessard, J. P., R. R. Dunn, and N. J. Sanders. 2007. "Rarity and diversity in forest assemblages of the Great Smoky Mountain National Park." *Southeastern Naturalist,* Special Issue 1: 215–228.

Lester, P. J. 2005. "Determinants for the successful establishment of exotic ants in New Zealand." *Diversity and Distributions* 11: 279–288.

Loarie, S. R., P. H. Duffy, H. Hamilton, G. P. Asner, C. B. Field, and D. D. Ackerly. 2009. "The velocity of climate change." *Nature* 462: 1052–1055.

Longino, J. T., J. Coddington, and R. K. Colwell. 2002. "The ant fauna of a tropical rain forest: Estimating species richness three different ways." *Ecology* 83: 689–702.

Lundholm, J., and A. Marlin. 1996. "Habitat origins and microhabitat preferences of urban plant species." *Urban Ecosystems* 9: 139–159.

Malcolm, J. R., A. Markham, R. P. Neilson, and M. Garaci. 2002. "Estimated migration rates under scenarios of global climate change." *Journal of Biogeography* 29: 835–849.

May, R. M. 1992. "How many species inhabit the Earth?" *Scientific American* October: 18–24.

Menke, S. B., B. Guénard, J. O. Sexton, M. D. Weiser, R. R. Dunn, and J. Silverman. 2011. "Urban areas may serve as habitat and corridors for dry-adapted, heat tolerant species; An example from ants." *Urban Ecosystems* 14 (2): 135–163.

Mouquet, N., J. A. Thomas, G. W. Elmes, R. T. Clarke, and M. E. Hochberg. 2005. "Population dynamics and conservation of a specialized predator: A case study of *Maculinea arion*." *Ecological Monographs* 75: 525–542.

New, T. R. 2008a. "Conserving narrow range endemic insects in the face of climate change: Options for some Australian butterflies." *Journal of Insect Conservation* 12: 585–589.

New, T. R. 2008b. "Insect conservation on islands: Setting the scene and defining the needs." *Journal of Insect Conservation* 12: 197–204.

Novotny, V., Y. Basset, S. E. Miller, G. D. Weiblen, B. Bremer, L. Cizek, and P. Drozd. 2002. "Low host specificity of herbivorous insects in a tropical forest." *Nature* 416: 841–844.

Novotny, V., P. Drozd, S. E. Miller, M. Kulfan, M. Janda, Y. Basset, and G. D. Weiblen. 2006. "Why are there so many species of herbivorous insects in tropical forests?" *Science* 313: 1115–1118.

Novotny, V., S. E. Miller, J. Hulcr, R. A. I. Drew, Y. Basset, M. Janda, G. P. Setliff, et al. 2007. "Low beta diversity of herbivorous insects in tropical forests." *Nature* 448: 692–698.

Odegaard, F. 2000. "How many species of arthropods? Erwin's estimate revised." *Biological Journal of the Linnean Society* 71: 583–597.

Odegaard, F., O. H. Diserud, S. Engen, and K. Aagaard. 2000. "The magnitude of local host specificity for phytophagous insects and its implications for estimates of global species richness." *Conservation Biology* 14: 1182–1186.

Parmesan, C. 2006. "Ecological and evolutionary responses to recent climate change." *Annual Review of Ecology, Evolution, and Systematics* 37: 637–669.

Phillips, S. J., R. P. Anderson, and R. E. Schapire. 2006. "Maximum entropy modeling of species geographic distributions." *Ecological Modelling* 190: 231–259.

Rangel, T., J. A. F. Diniz-Filho, and R. K. Colwell. 2007. "Species richness and evolutionary niche dynamics: A spatial pattern-oriented simulation experiment." *American Naturalist* 170: 602–616.

Rentz, D. C. F. 1977. "New and apparently extinct katydid from Antioch Sand Dunes (Orthoptera-Tettigoniidae)." *Entomological News* 88: 241–245.

Samways, M. J. 2006. "Insect extinctions and insect survival." *Conservation Biology* 20: 245–246.

Sanders, N. J., J. Moss, and D. Wagner. 2003. "Patterns of ant species richness along elevational gradients in an arid ecosystem." *Global Ecology and Biogeography* 12: 93–102.

Sekercioglu, C. H., S. H. Schneider, J. P. Fay, and S. R. Loarie. 2008. "Climate change, elevational range shifts, and bird extinctions." *Conservation Biology* 22: 140–150.

Shochat, E., P. S. Warren, S. H. Faeth, N. E. McIntyre, and D. Hope. 2006. "From patterns to emerging processes in mechanistic urban ecology." *Trends in Ecology & Evolution* 21: 186–191.

Stork, N. E. 1988. "Insect diversity: Facts, fiction and speculation." *Biological Journal of the Linnean Society* 353: 321–337.

Suarez, A. V., D. A. Holway, and P. S. Ward. 2005. "The role of opportunity in the unintentional introduction of nonnative ants." *Proceedings of the National Academy of Sciences, USA* 102: 17032–17035.

Thomas, J. A., D. J. Simcox, J. C. Wardlaw, G. W. Elmes, M. E. Hochberg, and R. T. Clarke. 1998. "Effects of latitude, altitude and climate on the habitat and conservation of the endangered butterfly *Maculinea arion* and its *Myrmica* ant hosts." *Journal of Insect Conservation* 2: 39–46.

Walther, G. R., S. Berger, and M. T. Sykes. 2005. "An ecological "footprint" of climate change." *Proceedings of the Royal Society B-Biological Sciences* 272: 1427–1432.

Williams, J. W., and S. T. Jackson. 2007. "Novel climates, no-analog communities, and ecological surprises." *Frontiers in Ecology and the Environment* 5: 475.

Williams, J. W., S. T. Jackson, and J. E. Kutzbacht. 2007. "Projected distributions of novel and disappearing climates by 2100 AD." *Proceedings of the National Academy of Sciences, USA* 104: 5738–5742.

Williams, S. E., E. E. Bolitho, and S. Fox. 2003. "Climate change in Australian tropical rainforests: An impending environmental catastrophe." *Proceedings of the Royal Society of London* 270: 1887–1892.

Wilson, R. J., D. Gutierrez, J. Gutierrez, D. Martinez, R. Agudo, and V. J. Monserrat. 2005. "Changes to the elevational limits and extent of species ranges associated with climate change." *Ecology Letters* 8: 1138–1146.

# Chapter 14

## *Extinction Risk from Climate Change in Tropical Forests*

YADVINDER MALHI

Tropical forests are biologically the richest biomes on Earth, home to half of global biodiversity and most of the insects described in the previous chapter. The prospects for the earth's treasure of living organisms over this century are thus inevitability tied to the prospects of its greatest treasure houses, the tropical rain forest regions, whether influenced by deforestation or climate change. The extinction risk in tropical forests will have a large effect on total global extinction risk, but like that of insects is currently difficult to quantify because of several key unknowns. This chapter explores the nature of contemporary and likely future climate change in the tropics, and possible implications for the biodiversity and functioning of tropical ecosystems. It begins by reviewing the likely nature of tropical climate change, then explores the likely response of tropical organisms to such change. It highlights key uncertainties in estimation of extinction risk from climate change, including the likely pattern of precipitation change, the influence of carbon dioxide on forest persistence, the upper thermal tolerance and adaptation/acclimation ability of tropical organisms, and the relationship between habitat restriction and extinction risk.

### The Nature of Climate Change in the Tropics

To assess global extinction risk due to climate change, it is essential to understand the interaction of global climate and the earth's tropical ecosystems. This section explores evidence of tropical climate change in the past as well as projections of climate change in the twenty-first century.

### *The Climate of Tropical Forest Regions over Earth's History*

Angiosperm-dominated tropical forests appear to have first appeared in the mid-Late Cretaceous (about 100 million years ago; Wang et al., 2009), and to have persisted throughout the Cenozoic (last 65 million years), albeit with periods of sharp turnover in species (Morley, 2000). Cenozoic temperatures broadly peaked in the Early Eocene climate optimum (50–53 million years ago), although this was preceded by a sharp warming fluctuation at the Paleocene-Eocene Thermal Maximum (55 million years ago), which occurred in less than 10,000 years and from which the recovery period lasted 200,000 years (Chapter 10, this volume; Zachos et al., 2001; Jaramillo et al., 2010). In both these events tropical temperatures may have been up to 10 degrees Celsius higher than the present (Huber, 2008).

Hence, the paleoclimatic evidence suggests that there is no tropical thermostat (such as cloud formation feedbacks) that limits tropical warming, at least within a range up to 10 degrees Celsius warmer than present (Huber, 2008). Given the likely persistence (and expansion) of recognizable tropical forests throughout this period, this suggests that in the long term the magnitude of projected twenty-first-century warming does not present a high risk of the fundamental physiological breakdown of the tropical forest biome, because taxa would acclimate and adapt to higher tropical temperatures. However, the rapidity of twenty-first-century warming compared to much of paleoclimatic history suggests that transient responses will be very important, as individuals, species, and communities race to adapt within and across generations to a rapidly increasing mean temperature (see below).

### *Future Change in Tropical Climates: Carbon Dioxide and Temperature*

Since the mid-1970s tropical land regions have been warming, fairly uniformly worldwide, at a rate of about 0.31 degree Celsius per decade (fig. 14-1B). This rate is only slightly less than the rate observed in northern temperate land regions (0.39 degree Celsius per decade);

southern temperate regions are warming at slower rates (0.09 degree Celsius per decade) because they are generally more maritime. This fact is often missed when global (land and ocean) warming rates are considered because a larger proportion of tropical latitudes is ocean (compared to the Northern Hemisphere) and tropical oceans on aver-

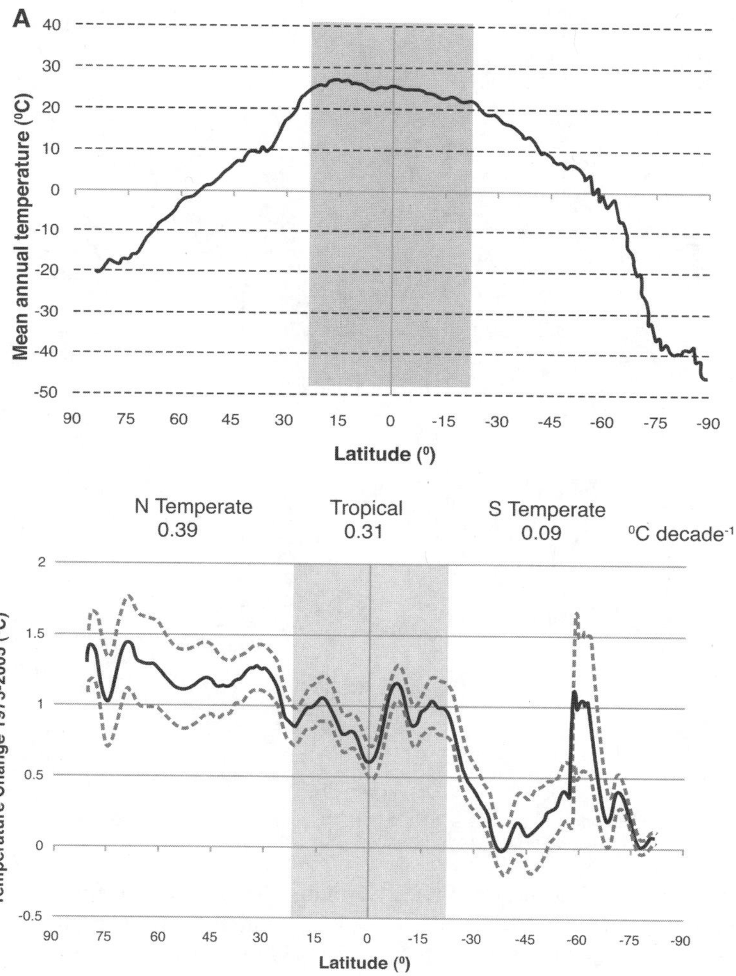


FIGURE 14-1. (A) Mean annual temperatures of land surface regions over the period 1975–2005, plotted against latitude. (B) Mean rate of warming of tropical land regions over the period 1975–2005. The tropical latitude band is shaded. Data are derived from the Climate Research Unit observational data set. Analysis updated from Malhi and Wright, 2004.

age warm at a slower rate than temperate oceans because they have a deeper surface mixing layer. Tropical land regions have been warming at rates broadly similar to temperate land regions.

Under the A2 scenario, a mean of fifteen Intergovernmental Panel on Climate Change global climate models (fig. 14-2A) suggests that

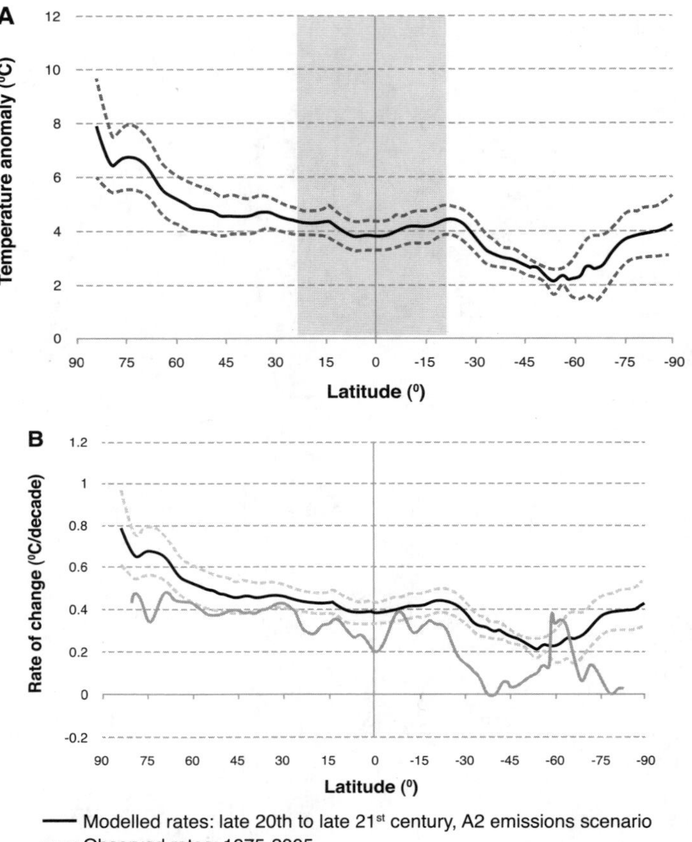


FIGURE 14-2. (A) The mean projected warming of land regions by the late twenty-first century (2070–2099), relative to the late twentieth century (1970–1999), as projected by fifteen IPCC Global Climate Models under the A2 greenhouse gas emissions scenario. The means are zonally averaged and plotted against latitude. Dotted lines indicate +/- one standard error of the mean. The tropical latitude band is shaded (B) a comparison of observed rates of warming of land regions in the late twentieth century (from fig. 14-1B) against projected rates over the twentieth century (from fig. 14-2A).

tropical land regions will undergo net warming of 4.1 degrees Celsius (range 3.2–5.0 degrees Celsius) by 2100. Hence, the mean projected rate of warming over the twenty-first century is 0.41 degree Celsius per decade (range 0.32–0.50 degree Celsius per decade), slightly higher than the observed rate from 1975 to 2005 of 0.31 degree Celsius per decade (fig. 14-2B). Most of the climate models do not incorporate possible consequences or feedbacks caused by tropical forest loss (whether from deforestation or climate change); those that do suggest greater surface warming of land regions because of reduction of the evaporative cooling service provided by forests.

It is likely that the twenty-first century will see the emergence of local climates that will include conditions not experienced at present or even in the past several millennia ("novel" climates), and that some present climates may completely disappear from the earth's surface (Williams et al., 2007). Disappearing climates are primarily concentrated in tropical mountains and the poleward sides of continents, and are particularly concentrated in the tropics. Specific areas in the tropics and near-tropics with disappearing climates include the Columbian and Peruvian Andes, Central America, African Rift Mountains, the Zambian and Angolan Highlands portions of the Himalayas, and the Indonesian and Philippine Archipelagos.

The absolute magnitude of warming is likely to have consequences on some taxa, but there have been periods of warmth in the past where tropical forests seem to have thrived (see above). The most important and alarming feature of twenty-first-century warming is the *rate* of warming. The current rates of warming and projected rates over this century may be thirty to forty times greater than those experienced by, say, Amazonia in the last glacial-interglacial transition (Bush et al., 2004). It is conceivable that such rates have not been experienced since at least the Paleocene-Eocene Thermal Maximum.

### *Future Change in Tropical Climates: Precipitation*

Climate models are fairly consistent in their predictions of temperature change over the twenty-first century, but regional changes in precipitation may have the greatest impact on the viability of existing tropical forest biomes. However, precipitation changes are more challenging to predict, with less agreement among models (Malhi et al., 2009).

Within this variability, the consequences for Amazonia are particularly important. Climate model predictions suggest that southeast Amazonia is among the most vulnerable to drying of the major tropical rain forest regions (Zelazowski et al., 2011), although paleoecological studies have clearly shown that Central Africa has been the major region with the most fluctuating rain forest cover (Maley, 2001). Cox et al. (2008) described a mechanism where asymmetric warming of the tropical North Atlantic relative to the tropical South Atlantic causes drought across much of Amazonia (as occurred in 2005), and suggested that 2005-style droughts may occur every other year by 2050. The high profile U.K. Hadley Centre model tends to be the most pessimistic model in terms of precipitation projections for Amazonia. Output from this model should be taken seriously, but its position at one extreme of the spectrum of possible responses of Amazonian climate should be kept in mind when evaluating extinction risk from this model (Malhi et al., 2009), as was done in Thomas et al. (2004).

Total global rainfall is likely to increase because of increased evaporation from warmer oceans, but regionally there will be winners and losers. Regional rainfall shifts will be driven by changes in general atmospheric circulation. Subtropical and Mediterranean regions are most likely to experience drought, as equator-pole temperature gradients weaken and seasonal rainfall shifts poleward. For tropical forest regions in the Northern Hemisphere the most vulnerable regions include Central America and the Caribbean and the Guyanas. In the Southern Hemisphere these include Eastern Amazonia, southern Africa, and Madagascar. In addition to possible regional shifts in absolute precipitation driven by shifting sea-surface temperature distribution, there is a general tendency for intensifying and narrowing tropical convection zones as the global atmospheric circulation increases in energy. This will likely lead to increased rainfall seasonality with more intense but shorter wet seasons, and more intense and longer dry seasons (Malhi et al., 2009).

Even if there were no change in precipitation, the changes in evapotranspiration rates caused by warming surface temperatures and increased carbon dioxide concentration will result in changes in water use by plants, although the exact direction of change is still not clear (Malhi et al., 2009).

In addition to the trends in temperature, rainfall, and carbon dioxide outlined above, there are a number of other contemporaneous trends in the abiotic environment of tropical forests that could influence forest functioning. These include possible changes in cloud regime driven by climate change (Butt et al., 2009) and numerous consequences from biomass burning, including increasing atmospheric haze (which changes light penetration in forests by increasing the fraction of diffuse radiation), aerosols causing intensification of tropical storms and more frequent extreme air downbursts, deposition of nitrogen and other compounds, and acid rain. Our understanding of almost all these processes in the context of the tropical rain forest regions is very limited.

## The Response of Tropical Species to Climate Change

### *Recent Observed Changes in Tropical Forests*

One line of evidence bearing on how tropical forests will respond to global atmospheric change comes from longtime monitoring of tropical ecological communities in situ in the context of the warming and rising carbon dioxide of the past three decades. The best and most spatially extensive studies in this context have been for tropical trees, which have the advantage of being easy to measure and trace over time. Large-scale international networks monitoring tropical trees include the Centre for Tropical Forest Science network (http://www.ctfs.si.edu/), which employs large sampling plots of 25–50 hectares (62–124 acres) worldwide, and the RAINFOR network in Amazonia (http://www.rainfor.org), which focuses on a large number of 1-hectare (2.5-acre) plots, together with its newly established sister network, AFRITRON, in Africa (Lewis et al., 2009).

Results from RAINFOR and AFRITRON suggest that old-growth tropical forests have increased in biomass over recent decades (Baker et al., 2004; Lewis et al., 2009) and have accelerated in both growth and death rates, possibly as a response to carbon dioxide fertilization, although this trend may be beginning to reverse, and the net biomass gain stopped during the Amazon drought of 2005 (Phillips et al., 2009). Other studies have documented a potential sensitivity to temperature (Clark et al., 2003). It is quite plausible that both factors are at play: carbon dioxide fertilization may be causing a (possibly transient) increase in growth rates and biomass, but there may also be a sensitivity to temperature and drought, which could cause a longer term decline in growth rates, an increase in mortality, and a decline in biomass. Any changes in dynamism are almost certainly accompanied by shifts in composition, but these are hard to pick out from the natural "noise" of ecological dynamics. Phillips et al. (2002) reported an increase of liana abundance in western Amazonia, a result that was corroborated by long-term studies of liana growth, flowering, and fruiting from Barro Colorado Island in Panama (Wright et al., 2006).

### *The Response of Tropical Ecosystems to Drought*

As outlined above, rapid changes in temperature are likely to affect species fitness, and this will cascade down to changes in community composition and possible extinction. Thus temperature change may strongly affect community ecology and diversity, but is less likely to cause large-scale dieback of tropical forests. Intensified drought, on the other hand, has the potential to lead to the retreat of moist tropical forest in favor of dry, fire-adapted communities.

A number of field studies have looked at the effects of drought. Some have exploited natural interannual variability, such as El Niño droughts, or the 2005 drought in Amazonia (e.g., Condit et al., 2004; Phillips et al., 2009). These studies tend to demonstrate a reduction in growth and an increase in mortality in drought years, particularly at inherently drier sites. The forests appear to be sensitive to drought, but also resilient enough to recover if drought conditions do not persist for several years. This resilience has been explored in greater mechanistic detail in two drought ("throughfall exclusion") experiments in eastern Amazonia, where plastic panels have excluded about 50 percent of rainfall from 1-hectare patches of forest for several years (Nepstad et al., 2007; Fisher et al., 2007). In the first few years of drought the forest showed some reduction in photosynthesis and physiological performance, but overall the forest plots seemed to prove resilient by tapping the substantial reserves of soil water throughout the dry season. However, if these reserves are not sufficiently replenished in subsequent seasons, eventually a threshold is reached where tree mortality rates rise, with large canopy trees and lianas appearing most vulnerable. Thus resilience is dependent on total annual rainfall and the reserve of soil water that can be maintained in the dry season. Although the forest as a whole may persist during intensified seasonal drought, there are likely to be long-term shifts in community composition in favor of drought-adapted taxa, at the expense of drought-averse taxa, which tend to account for a greater proportion of tropical tree diversity.

Although forests may demonstrate some physiological resilience to intensified drought, their litter layer does become seasonally flammable as it dries out, allowing for the spread of ground fires. In drought years in Amazonia, there can be substantial leakage of fires from agricultural zones into surrounding, temporarily flammable forests (Alencar et al., 2004; Aragão et al., 2007). Each fire opens up the forest further, and if fires occur frequently enough these forests degrade to a fire-adapted scrub with substantial decline in biodiversity (Barlow and Peres, 2008).

Hence, it is this interaction between climate change and land use change that perhaps represents the greatest threat to tropical forest species. Land use change is directly reducing forest cover; fragmenting remaining forests; reducing the potential for migration or dispersal to cooler or wetter regions; increasing the length of dry, fire-prone forest edge; and opening up forests to logging and bushmeat hunting, which can trigger local extinction of large fauna and trigger trophic cascades that affect the whole ecosystem. It also has effects on local climate by potentially changing local cloud climatology, reducing dry-season precipitation, increasing atmospheric aerosol content and the nature and intensity of storm events, and warming surface temperatures by reducing evaporative cooling. Conversely, the maintenance of sufficient and intact forest area would be a strategy to lessen the impacts of global climate change, by ameliorating local effects on temperature and rainfall (Malhi et al., 2008). Maintaining forest area must also be coupled with maintaining forest connectivity, to maximize the potential for migration or dispersal to potential highland and moist refugia. Hence the threat of climate change also provides an opportunity for conservation of tropical forests, by placing direct value on their potential role in global mitigation and regional adaptation. Mechanisms for carbon payment, such as Reduced Emissions from Deforestation and Degradation (Grainger et al., 2009), are but the first step in such a process of valuing the ecosystems services that tropical forests provide.

### *The Response of Tropical Organisms to Warming Temperatures*

Given a world in which tropical land regions may be about 4 degrees Celsius warmer within a century, how will tropical organisms, ecology, and ecosystems fare? The critical features of current and projected tropical warming are (i) its large projected magnitude relative to the usual range of diurnal, seasonal, and interannual temperature variability experienced by most tropical organisms; and (ii) its rapid rate—in many regions much more rapid than the warming experienced at glacial-interglacial transitions. This question may be considered in three parts: first the current thermal tolerance ranges of tropical organisms in contrast to temperate organisms, then how these thermal tolerances will respond to the overall magnitude of warming, and finally how the rapidity of contemporary warming will influence the outcome.

### *Thermal Tolerances to Temperature Increase*

The low variability of tropical temperatures, whether on diurnal, seasonal, or interannual timescales, has resulted in most tropical organisms having a narrower range of thermal tolerance compared to higher latitude organisms, and also being closer to climatic optima or critical upper limits (fig. 14-3). This has been demonstrated for ectotherms (frogs and toads, lizards and turtles; Deutsch et al., 2008). One method for deriving thermal performance for candidate species is to measure populations' intrinsic growth rates under laboratory conditions at various fixed temperatures; for insects, this is a direct measure of Darwinian fitness. It is unclear how narrow the thermal tolerance may be for other organisms, such as plants or endotherms. In plants the fundamental ecophysiological processes of photosynthesis and respiration appear to acclimate rapidly to warming, in situ either in existing leaves or in new leaves (Campbell et al., 2007; Atkin and Tjoelker, 2003). When the whole life cycle of the plant is considered, there may conceivably be bottlenecks at critical stages, such as reproduction or germination, where a narrower thermal tolerance regime pervades. As daytime ambient tropical temperatures approach body temperature for mammals and birds, while atmospheric relative humidity remains fairly invariant, it seems likely that the difficulty of cooling will also cause a net decline in performance in endotherms. It is costly to maintain broad thermal tolerance regimes, and these would likely have been selected against by natural selection in tropical regions where ambient temperatures show little variation. The high competitive pres-

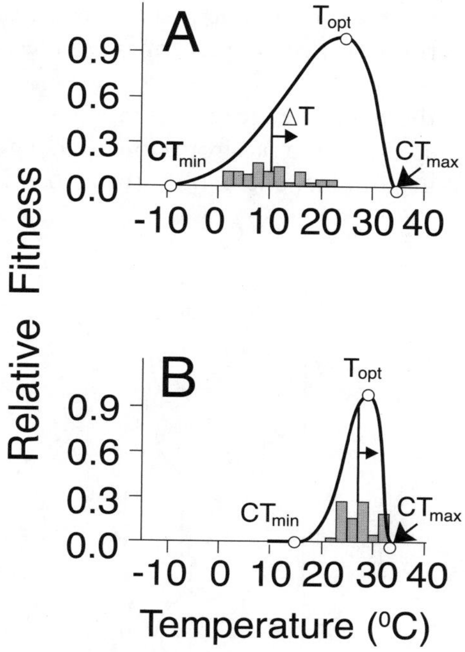


FIGURE 14-3. Fitness curves for representative insect taxa from temperate (A) and tropical (B) locations. Fitness curves are derived from measured intrinsic population growth rates versus temperature for 38 species, including *Acyrthosiphon pisum* (Hemiptera), from 52°N (England) (A), and the same for *Clavigralla shadabi* (Hemiptera) from 6°N (Benin) (B). $CT_{min}$ is the critical minimum temperature, $T_{opt}$ is the optimum temperature, and $CT_{max}$ is the critical maximum temperature. Climatological mean annual temperature from 1950 to 1990 ($T_{hab}$, drop lines from each curve), its seasonal and diurnal variation (gray histogram), and its projected increase because of warming in the next century (T, arrows) are shown for the collection location of each species. Reproduced from Deutsch et al., 2008; © (2008) National Academy of Sciences, USA.

sure in the tropical forest environment may have driven these thermal niches to be very narrow.

Given these narrower tolerance regimes, Deutsch et al. (2008) modeled the performance of ectotherms (in their study, insects, frogs and toads, lizards and turtles) under an A2 emissions scenario). They found that almost all projected declines in thermal performance were located in the tropics, with a typical 15 percent decline in tropical ecotherm performance over the twenty-first century, whereas most temperate ectotherms improved in performance in response to warming (fig. 14-4).

The issue of thermal tolerance is perhaps the most critical unknown in evaluating extinction risk from climate change. Colwell et al. (2008) explored the potential effects of warming on Costa Rican rain

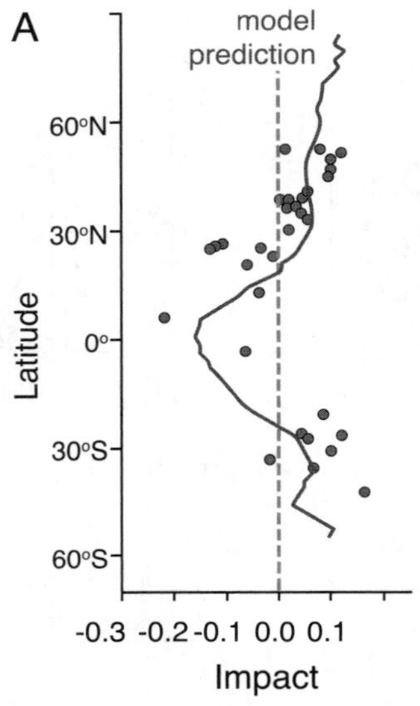


FIGURE 14-4. Predicted impact of warming on the thermal performance of ectotherms in 2100. (A) Impact versus latitude for insects using thermal performance curves fit to intrinsic population growth rates measured for each species and for a global model (solid line) in which performance curves at each location are interpolated from empirical linear relationships between seasonality and both warming tolerance and thermal safety margin. (B and C) Results from the simplified conceptual model are shown globally for insects (B), for which performance data are most complete, and versus latitude for three additional taxa of terrestrial ectotherms: frogs and toads, lizards, and turtles (C), for which only warming tolerance was available. In general darker regions on the map indicate negative impact and lighter regions show positive impact of rising temperature. On the basis of patterns in warming tolerance, climate change is predicted to be most deleterious for tropical representatives of all four taxonomic groups. Performance is predicted to increase in mid- and high latitudes because of the thermal safety margins observed there for insects, and provisionally attributed to other taxa. Reproduced from Deutsch et al., 2008, © (2008) National Academy of Sciences, USA.

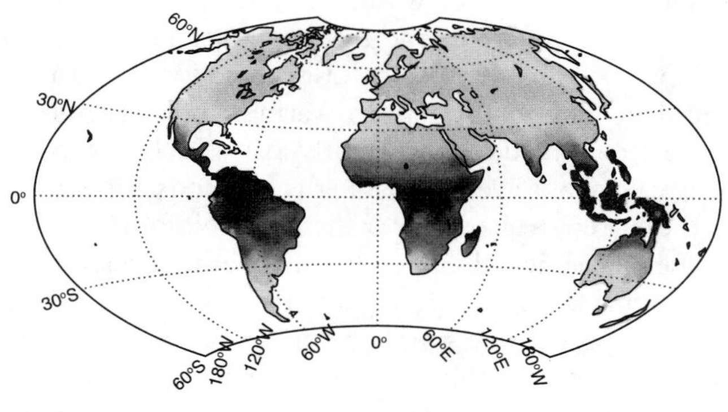


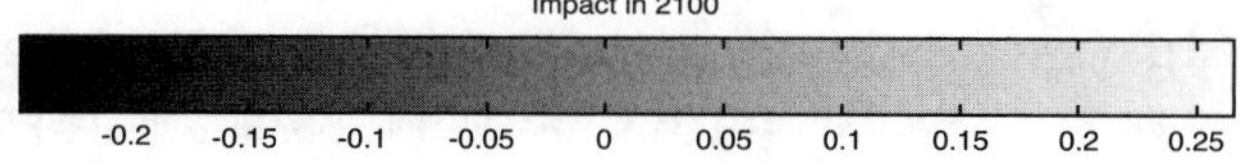


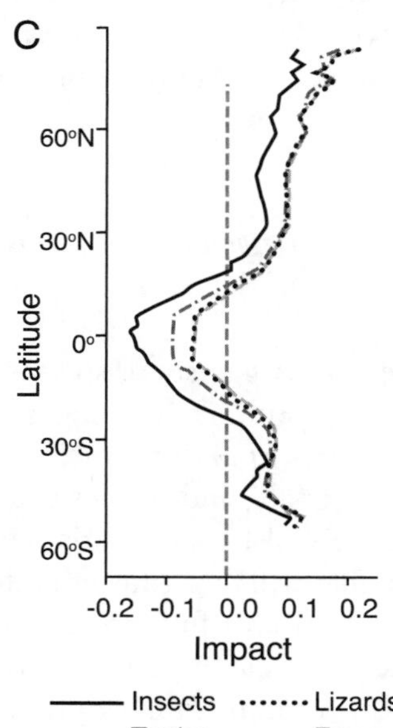


FIGURE 14-4. Continued.

forest species by looking at the current thermal range of species, and then modeling the effects of warming. They described a process of "attrition" of biodiversity in the tropical lowlands, as temperature-intolerant species emigrate upslope or become locally extinct, but are not replaced by immigration from warmer moist regions, because warmer moist regions do not currently exist. Such a scenario would likely see extensive biodiversity loss in the lowlands. However, it is not clear that the current realized upper thermal limit for tropical species is equal to the fundamental upper thermal limit. Feeley and Silman (2010) argue that the apparent narrow (field-observed) thermal niche of Amazonian species may simply be an artifact because warmer climates do not exist, but this does not imply that particular species would necessarily retreat in those warmer climates. They note that tropical montane tree species typically have a mean thermal niche width of 10 degrees Celsius, and that a similar niche width may apply in the lowlands. In such a case, the rate of projected biodiversity attrition in the lowlands would be much lower. Of course, the reality will be that some taxa do have a lower thermal threshold than others, likely leading to a shift on lowland community composition even if extinction rates are low.

### *Medium and Long-Term Responses of Organisms and Ecosystems to Warming*

The performance curve of an organism or species under current conditions does not map directly onto likely changes in performance in the medium term. How changes in performance will cascade through to changes in abundance, diversity, and species viability is unclear, but it is more certain that there will be substantial changes in the relative competitiveness of taxa that will feed through into changes in ecological community composition and functioning. Individual organisms and species will react to a decline in performance in a variety of ways, including (i) acclimation (modification of physiological performance curves, if possible), (ii) behavioral and phenological adaptation (modification of behavior, such as shifts in timing of activity in the diurnal and seasonal cycle), (iii) evolutionary adaptation over generations as natural selection favors heat- or drought-tolerant traits, and (iv) dispersal or migration to cooler or wetter climes where possible. All these responses will be in the context of complex ecological communities with intense interorganism competition, mutualism, and predator-prey-herbivory networks. Hence, for example, the viability of a particular tropical tree species under rapid warming may depend as much on the climate sensitivity of pollinators, herbivores, and pathogens as it does on the climate sensitivity of plant physiology. The specific viability of most organisms will be hard to predict, but, given the overall decline in performance for many organisms, it seems plausible that there may be an overall decline in abundance and an increase of extinction rates for many tropical organisms.

Perhaps the most critical question here is how these rates of biological response and adaptation compare to the extremely rapid changes in surface temperature projected. The fact that tropical forests have been warmer in the past and rich in life-forms suggests that there is no immediate upper limit to the viability of many tropical organisms (although some would decline at the expense of others), but the rate of temperature change may represent a real bottleneck in population viability. Organisms with the greatest risk of species decline and extinction are those with a low current tolerance for warming (i.e., potentially many tropical organisms), limited inherent ability to acclimate (little plasticity in the upper thermal tolerance limit), low dispersal or migration ability, a small species range, and location in a fragmented landscape, whether artificial (e.g., forest fragments) or natural (e.g., mountaintops or islands). Range sizes tend to be smaller in the tropics (Rapoport's Rule: Stevens, 1989), again potentially increasing the vulnerability of tropical organisms. It is unclear how generation life span affects the risk level: shorter lived organisms may be exposed to greater pressure if the temperature threshold acts most on a particular stage (e.g., juvenile) of the life cycle, but at the same time may have more capacity for rapid evolutionary adaptation. On the other hand, longer lived organisms (e.g., trees) may simply be able to "ride out" the coming century as mature adults even if juvenile viability is affected, but may have less ability to adapt and acclimate at suitable speed across generations.

### *Challenges to Migration and Dispersal*

If the potential for acclimation or adaptation is limited for some tropical organisms, the only alternative to maintain species fitness may be to migrate or disperse seeds to a cooler climate. Here, the tropical lowlands present another challenge. Tropical regions have shallow spatial gradients in temperature, on the order of 1 degree Celsius per 380 kilometers (236 miles) over the tropical belt, in contrast to 1 degree Celsius per 140 kilometers (87 miles) over both northern and southern temperate regions (fig. 14-5). Hence it is far more challenging and less likely for species to migrate across continents in tropical regions to remain close to a thermal optimum. For example, the recent rate of warming of tropical land regions (0.31 degree Celsius per decade; fig. 14-2A) would require poleward migration rates of 116 kilometers (72 miles) per decade (equivalent to 32 meters [105 feet] per day!). This is more than twice the required migration rate in northern temperate land regions (warming at 0.39 degree Celsius per decade; required migration rate of 55 kilometers [34 miles] per decade), and ten times that in southern temperate land regions (warming at 0.09 degree Celsius per decade; required migration rate of 13 kilometers [8 miles] per decade). In practice, local gradients tend to be steeper than this because

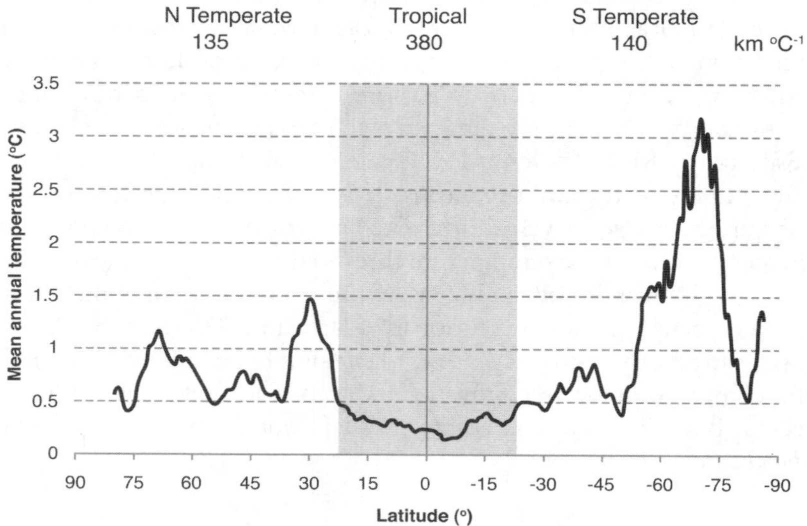


FIGURE 14-5. The latitudinal gradient in surface temperature, derived as the slope of figure 14-1A. This provides an approximate indicator of lowland spatial temperature gradients, but averaging over latitude zone misses some aspects of horizontal temperature gradients, such as temperature contrasts between continental interior and coastal regions.

of topography and proximity to large water bodies, but the relative comparison between tropical and temperate regions is still relevant.

It will be a significant challenge, if not impossible, for many heat-intolerant taxa to migrate or disperse at this rate. This challenge makes the potential for migration to higher elevation (lapse rate about 1 degree Celsius per 180 meters [591 feet]; 2,000 times steeper than the tropical horizontal gradient) far more important, and tropical montane and highland environments a particularly critical component of biodiversity resilience in the face of rapid tropical lowland warming. There is evidence from elevation gradient studies that tropical tree and insect communities are indeed migrating upslope (Chen et al., 2009; Feeley et al., 2011). Mountains and highland environments have a number of other features that favor adaptation and resilience, in particular the high diversity of microhabitats that are created by variation in aspect and slope. With too much warming, however, lower mountains and hills risk becoming traps, as species become isolated on warming hilltops.

Even if such rates of horizontal migration and dispersal are theoretically possible for some taxa, they will be a challenge in a fragmented landscape. This can include natural fragmentation, such as isolation on mountains or islands, or in swamps or riverine forests, but, of course, in many regions of the tropics, the rate of anthropogenic fragmentation is very high, with clearance of forests for farms, plantations, and cattle ranches. Deforestation in piedmont regions is a particular threat, as it isolates the warming lowlands from potential refugia in the highlands. Maintaining this connectivity to the highlands (low hill regions as well as high mountains) must be a conservation priority in the face of rapid warming of the lowlands.

Small-scale variation in microclimate has the potential to lessen extinction risk. Local microrefugia created by topographic or hydrological variation will lessen the need for extremely high dispersion rates. However, these microrefugia are useful only if the degree of local variation exceeds the larger scale variation over time; if it does not, microrefugia have the potential to become traps.

Estimates of extinction risk tend to include the two extreme migration/dispersion scenarios that are simplest to model: perfect dispersion and no dispersion. Dispersion rates will vary substantially by taxon (e.g., wind-dispersed vs. locally dispersed seeds), but from the evidence presented above it seems likely that the perfect migration rates required in the lowland tropics will be too high for most taxa.

Hence the no-dispersion assumption may be closer to reality, especially if temperature acts as a major limiting factor.

## Conclusions

In this chapter it has been argued that tropical organisms, species, and ecosystems face a number of acute threats from climate change that have perhaps been underestimated. This vulnerability in the tropics arises from a number of factors: instrinsic high diversity, shallow spatial gradients in temperature, limited history of temperature fluctuations on all timescales (diurnal, seasonal, interannual, and long-term) potentially resulting in narrow thermal tolerance, and the high pressure from ongoing logging, deforestation, fragmentation and fire use.

There is much we do not know about the response of tropical organisms to climate change. There is clearly a need for greater research into the regional patterns and mechanisms of precipitation change, the thermal niches of tropical taxa, the ability and potential speeds of acclimation and adaptation, especially to warming temperatures, the potential rates of migration and dispersal, and the interactions with fragmentation. There is also a need for systematic monitoring to look for early signs of change and clues to vulnerable taxa, extending beyond trees to other key taxa such as vertebrates or insects. However, the tropics are already warming rapidly, and research should not be a substitute for action. It is a time for immediate action, both in stabilizing and reversing the global rise in greenhouse gas concentrations that is the fundamental problem, but also in designing and implementing tropical biome protection of migration corridors to maximize the resilience of the biomes to the rapid change that is already under way. This will almost certainly be the most challenging century that tropical biodiversity has faced for a long, long time.

## Acknowledgments

Carol MacSweeney extracted the data on zonal climate trends, and the chapter benefited from discussions with Kathy Willis. The author is supported by the Jackson Foundation.

## REFERENCES


Alencar, A. A. C., L. A. Solorzano, and D. C. Nepstad. 2004. "Modeling forest understory fires in an Eastern Amazonian landscape." *Ecological Applications* 14: S139–S149.

Aragão L. E. O. C., Y. Malhi, R. M. Roman-Cuesta, S. Saatchi, L. O. Anderson, and Y. E. Shimabukuro. 2007. "Spatial patterns and fire response of recent Amazonian droughts." *Geophysical Research Letters* 34: L07701.

Atkin, O. K., and M. G. Tjoelker. 2003. "Thermal acclimation and the dynamic response of plant respiration to temperature." *Trends in Plant Science* 8: 343–351.

Baker, T. R., O. L. Phillips, Y. Malhi, S. Almeida, L. Arroyo, A. DiFiore, T. Erwin, et al. 2004. "Increasing biomass in Amazonian forests." *Philosophical Transactions of Royal Society of London-B Biological Sciences* 359: 353–365.

Barlow, J., and C. A. Peres. 2008. "Fire-mediated dieback and compositional cascade in an Amazonian forest." *Philosophical Transactions of Royal Society of London-B Biological Sciences* 363: 1787–1794.

Bush, M. B., M. R. Silman, and D. H. Urrego. 2004. "48,000 years of climate and forest change in a biodiversity hot spot." *Science* 303: 827–829.

Butt, N., M. New, G. Lizcano, and Y. Malhi. 2009. "Spatial patterns and recent trends in cloud fraction and cloud-related diffuse radiation in Amazonia." *Journal of Geophysical Research-Atmospheres* 114.

Campbell, C., L. Atkinson, J. Zaragoza-Castells, M. Lundmark, O. Atkin, and V. Hurry. 2007. "Acclimation of photosynthesis and respiration is asynchronous in response to changes in temperature regardless of plant functional group." *New Phytologist* 176: 375–389.

Chen, I. C., H. J. Shiu, S. Benedick, J. D. Holloway, V. K. Cheye, H. S. Barlow, J. K. Hill, and C. D. Thomas. 2009. "Elevation increases in moth assemblages over 42 years on a tropical mountain." *Proceedings of the National Academy of Sciences, USA* 106: 1479–1483.

Clark, D. A., S. C. Piper, C. D. Keeling, and D. B. Clark. 2003. "Tropical rain forest tree growth and atmospheric carbon dynamics linked to interannual temperature variation during 1984–2000." *Proceedings of the National Academy of Sciences* 100: 5852–5857.

Colwell, R. K., G. Brehm, C. L. Cardelus, A. C. Gilman, and J. T. Longino. 2008. "Global warming, elevational range shifts, and lowland biotic attrition in the wet tropics." *Science* 322: 258–261.

Condit, R., S. Aguilar, A. Hernandez, R. Perez, S. Lao, G. Angehr, S. Hubbell, and R. Foster. 2004. "Tropical forest dynamics across a rainfall gradient and the impact of an El Nino dry season." *Journal of Tropical Ecology* 20: 51–72.

Cox, P. M., P. P. Harris, C. Huntingford, R. A. Betts, M. Collins, C. D. Jones, T. E. Jupp, J. A. Marengo, and C. A. Nobre. 2008. "Increasing risk of

Amazonian drought due to decreasing aerosol pollution." *Nature* 453 (7192) (May): 212–215.

Deutsch, C. A., J. J. Tewksbury, R. B. Huey, K. S. Sheldon, C. K. Ghalambor, D. C. Haak, et al. 2008. "Impacts of climate warming on terrestrial ectotherms across latitude." *Proceedings of the National Academy of Sciences, USA* 105: 6668–6672.

Feeley, K. J., and M. R. Silman. 2010. "Biotic attrition from tropical forests correcting for truncated temperature niches." *Global Change Biology* 16: 1830–1836.

Feeley, K. J., M. R. Silman, M. B. Bush, W. Farfan, K. G. Cabrera, Y. Malhi, P. Meir, N. S. Revilla, M. N. R. Quisiyupanqui, and S. Saatchi. 2011. "Upslope migration of Andean trees." *Journal of Biogeography* 38: 783–791.

Fisher, R. A., M. Williams, A. L. da Costa, Y. Malhi, R. F. da Costa, et al. 2007. "The response of an E. Amazonian rain forest to drought stress: Results and modelling analyses from a throughfall exclusion experiment." *Global Change Biology* 13: 2361–2378.

Grainger, A., D. H. Boucher, P. C. Frumhoff, W. F. Laurance, T. Lovejoy, J. McNeely, M. Niekisch, et al. 2009. "Biodiversity and REDD at Copenhagen." *Current Biology* 19: R974–R976.

Huber, M. 2008. "A hotter greenhouse?" *Science* 321: 353–354.

Jaramillo, C., D. Ochoa, L. Contreras, M. Pagani, H. Carvajal-Ortiz, L. M. Pratt, S. Krishnan, et al. 2010. "Effects of rapid global warming at the Paleocene-Eocene boundary on neotropical vegetation." *Science* 330: 957–961.

Lewis, S. L., G. Lopez-Gonzalez, B. Sonke, K. Affum-Baffoe, T. R. Baker, L. O. Ojo, et al. 2009. "Increasing carbon storage in intact African forests." *Nature* 457: 1003–1006.

Maley, J. 2001. "A catastrophic destruction of African forests around 2,500 years ago still exerts a major influence on present vegetation formations." *IDS-Bulletin of Development Studies* 33: 13–20.

Malhi, Y., and J. Wright. 2004. "Spatial patterns and recent trends in the climate of tropical rainforest regions." *Philosophical Transactions of Royal Society of London-B Biological Sciences* 359: 311–329.

Malhi, Y., J. T. Roberts, R. A. Betts, T. J. Killeen, W. H. Li, et al. 2008. "Climate change, deforestation, and the fate of the Amazon." *Science* 319: 169–172.

Malhi, Y., L. E. O. C. Aragão, D. Galbraith, C. Huntingford, R. Fisher, P. Zelazowski, et al. 2009. "Evaluating the likelihood and mechanism of a climate-change-induced dieback of the Amazon rainforest." *Proceedings of the National Academy of Sciences, USA* 106: 20610–20615.

Morley, R. J. 2000. *Origin and Evolution of Tropical Rain Forests*. Chichester: Wiley Blackwell.

Nepstad, D. C., I. M. Tohver, D. Ray, P. Moutinho, and G. Cardinot. 2007. "Mortality of large trees and lianas following experimental drought in an Amazon forest." *Ecology* 88: 2259–2269.

Phillips, O. L., R. V. Martinez, L. Arroyo, T. R. Baker, T. Killeen, S. L. Lewis, et al. 2002. "Increasing dominance of large lianas in Amazonian forests." *Nature* 418: 770–774.

Phillips, O. L., L. E. O. C. Aragão, S. L. Lewis, J. B. Fisher, J. Lloyd, G. Lopez-Gonzalez, et al. 2009. "Drought sensitivity of the Amazon forest." *Science* 323: 1344–1347.

Stevens, G. C. 1989. "The latitudinal variations in geographical range—how so many species coexist in the tropics." *American Naturalist* 133: 240–256.

Thomas, C. D., A. Cameron, R. E. Green, M. Bakkenes, L. J. Beaumont, Y. C. Collingham, B. F. N. Erasmus, et al. 2004. "Extinction risk from climate change." *Nature* 427: 145–148.

Wang, H. C. , M. J. Moore, P. S. Soltis, C. D. Bell, S. F. Brockington, R. Alexandre, et al. 2009. "Rosid radiation and the rise of angiosperm-dominated forests." *Proceedings of the National Academy of Sciences, USA* 106: 3853–3858.

Williams, J. W., S. T. Jackson, and J. E. Kutzbach. 2007. "Projected distributions of novel and disappearing climates by 2100 AD." *Proceedings of the National Academy of Sciences, USA* 104: 5738–5742.

Wright, S. J., O. Calderon, A. Hernandez, and S. Paton. 2006. "Are lianas increasing in importance in tropical forests? A 17-year record from Panama." *Ecology* 85: 484–489.

Zachos, J., et al. 2001. "Trends, rhythms and aberrations in global climate 65 Ma to present." *Science* 292: 686–693.

Zelazowski, P., Y. Malhi, C. Huntingford, S. Sitch, and J. B. Fisher. 2011. "Changes in the potential distribution of humid tropical forests on a warmer planet." *Philosophical Transactions of the Royal Society A-Mathematical, Physical and Engineering Sciences* 369: 137–160.

# Chapter 15

## *Coral Reefs, Climate Change, and Mass Extinction*

OVE HOEGH-GULDBERG

The term "extinction" is a loaded phrase when applied to coral reefs. Often, it is applied to coral reefs generically, with the phrase "extinction of coral reefs" appearing regularly in the scientific and popular press (e.g., it appeared more than 15,000 times when typed into the Google search engine, June 4, 2011). This phrase is generally used to describe the disappearance of coral reefs as an ecosystem, which is very distinct from the extinction of a particular coral reef species. This distinction becomes increasingly important when considering the likely outcome for coral reefs and their biodiversity under rapid global change.

Discussions of the characteristics that are likely to influence the vulnerability of species to extinction have appeared several times already in this book. Currently, many marine species are often abundant with large dispersal ranges. These characteristics are such that many marine species are relatively resistant to extinction, especially when compared to many terrestrial species, which are often narrowly distributed with limited dispersal ability. Roberts and Hawkins (1999) list a number of characteristics associated with marine species that make them more or less vulnerable to extinction. In this respect, they explored the influence of parameters that affect population turnover, reproduction, recovery capacity, range and distribution, trophic level, and whether or not a species is common or rare. Under their analysis, long-lived and slow-growing organisms that have short dispersal distances and narrow nearshore ranges are at greater risk of extinction than species that are short-lived, fast-growing, dispersed widely, and which have broad geographical ranges (table 15-1).

Reef-building corals tend to be long-lived, slow-growing, reproductive at large size, and immobile as adults (table 15-1). Together with their narrow depth range and nearshore distribution, and their sensitivity to the activities of humans, corals make ideal candidates for extinction. On the other hand, they often have sizable reproduction biomasses and reproduce many times in their lives, with the ability to regenerate and grow asexually from fragments. When in good health, corals are also effective competitors and colonizers with strong larval recruitment. They also have some of the largest geographical ranges of any known animal. For these reasons, they should be relatively less vulnerable to extinction.

Typical coral-dwelling fish, on the other hand, differ from corals in that they are relatively short-lived, mature at small sizes, reasonably mobile but unable to reproduce asexually. According to the Roberts and Hawkins' (1999) framework, typical coral-dwelling fish will also have a degree of vulnerability to extinction but for different reasons. Across the range of organisms that live on coral reefs, there are clearly those that are relatively resistant to extinction versus those that are highly vulnerable. Under a rapidly changing climate, one of the interesting questions becomes: Can we predict which organisms are going to be "winners" and which organisms are going to be "losers" with respect to extinction?

## Rapidly Changing Ocean Environments

One of the defining characteristics of our times is that there is no longer anywhere on our planet that humans have not significantly impacted (Halpern et al., 2008). A depressing yet sobering fact, the signature of human activities can be found at the bottom of the deepest oceanic trench and at the top of the highest mountain. For coral reefs, the signature has been unambiguous for some time, with a 1–2 percent decrease in coral cover each year and the loss of 40 percent of coral reefs over the past three decades (Bruno and Selig, 2007).

Both local and global stresses are driving this decline. Local factors include deteriorating water quality (i.e., too many nutrients and

TABLE 15-1. Characteristics that are likely to render marine species vulnerable to extinction (adapted from Roberts and Hawkins 1999).

| | *Vulnerability* | | | |
|---|---|---|---|---|
| *Characteristics* | *High* | *Low* | | |
| *Population turnover* | | | *Corals* | *Fish* |
| Longevity | Long | Short | Long | Short |
| Growth rate | Slow | Fast | Slow | Moderate |
| Natural mortality rate | Low | High | Moderate | Moderate |
| Production biomass | No | High | High | Moderate |
| *Reproduction* | | | | |
| Reproductive efforts | Low | High | Low | Moderate |
| Reproductive frequency[1] | Semelparity | Iteroparity | Iteroparity | Iteroparity |
| Age or size at sexual maturity | Old or large | Young or small | Old or large | Moderate |
| Sexual dimorphism | Large difference in size between sexes | Does not occur | Does not occur | Does not occur |
| Sex change | Occurs (protandry[2] in particular) | Does not occur | rare | Occurs (but many protogynous) |
| Spawning | In aggregation is at predictable locations | Not in aggregations | Not in aggregations | In aggregation is at predictable locations |
| Allee effects[3] at reproduction | Strong | Weak | Strong | Weak |
| *Capacity for recovery* | | | | |
| Regeneration from fragments | Does occur | Occurs | Occurs | Does occur |
| Dispersal | Short distance | Long-distance | Long distance | Long distance |
| Competitive ability | Poor | Good | Good | Good |
| Colonizing ability | Poor | Good | Good | Good |

TABLE 15-1. Continued.

| Characteristics | Vulnerability | | | |
|---|---|---|---|---|
| | *High* | *Low* | | |
| Adult mobility | Low | High | Low | Moderate |
| Recruitment by larval settlement | Irregular and/or low | Frequent and intense | Frequent and intense | Frequent and intense |
| Allee effects at settlement | Strong | Weak | Weak | Weak |
| *Range and distribution* | | | | |
| Horizontal distribution | Nearshore | Offshore | Nearshore | Nearshore |
| Vertical depth range | Narrow | Broad | Narrow | Narrow |
| Geographic range | Small | Large | Large | Large |
| Patchiness of population within range | High | Low | Low | Low |
| Habitat specific | High | Low | High | High |
| Habitat vulnerable to destruction by people | High | Low | High | High |
| *Commonness and/or rarity* | Rare | Abundant | Abundant | Abundant |
| *Trophic level* | High | Low | Low | Moderate |

[1]Semelparity: Organisms in which the life cycle is characterized by a single reproductive episode before death.

Iteroparity: Organisms in which the life cycle is characterized by multiple reproductive cycles over the course of the life history.

[2]Protandry: Organisms that start off life as males and transform into females. Opposite is protogyny, in which organisms start off life as females and transform later into males.

[3]Allee effects: Situation in which there is a positive correlation between population density and the per capita population growth rate in very small populations.

sediments from disturbed coastal areas adjacent to coral reefs), pollution, destructive fishing using dynamite and cyanide, impacts from shipping, and the overexploitation of key fisheries species (Bryant, 1998; Halpern et al., 2008). These factors have played a dominant role in the the destruction of coral reefs, with the role of climate change being relegated to a secondary and long-term threat. Evidence over the past decade (reviewed by Hoegh-Guldberg et al., 2007), however, reveals that climate change has already affected reefs heavily, and that current trends in the warming and acidification of the oceans will almost certainly destroy coral reefs unless we take action to reduce greenhouse gas emissions.

## Impacts of Global Warming

The atmospheric concentration of carbon dioxide in the earth's atmosphere now exceeds 390 parts per million, which is approximately 100 parts per million higher than the maximal values seen over the past 740,000 years (Petit et al., 1999; Augustin et al., 2004), if not 20 million years (Raven et al., 2005). These changes have driven the average temperature of the ocean up by 0.74 degree Celsius and sea levels by 17 centimeters (7 inches) (IPCC, 2007). Second-order effects have also begun to occur, with evidence of shifting patterns of rainfall, drought, and storm intensity across the tropics leading to complex changes in the conditions surrounding natural ecosystems (Walther et al., 2002; IPCC, 2007).

Even though these changes appear subtle, they have already had a huge impact on coral reef ecosystems. Reef-building corals form intimate mutualistic endosymbioses (in which one organism lives inside the cells of another) with tiny plantlike organisms called dinoflagellates from the genus *Symbiodinium*. The partnership that corals form with these primary producers enables the symbiosis to trap large amounts of solar energy, allowing corals access to an abundant source of energy. As a result of this rich source of photosynthetic energy, reef-building corals are able to precipitate copious quantities of calcium carbonate, which ultimately lead to the formation of the three-dimensional framework of coral reefs (Muscatine, 1990).

The symbiosis between *Symbiodinium* and corals is relatively stable within the normal range of environmental variability. When conditions change too rapidly, or exceed the natural range of environmental

FIGURE 15-1. Small increases in sea temperature can destabilize the symbiosis between reef-building corals and their dinoflagellate symbionts (*Symbiodinium*). A mass bleaching event that affected more than 60 percent of the Great Barrier Reef occurred in 2002, leading to the death of about 5–10 percent of reef-building corals within the Great Barrier Reef Marine Park. Photo: O. Hoegh-Guldberg.

variability, the symbiosis breaks down, with the brown *Symbiodinium* moving rapidly out of the tissues of the coral host (fig. 15-1). This phenomenon is referred to as coral bleaching (Hoegh-Guldberg, 1999). In the early 1980s, coral reefs around the world began to bleach across large areas of the tropics (Glynn, 1983; Lasker et al., 1984; Roberts, 1987), with no precedent in the scientific literature. Work over a number of years revealed that these mass bleaching events were being driven by small increases in sea temperature (Hoegh-Guldberg and Smith, 1989; Glynn and D'Croz, 1990; Strong et al., 1996) above the summer maximal temperatures for a particular region. This led some to speculate on whether this was connected to climate change (Glynn et al., 1988). These early suspicions were eventually confirmed, with projections of how sea temperature would change under greenhouse forcing revealing that future conditions were likely to be extremely hostile to coral communities (Hoegh-Guldberg, 1999).

## Ocean Acidification and Marine Calcification

The increase in atmospheric carbon dioxide has resulted in a greater amount of carbon dioxide entering the world's oceans (fig. 15-2A). When carbon dioxide enters the ocean, it reacts with water to produce a weak acid, carbonic acid. Carbonic acid dissociates producing a bicarbonate ion and a proton, the latter of which reacts with carbonate

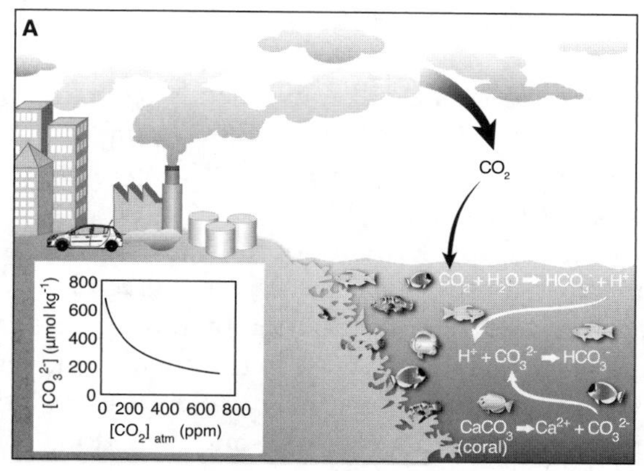


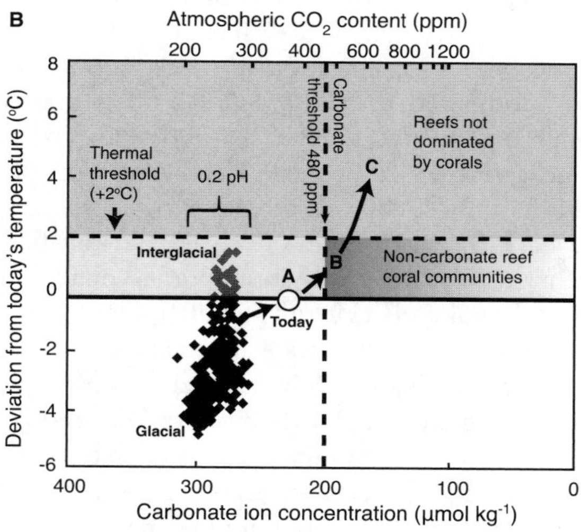


FIGURE 15-2. (A) Ocean acidification is occurring due to the increased amount of carbon dioxide that is entering the ocean. In the ocean, carbon dioxide combines with water to create carbonic acid, which releases a proton that reacts in turn with carbonate ions. The net effect of this is that rising atmospheric carbon dioxide has driven down the concentration of carbonate ions in the ocean. This has already slowed the calcification of corals by around 15 percent since 1990. (B) Reconstruction of the temperature and carbonate ion concentrations of a typical tropical ocean over the past 420,000 years (reconstructed using data from the Vostok ice core series, Petit et al. 1999. See Hoegh-Guldberg et al. 2007 for a full explanation of the methods). Three important features are shown: (i) Present-day values are outside the conditions seen over at least the past 420,000 years. (ii) Rates of change dwarf anything seen over the same period. (iii) Two thresholds, one with temperature and the other with carbonate ion concentrations, will be crossed within the next 40 years. Figures 15-2 and 15-3 reprinted from Hoegh-Guldberg et al. (2007) with permission from *Science Magazine*.

ions, producing bicarbonate and consequently reducing their concentration and availability. The net effect of increasing carbon dioxide in the atmosphere is that it causes a precipitous drop in the concentration of carbonate ions (Raven et al., 2005).

These changes in the concentration of carbonate ions have huge implications for organisms such as corals that precipitate calcium carbonate. A doubling of the concentration of atmospheric carbon dioxide, for example, results in a decrease in the calcification rate of 15–45 percent across a wide range of marine calcifiers such as red calcareous algae to reef-building corals (Kleypas and Langdon, 2006). These changes are now being reported from corals in the field, with studies done on the Great Barrier Reef (De'ath et al., 2009) reporting an unprecedented decrease in calcification of corals in both regions by 15 percent when compared to 1990. Similar observations have been made for corals in Thailand (Tanzil et al., 2009). Given that reef structures are a delicate balance between calcification on one hand and physical and biological erosion on the other, these changes in the calcification rate have the potential to decrease the calcification rate of corals and other calficiers below that required to maintain the carbonate structures of coral reefs (Hoegh-Guldberg et al., 2007). A direct implication of this is that coral reefs will soon reach a point where they are likely to erode and dissolve (Silverman et al., 2009).

When the current geographic distribution of carbonate coral reefs is plotted relative to the aragonite saturation state of seawater, it is clear that coral reefs require a certain concentration of calcium and carbonate ions to calcify at the rate required to maintain coral reefs. The aragonite saturation constant ($\Omega_{aragonite}$) is the ratio of the calcium and carbonate ion concentrations relative to the solubility product of aragonite (the crystal form of calcium carbonate that corals preferentially deposit in their skeletons). Figure 15-3 illustrates how changing the carbon dioxide concentration will essentially shrink the distribution of areas where water contains enough calcium and carbonate for the formation of carbonate coral reefs (i.e., have aragonite saturation in excess of 3.3) to a small band around the equator (at atmospheric carbon dioxide concentrations of 450 parts per million). These conditions are largely eliminated when concentrations rise to 550 parts per million or more. This modelling study illustrates the extreme sensitivity of carbonate coral reef ecosystems to the atmospheric concentration of carbon dioxide (Hoegh-Guldberg et al., 2007).

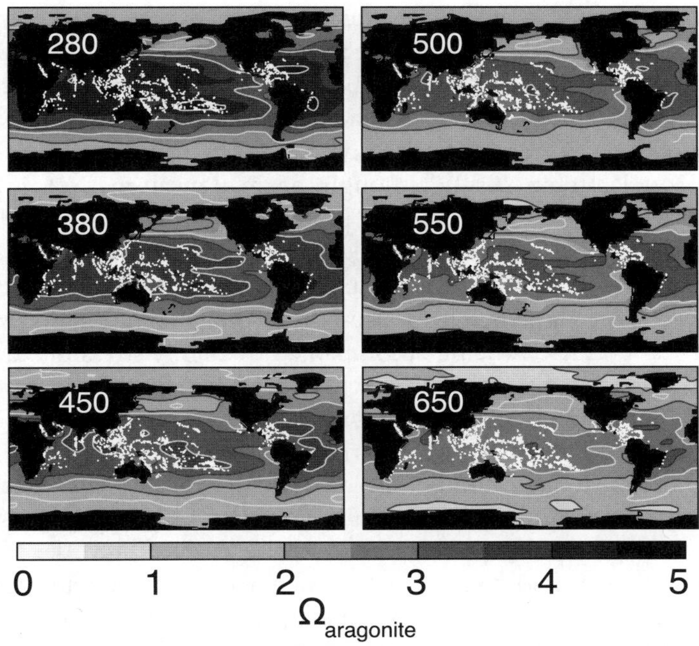


FIGURE 15-3. Distribution of carbonate coral reefs (white dots) relative to the aragonite saturation ($\Omega_{aragonite}$ = $[Ca^{2+}].[CO_3^{2-}]$/Ksp aragonite) calculated for different atmospheric concentrations of carbon dioxide (white number in each box). The aragonite saturation is a measure of the concentration of calcium and carbonate ions relative to solubility of aragonite (the chief form of calcium carbonate deposited by corals and other marine calcifiers). The distribution of today's coral reefs relative to the current aragonite saturation is shown in the panel labeled 380, revealing the association of carbonate coral reefs with waters that have an aragonite saturation of more than 3.3 (darker gray areas). As atmospheric carbon dioxide increases, the distribution of these waters contracts to the equator and more or less disappears when concentrations of atmospheric carbon dioxide rise above 550 parts per million.

## Sea-Level Rise and Other Factors

Global sea level is currently rising at the rate of 3.3 millimeters (0.13 inch) per year, and sea level is conservatively expected to rise between 11 and 77 centimeters (4 and 30 inches) by the end of the century (IPCC, 2007). These estimates of sea-level rise are widely considered to be conservative, especially in the light of the recent rapid decreases in land-locked glacial and polar ice mass in Greenland (Cressey, 2007; Steffensen et al., 2008; Zhang et al., 2008) and Antarctica (Barnes and Peck 2008). Under conditions where coral growth is maximal, coral reefs are able to keep pace with comparatively high rates of sea-level rise (Douglas et al., 2000), although this may mean that reef growth may "back-step" in cases of catastrophic rises such as that seen 121,000 years ago (Blanchon et al., 2009). In past periods of rapid sea-level change, corals have been healthy due to the relatively small or nonexistent impact of humans. These assumptions may be invalidated today, when sea-level rise is being accompanied by highly stressful sea temperatures and acidities that are likely to dramatically slow the growth of corals and coral reefs, thereby presenting the specter of deteriorating reefs that "drown" as sea level rises.

There are a large number of other factors that will influence the health of coral reefs into the future. Warmer sea temperatures will drive more intense storm activity (Webster et al., 2005), leading to greater damage to some coral reefs, which when coupled with slowing growth and reduced coral survivorship, may mean that the balance is tipped away from coral-dominated reefs to reefs that are substantially different.

## Prognosis for the Future of Coral Reefs

In exploring the future of coral reefs, it is instructive to look at the conditions that coral reefs have experienced over the past several hundred thousand years. This type of analysis allows one to understand the range of variability in the past relative to the conditions projected to occur in the future. We recently analyzed (Hoegh-Guldberg et al., 2007) sea temperature (from proxies) and carbonate ion concentrations (from carbon dioxide concentrations) for typical coral reef environments over the past 420,000 years using measurements of ancient atmospheric carbon dioxide and global temperatures derived from the Vostok ice core data series (Petit et al., 1999). This analysis revealed that the key conditions for coral reefs (i.e., temperature and carbonate ion concentration; Kleypas et al., 1999) varied significantly over the past 420,000 years (fig. 15-2) but that current conditions on coral reefs were well outside the envelope in which this variability had occurred. Perhaps even more significant is that fact that the changes that are occuring over decade and century timescales today used to occur over thousands of years in the past (Hoegh-Guldberg et al., 2007).

The trajectory that coral reefs are currently on is rapidly approaching two significant thresholds (fig. 15-2B). The first occurs when tropical seas become 2 degrees Celsius warmer than they were prior to the Industrial Revolution, a condition that we know will cause unsustainable coral bleaching and mortality on an annual basis. The second, which occurs more or less simultaneously, is that oceans will be acidified to the point where they will have carbonate ion concentrations of less than 200 micromoles per kilogram of water. Both field and laboratory studies reveal that the latter is around the point at which coral calcification struggles to keep up with erosion, and hence maintain carbonate reef structures. As a result, many of these all-important structures will crumble and slowly disappear. Hopefully, decisive and effective action to reduce emissions of carbon dioxide will stabilize and eventually reverse the upward trend in atmospheric carbon dioxide.

As a final point in this discussion, it is instructive to consider how coral communities will change over the coming decades. The very rapid rate of rise in atmospheric carbon dioxide is already driving periodic mass mortality events and a slowing of reef calcification. Given that not all corals have the same sensitivity to thermal stress (Hoegh-Guldberg and Salvat, 1995; Marshall and Baird, 2000; Loya et al. 2001; McClanahan, 2007) or even reduced carbonate concentrations (Kleypas and Langdon, 2006), some are likely to be more persistent than others as conditions change. For this reason, massive and encrusting corals (e.g., *Porites, Favia*) may be more prominent on future reefs than branching corals (e.g., *Acropora, Stylophora*), leading to changes in the community structure of reef-building corals and the extinction of some branching coral species. Naturally, given that even these tougher species have their limits, reefs will eventually become largely devoid of corals as ocean temperatures and acidities continue to rise.

These futures are not distant. Three decades ago, the potential extinction of a reef-building coral species would have been unthinkable given the stability of tropical environments and the vast geographic distributions of most species (Veron and Stafford-Smith, 2000). In 2004, however, the US National Marine Fisheries Service received a request from the Center for Biological Diversity to place the Caribbean species of *Acropora* (*Acropora palmata* and *Acropora cervicornis*) on the US Endangered Species list on account of the precipitous decrease in the distribution and abundance of these once dominant coral species. This was finally granted on May 4, 2006. Although the circumstances of a restricted and highly stressed ocean basin such as the Caribbean probably predispose coral species in the Caribbean to extinction (when compared to the vast, moderately stressed Pacific ocean), the listing of these two species represented a wake-up call for coral reef biologists and conservations, who may have been concerned about the loss of functional coral reefs as opposed to the extinction of coral species (Bruckner et al., 2002; Precht et al., 2002; Precht et al., 2004). According to Carpenter (2008), almost one-third of coral species are vulnerable to extinction by climate change and local threatening processes.

## The Fate of Coral Reef–Associated Organisms

Rapid changes in the environmental factors associated with climate change, such as sea temperature and acidity, are likely to eliminate coral-dominated reefs (fig. 15-4A) and transform them into vastly different systems that are occupied by very different organisms such as seaweeds. Today, there are a growing number of examples that illustrate what these ecosystems might look like in the future. For example, many areas across the Caribbean have lost almost all of their coral-dominated reefs, leaving behind communities that are vastly different from those of 50 years ago. Naturally occurring coral reefs at high latitudes also give us clues as to what future communities might look like. These areas, like those of the southwest Australia (fig. 15-4B), typically have lower coral cover, which does not contribute enough calcium carbonate to maintain significant reef frameworks. This situation will expand in warmer, more acidic seas, however, with carbonate reef structures beginning to crumble and disappear, as is already happening in some heavily affected areas (fig. 15-4C). As a result of these changes, ecological processes and resources will change radically, leading to changes in the species composition of the organisms that associate with these new reef communities and structures. With this change comes the question of what happens to all the other species that normally associate with coral reefs. Perhaps surprisingly, this question has not been answered adequately for anything more than reef fish at this point. This said, there are a number of "commonsense rules" that appear to apply in terms of which species on the coral reef are more sensitive to global change than others.

FIGURE 15-4. Coral reefs are sensitive to climate change. (A) Healthy coral-dominated reef on the Great Barrier Reef in Australia; (B) high-latitude coral reef of Jurien Bay, western Australia, and (C) heavily impacted reef in Karimunjawa, Indonesia, showing the devastated reef framework and dead coral. Photos: O. Hoegh-Guldberg.

Returning to table 15-1, we see that a high level of habitat specificity is associated with increased vulnerability to extinction. This characteristic varies quite significantly across the various organisms that occupy coral reefs. Some organisms have highly specific requirements for habitat and will not be found on reefs that no longer have that habitat. For example, many butterfly fish (Chaetodontidae) require living colonies of corals from the family Acroporidae as habitat for recruitment, protection from predators, and as food (fig. 15-5A). These species tend not to be found on reefs where corals such as *Acroporid* corals have disappeared or are naturally absent (Pratchett et al., 2006). On the other hand, fish such as surgeon fish (fig. 15-5B), scarids, and siganids appear to be somewhat independent or even positively correlated with the loss of coral (Wilson et al., 2008). In this case, the fish involved do not have to eat coral (if they do at all), and have ecological requirements that don't require corals to be present. Not surprisingly, these fish are often found in areas within coral reefs that often do not have high levels of coral cover (e.g., inshore and back-reef areas).

The general principles underlying which species are more vulnerable appear to apply across the board for fish on coral reefs. Wilson and colleagues (2008) visited twenty-one sites across the Indo-Pacific that had been affected by coral bleaching, mass coral bleaching, and other stresses, and which had recovered to varying extents. Using this information, they related fish community composition to the amount of coral cover at the various sites. Fish that were highly tied to coral declined dramatically, while fish that were either omnivorous or herbivorous (and had no specific requirements for coral) did not change or increased (fig. 15-6). Often in these cases, the supply of grazing surfaces and provision of a place to hide from predators will be sufficient. However, one has to be careful with assumptions based on the apparent requirements of the adult phase given that specific requirements can arise in the juvenile stages which are not met by the adult habitat. Munday et al. (2004) has also pointed out that specialist species are far more vulnerable than generalist species, not only because they are specific in their requirements by definition, but they are often in lower abundance and distributed more patchily. Roberts and Hawkins (1999) identified both of these as characteristics conferring vulnerability to extinction (table 15-1). Understanding the direction of change in terms of which species will become vulnerable to extinction will require careful consideration of the full set of ecological requirements of any particular species or group of organisms.

FIGURE 15-5. Coral fish show a broad range of associations with coral communities, from those that are obligatory to those that are not. The requirement for coral may affect their extinction vulnerability. (A) The butterfly fish, *Chaetodon plebius,* has an obligatory relationship with *Acroporid* corals—eating, recruiting into, and hiding among them. (B) Generalists such as the herbivorous surgeon fish, *Acanthurus lineatus,* have a far more facultative relationship with corals, and may show increases in abundance when coral cover decreases. Photos: O. Hoegh-Guldberg.

There is probably little doubt that similar results would be obtained for coral-dependent invertebrates and the many other organisms that associate with coral reefs. Information beyond that about fish and corals, however, is extremely limited (Przeslawski et al., 2008). Several previous studies have focused on benthic crustacea associated with living coral colonies. These studies have revealed that the species composition of decapod crustacea (shrimps and crabs)

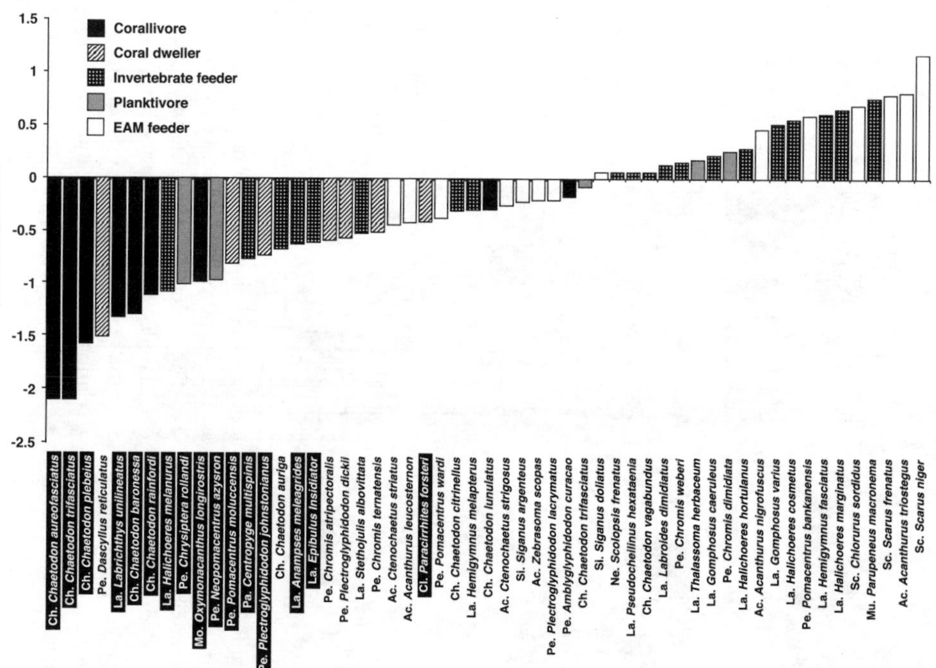


FIGURE 15-6. Response of coral fish species to a decline in coral cover measured for Indo-Pacific coral reefs. Bars that drop below the *x*-axis represent fish whose abundance decreased as coral cover decreased, while those above the *x*-axis represent fish species that increase their abundance in situations of low coral cover. Key ecological characteristics of each fish species are depicted by different patterns on the bars. Species names in black indicate differences to zero (95% confidence intervals). EAM = epilithic algal matrix. Reprinted with permission of Blackwell Publishing Ltd.

responds to the coral colony size, the surrounding habitat, and behavioral interactions between species (Abele and Patton, 1976; Black and Prince, 1983; Gotelli et al., 1985). When the host coral colony dies, the infaunal community shifts from a dominance of predominantly symbiotic species to small facultatively associated normal symbionts (Coles, 1980; Glynn, 1993). Preston and Doherty (1990) analyzed the agile shrimp fauna living in live and dead colonies of a single species of reef-building coral, *Pocillopora verrucosa,* on the central Great Barrier Reef in Australia. Their study highlights the potentially enormous diversity of coral-dependent organisms, with the identification of more than twenty species of coral-dwelling shrimp that regularly inhabited colonies of *P. verrucosa,* with three species being coral specialists (obligates). Clearly this complexity within complexity is just the tip of the iceberg with a very clear message that the loss of corals and the structures that they build are very clear threats to a poorly known yet sizable component of marine biodiversity.

Until advances are made in the taxonomy of the multitude of species that live in and around coral reefs, estimates of the number of species that are highly vulnerable to extinction through the loss of coral-dominated reef systems will be speculative at best. However, one might list the following characteristics of coral reef species that would make them a high risk for extinction.

- Obligatory dietary requirement for coral or coral-dependent organisms. Many fish and invertebrates depend on living coral as part of their diets. The loss of corals as a food source would directly threaten these organisms with extinction (Moran, 1986; Rinkevich et al., 1991; Turner, 1994; Graham et al., 2007; Pratchett et al., 2008; Wilson et al., 2008).
- Commitment to a settlement cue that is dependent on coral or involves a requirement for live coral by new recruits. This is widespread on coral reefs with highly specific settlement preferences (Sale et al., 1984; Feary et al., 2007) and often involves homing in on particular coral species or genera (Gotelli and Abele, 1983; Gotelli et al., 1985; Öhman et al., 1998).
- Requirement for complex three-dimensional reef structure to hide from predators. Many organisms on coral reefs require their three-dimensional structures, which provide a complex set of hiding places for coral reef organisms (Pratchett et al., 2008). Where corals have disappeared due to human stresses

(fig. 15-5C) or storms, much of this three-dimensional topology and complexity also disappears.

- Need for close spacing of coral habitat to be able to access mates. Although corals can reproduce asexually by fragmentation (Highsmith, 1982; McKinney, 1983), and could theoretically exist for centuries without undergoing sexual reproduction, many organisms require access to the opposite sex. In order to be able to persist, these organisms require coral colonies to be within close proximity to each other to have access to mates (Abele and Patton, 1976; Black and Prince, 1983; Gotelli et al., 1985; Roberts and Hawkins, 1999; Cowen et al., 2000; Sale, 2002; Cowen et al., 2006).

The shaping of coral reef ecosystems by reef-building corals necessarily dictates that many species are dependent on the health and presence of these organisms. Although similar relationships hold for other marine communities such as rocky shores, kelp forests, and sea-grass beds, the sheer number of species involved in coral reefs (i.e., 1 million–9 million species) means that the projected loss of corals and the reef structures due to climate change will have huge consequences for tropical and indeed global biodiversity.

## Species Extinction Versus Functional Extinction

As pointed out earlier, the phrase "extinction of coral reefs" is used frequently in both public and academic discourse as part of the concern for the future of coral reefs with rapid climate change. The phrase implies that coral reefs and the species that live on them will disappear if we continue down the current enhanced greenhouse pathway. Many coral reef organisms, however, have characteristics that make them relatively invulnerable to extinction. This suggests that many coral reef organisms will survive even if coral reefs as the functional ecosystem do not. What will be fascinating (though disturbing) for future generations of biologists is how these new assemblages will take shape and function. Beyond this, the functional extinction of coral reef ecosystems remains a serious crisis for humanity. Coral reef ecosystems support enormous numbers of relatively disadvantaged people living along tropical coastlines through the provision of subsistence food sources, income, and coastal protection. The loss of functional coral reef ecosystems represents a severe threat to these people and their societies. It is hoped that there will be a global commitment not to let this disturbing future unfold.

## REFERENCES

Abele, L., Patton, W. 1976. "The size of coral heads and the community biology of associated decapod crustaceans." *Journal of Biogeography* 3: 35–47.

Augustin, L., Barbante, C., Barnes, P. R. F., Barnola, J. M, Bigler, M., Castellano, E., Cattani, O., et al. 2004. "Eight glacial cycles from an Antarctic ice core." *Nature* 429: 623–628.

Barnes, D. K. A., Peck, L. S. 2008. "Vulnerability of Antarctic shelf biodiversity to predicted regional warming." *Climate Research* 37: 149–163.

Black, R., Prince, J. 1983. "Fauna associated with the coral *Pocillopora damicornis* at the southern limit of its distribution in Western Australia." *Journal of Biogeography* 10: 135–152.

Blanchon, P., Eisenhauer, A., Fietzke, J., Liebetrau, V. 2009. "Rapid sea-level rise and reef back-stepping at the close of the last interglacial highstand." *Nature* 458: 881–884.

Bruckner, A., Hourigan, T., Moosa, M., Soemodihardjo, S., Soegiarto, A., Romimohtarto, K., Nontji, A., and Suharsono, S. 2002. "Proactive management for conservation of *Acropora cervicornis* and *Acropora palmate*: Application of the U.S. Endangered Species Act." *Proceedings of the Ninth International Coral Reef Symposium* 2: 661–665; Bali, 23–27 October 2000.

Bruno, J. F., Selig, E. R. 2007. "Regional decline of coral cover in the Indo-Pacific: Timing, extent, and subregional comparisons." *PLoS ONE* 2: e711.

Bryant, D. 1998. *Reefs at Risk: A Map-Based Indicator of Threats to the World's Coral Reefs*. Washington, D.C.: World Resources Institute.

Carpenter, K., Abrar, M., Aeby, G., Aronson, R., Banks, S., Bruckner, A., Chiriboga, A., et al. 2008. "One-third of reef-building corals face elevated extinction risk from climate change and local impacts." *Science* 321: 560–563.

Coles, S. 1980. "Species diversity of decapods associated with living and dead reef coral *Pocillopora meandrina*." *Marine Ecology Progress Series* 2: 281–291.

Cowen, R., Lwiza, K., Sponaugle, S., Paris, C., Olson, D. 2000. "Connectivity of marine populations: Open or closed?" *Science* 287: 857.

Cowen, R., Paris, C., Srinivasan, A. 2006. "Scaling of connectivity in marine populations." *Science* 311: 522.

Cressey, D. 2007. "Arctic melt opens Northwest passage." *Nature* 449: 267.

De'ath, G., Lough, J. M., Fabricius, K. E. 2009. "Declining coral calcification on the Great Barrier Reef." *Science* 323: 116–119.

Douglas, B., Kearney, M., Leatherman, S. 2000. *Sea Level Rise: History and Consequences*. International Geophysics Series, Vol. 75. New York: Academic Press.

Feary, D., Almany, G., Jones, G., McCormick, M. 2007. "Coral degradation and the structure of tropical reef fish communities." *Marine Ecology Progress Series* 333: 243–248.

Glynn, P. 1993. "Coral reef bleaching: Ecological perspectives." *Coral Reefs* 12: 1–17.

Glynn, P. W. 1983. "Extensive bleaching and death of reef corals on the Pacific coast of Panama." *Environmental Conservation* 10: 149–154.

Glynn, P. W., D'Croz, L. 1990. "Experimental-evidence for high-temperature stress as the cause of El-Niño-coincident coral mortality." *Coral Reefs* 8: 181–191.

Glynn, P. W., Cortes, J., Guzman, H. M., Richmond, R. H. 1988. "El Niño (1982–1983) associated coral mortality and relationship to sea surface temperature deviations in the tropical eastern Pacific." In *Proceedings of the Sixth International Coral Reef Symposium,* edited by J. H. Choat, et al., 237–243. Townsville, Australia: 6th International Coral Reef Symposium Executive Committee. Vol. 11.

Gotelli, N., Abele, L. 1983. "Community patterns of coral-associated decapods." *Marine Ecology Progress Series. Oldendorf* 13: 131–139.

Gotelli, N., Gilchrist, S., Abele, L. 1985. "Population biology of *Trapezia* spp. and other coral-associated decapods." *Marine Ecology Progress Series. Oldendorf* 21: 89–98.

Graham, N. A., Wilson, S. K., Jennings, S., Polunin, N. V., Robinson, J., Bijoux, J. P., Daw, T. M. 2007. "Lag effects in the impacts of mass coral bleaching on coral reef fish, fisheries, and ecosystems." *Conservation Biology* 21: 1291–1300.

Halpern, B. S., Walbridge, S., Selkoe, K. A., Kappel, C. V., Micheli, F., D'Agrosa, C., Bruno, J. F., et al. 2008. "A global map of human impact on marine ecosystems." *Science* 319: 948–952.

Highsmith, R. 1982. "Reproduction by fragmentation in corals." *Marine Ecology Progress Series. Oldendorf* 7: 207–226.

Hoegh-Guldberg, O. 1999. "Climate change, coral bleaching and the future of the world's coral reefs." *Marine and Freshwater Research* 50: 839–866.

Hoegh-Guldberg, O., Salvat, B. 1995. "Periodic mass-bleaching and elevated sea temperatures—bleaching of outer reef slope communities in Moorea, French-Polynesia." *Marine Ecology Progress Series* 121: 181–190.

Hoegh-Guldberg, O., Smith, G. J. 1989. "The effect of sudden changes in temperature, light and salinity on the population density and export of zooxanthellae from the reef corals *Stylophora pistillata* and *Seriatopra hystrix*." *Journal of Experimental Marine Biology and Ecology* 129: 279–303.

Hoegh-Guldberg, O., Mumby, P. J., Hooten, A. J., Steneck, R. S., Greenfield, P., Gomez, E., Harvell, C. D., et al. 2007. "Coral reefs under rapid climate change and ocean acidification." *Science* 318(5857): 1737–1742.

IPCC. 2007. *Synthesis Report*. Contribution of Working Groups I, II and III to the Fourth Assessment Report of the Intergovernmental Panel on Climate Change. Geneva, Switzerland: Intergovernmental Panel on Climate Change.

Kleypas, J. A., Langdon, C. 2006. "Coral reefs and changing seawater chemistry." In *Coral Reefs and Climate Change: Science and Management. AGU Monograph Series, Coastal and Estuarine Studies,* vol. 61, edited by Phinney, J., Hoegh-Guldberg, O., Kleypas, J., Skirving, W., and Strong, A. E., 73–110. Washington, D.C.: Geophysical Union.

Kleypas, J. A., McManus, J. W., Menez, L. A. B. 1999. "Environmental limits to coral reef development: Where do we draw the line?" *American Zoologist* 39: 146–159.

Lasker, H., Peters, E., Coffroth, M. 1984. "Bleaching of reef coelenterates in the San Blas Islands, Panama." *Coral Reefs* 3: 183–190.

Loya, Y., Sakai, K., Yamazato, K., Nakano, Y., Sambali, H., van Woesik, R. 2001. "Coral bleaching: The winners and the losers." *Ecology Letters* 4: 122–131.

Marshall, P. A., Baird, A. H. 2000. "Bleaching of corals on the Great Barrier Reef: Differential susceptibilities among taxa." *Coral Reefs* 19: 155–163.

McClanahan, T. R. 2007. "Western Indian Ocean coral communities: Bleaching responses and susceptibility to extinction." *Marine Ecology Progress Series* 337: 1–13.

McKinney, F. 1983. "Asexual colony multiplication by fragmentation: An important mode of genetic longevity in the Carboniferous bryozoan *Archimedes*." *Paleobiology* 9 (1): 35–43.

Moran, P. 1986. "The *Acanthaster* phenomenon." *Oceanography and Marine Biology* 24: 379–480.

Munday, P., Jones, G., Sheaves, M., Williams, A., Goby, G. 2007. "Vulnerability of fishes of the Great Barrier Reef to climate change." *Climate Change and the Great Barrier Reef* 357–391.

Muscatine, L. 1990. "The role of symbiotic algae in carbon and energy flux in reef corals." In *Ecosystems of the World*, vol. 24, *Coral Reefs*, edited by Z. Dubinsky, 75–87. Amsterdam: Elsevier.

Öhman, M., Munday, P., Jones, G., Caley, M. 1998. "Settlement strategies and distribution patterns of coral-reef fishes." *Journal of Experimental Marine Biology and Ecology* 225: 219–238.

Petit, J. R., Jouzel, J., Raynaud, D., Barkov, N. I., Barnola, J. M., Basile, I., Bender, M., et al. 1999. "Climate and atmospheric history of the past 420,000 years from the Vostok ice core, Antarctica." *Nature* 399: 429–436.

Pratchett, M., Wilson, S., Baird, A. 2006. "Declines in the abundance of *Chaetodon* butterflyfishes following extensive coral depletion." *Journal of Fish Biology* 69: 1269–1280.

Pratchett, M. S., Munday, P. L., Wilson, S. K., Graham, N. A. J., Cinnerm, J. E., Bellwood, D. R., Jones, G. P., Polunin, N. V. C., McClanahan, T. R. 2008.

"Effects of climate-induced coral bleaching on coral-reef fishes—Ecological and economic consequences." *Oceanography and Marine Biology: An Annual Review* 46: 251–296.

Precht, W., Bruckner, A., Aronson, R., Bruckner, R. 2002. "Endangered acroporid corals of the Caribbean." *Coral Reefs* 21: 41–42.

Precht, W., Robbart, M., Aronson, R. 2004. "The potential listing of *Acropora* species under the US Endangered Species Act." *Marine Pollution Bulletin* 49: 534–536.

Preston, N., Doherty, P. 1990. "Cross-shelf patterns in the community structure of coral dwelling Crustacea in the central region of the Great Barrier Reef. 1. Agile shrimps." *Marine Ecology Progress Series* 66: 47–61.

Przeslawski, R., Ahyong, S., Byrne, M., Worheide, G., Hutchings, P. 2008. "Beyond corals and fish: The effects of climate change on noncoral benthic invertebrates of tropical reefs." *Global Change Biology* 14: 2773–2795.

Raven, J., Caldeira, K., Elderfield, H., Hoegh-Guldberg, O., Liss, P., Riebesell, U., Shepherd, J., Turley, C., Watson, A. 2005. *Ocean Acidification Due to Increasing Atmospheric Carbon Dioxide*. London: Royal Society.

Rinkevich, B., Wolodarsky, Z., Loya, Y. 1991. "Coral-crab association: A compact domain of a multilevel trophic system." *Hydrobiologia* 216: 279–284.

Roberts, C., Hawkins, J. 1999. "Extinction risk in the sea." *Trends in Ecology and Evolution* 14: 241–246.

Roberts, C., Ormond, R. 1987. "Habitat complexity and coral reef fish diversity and abundance on Red Sea fringing reefs." *Marine Ecology Progress Series. Oldendorf* 41: 1–8.

Roberts, L. 1987. "Coral bleaching threatens Atlantic reefs: Unexplained changes are occurring in some of the most productive ecosystems on the planet, the Caribbean coral reefs." *Science* 238: 1228–1229.

Sale, P., Douglas, W., Doherty, P. 1984. "Choice of microhabitats by coral reef fishes at settlement." *Coral Reefs* 3: 91–99.

Sale, P. F. 2002. *Coral Reef Fishes: Dynamics and Diversity in a Complex Ecosystem*. San Diego: Academic Press.

Silverman, J., Lazar, B., Cao, L., Caldeira, K., Erez, J. 2009. "Coral reefs may start dissolving when atmospheric $CO_2$ doubles." *Geophysical Research Letters* 36: L05606, doi:05610.01029/02008GL036282.

Steffensen, J. P., Andersen, K. K., Bigler, M., Clausen, H. B., Dahl-Jensen, D., Fischer, H., Goto-Azuma, K., et al. 2008. "High-resolution Greenland ice core data show abrupt climate change happens in few years." *Science* 321: 680–684.

Strong, A. E., Barrientos, C. S., Duda, C., Sapper, J. 1996. "Improved satellite technique for monitoring coral reef bleaching." *Proceedings of the Eighth International Coral Reef Symposium*, Panama, 1495–1497.

Tanzil, J., Brown, B., Tudhope, A., Dunne, R. 2009. "Decline in skeletal growth of the coral *Porites lutea* from the Andaman Sea, South Thailand between 1984 and 2005." *Coral Reefs* doi:10.1007/s00338-008-0457-5.

Turner, S. 1994. "The biology and population outbreaks of the corallivorous gastropod *Drupella* on Indo-Pacific reefs." *Oceanography and Marine Biology* 32: 461–530.

Veron, J. E. N., Stafford-Smith, M. 2000. *Corals of the World*. Townsville, Queensland: Australian Institute of Marine Science.

Walther, G. R., Post, E., Convey, P., Menzel, A., Parmesan, C., Beebee, T. J., Fromentin, J. M., Hoegh-Guldberg, O., Bairlein, F. 2002. "Ecological responses to recent climate change." *Nature* 416: 389–395.

Webster, P. J., Holland, G. J., Curry, J. A., Chang, H. R. 2005. "Changes in tropical cyclone number, duration, and intensity in a warming environment." *Science* 309: 1844–1846.

Wilson, S. K. 2006. "Multiple disturbances and the global degradation of coral reefs: Are reef fishes at risk or resilient?" *Global Change Biology* 12: 2220–2234.

Wilson, S. K., Burgess, S. C., Cheal, A. J., Emslie, M., Fisher, R., Miller, I., Polunin, N. V. C., Sweatman, H. P. A. 2008. "Habitat utilization by coral reef fish: Implications for specialists vs. generalists in a changing environment." *Journal of Animal Ecology* 77: 220–228.

Zhang, J., Lindsay, R., Steele, M., Schweiger, A. 2008. "What drove the dramatic retreat of arctic sea ice during summer 2007?" *Geophysical Research Letters* 35: L11505, doi:11510.11029/12008GL034005.

## Chapter 16

# *Extinction Risk in a Changing Ocean*

Benjamin S. Halpern and Carrie V. Kappel

Only a handful of marine species are known to have gone globally extinct in modern times (Carlton et al., 1999; Dulvy et al., 2003). Yet it is difficult to know if this low extinction rate is due to greater resilience of marine species relative to terrestrial species, less human impact on marine species, or simply an artifact of so little of the ocean having been explored and documented. Into this already uncertain extinction picture, climate change presents additional complication.

Climate change is already leaving its mark on the oceans, and future changes are expected to be much greater. There are four key drivers of change in oceans: rising sea level, increasing ultraviolet light exposure, changing ocean temperature, and acidification. Ocean acidification in particular has raised the specter of catastrophic species extinction, as any species with a calcified structure (e.g., corals, mollusks) may no longer be able to exist if the oceans become too acidic. Rapid loss of sea ice and the "squeeze" of coastal habitats caught between developed shorelines and rising sea level seriously threaten many species. Oceans are also expected to respond indirectly to climate change through shifts in circulation and upwelling dynamics, primary productivity, storm (disturbance) regimes, and patterns of precipitation that alter land-based inputs into oceans and likely lead to more frequent and severe hypoxia in coastal waters (Justic et al., 1996) and changing sediment regimes. The causes of these changes to the oceans are fairly well known, but we are just beginning to understand their consequences for marine species (Scavia et al., 2002; Harley et al., 2006).

Our aim with this chapter is to assess and predict how climate change will drive future exinctions in non-coral ocean organisms, using known extinctions and threatened species in the oceans as a point of reference. Coral reefs—the ocean's canary in the coal mine—have received most of the recent attention with respect to climate change impacts (e.g., chapters 7 and 15, this volume), yet plenty is known about species' vulnerabilities and patterns of threats for the rest of the ocean, and our focus here is on species that live beyond coral reefs. Instead of asking how many species are at risk of extinction due to each climate change driver—an important but difficult question to answer—we instead review what makes marine species vulnerable to extinction in general and evaluate how different aspects of climate change interact with those traits or ecosystem states to increase species' vulnerability to extinction. This predictive framework, which includes species' capacity to adapt to climate change, should apply to any species, including coral reef species. Additionally, we place this framework in the broader context of the full suite of human threats to marine species to evaluate how climate change interacts with other threats to affect species' vulnerability to extinction.

## Drivers of Extinction Risk in Marine Species

Below we focus on and evaluate seven categories of biological or physical factors that influence marine species' susceptibility to extinction. Similar efforts have been done by the International Union for Conservation of Nature (IUCN) in developing the Red List of threatened species (Baillie et al., 2004), and to a lesser extent in the United States as part of the Endangered Species Act (relatively few marine species have been listed to date). The literature addressing each individual category is vast, and so we seek to be illustrative, not comprehensive. Figure 16-1 provides a schematic overview of how different threats to species—climate change–related and otherwise—directly and indirectly affect these extinction risk factors; it builds on previous schema by others (Harley et al., 2006; Brook et al., 2008).

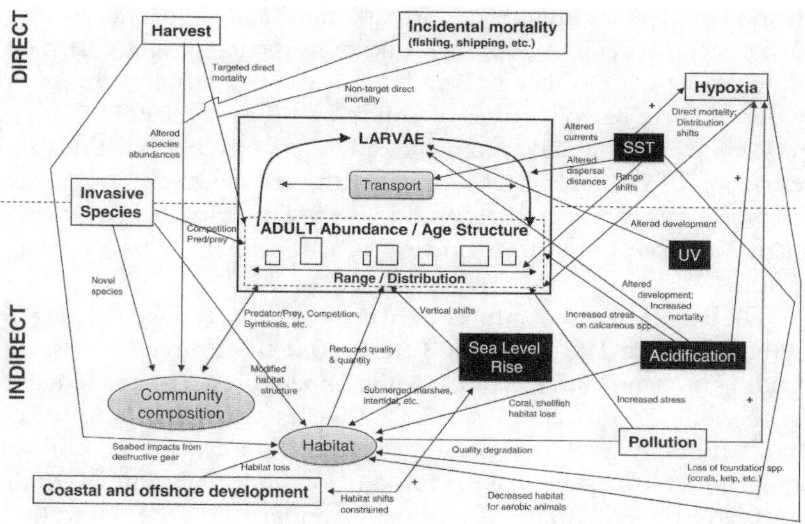


FIGURE 16-1. Schematic of the effects of human-derived stressors on marine populations and their extinction risk. The population is represented by the central box and divided into larval and adult life stages, with potential effects on larval and adult abundances and population range and distribution. Ecological factors that are (at least in part) a function of the species' environment (transport, community composition, and habitat) are represented by gray bubbles. Climate change drivers are highlighted in black rectangles with white text. Other drivers are in white boxes. Stressors can affect populations directly and indirectly, via interactions with community composition and habitat. Arrows highlight the paths by which drivers affect populations and are labeled with their potential outcomes.

## *Ecophysiology*

Organisms' physiologies dictate the environmental conditions under which they can survive, grow, and reproduce, yet direct physiological impacts seldom lead to species extinctions (Clarke, 1993). Increased ocean temperatures, ocean acidity, nutrients, ultraviolet radiation, and turbidity, as well as more frequent and severe hypoxia events, sea-level rise, and changing patterns of ocean circulation and weather all affect the environmental conditions experienced by organisms and, in turn, their distribution and ecology (fig. 16-1). The impact of these changes on species' physiology will depend on their tolerance for and adaptability to variation in physical and chemical conditions, where within their own tolerance range they currently exist, and the overall ecological context. Living in suboptimal conditions is energetically costly, because growth, reproduction, and survival are generally optimized within a fairly narrow range of conditions (Clarke, 1993; Pörtner, 2008). Such costs could eventually lead to loss of subpopulations and restriction of species' ranges. Species with very specialized or narrow tolerances for environmental conditions and limited ability to move or adapt are more prone to extinction (Carlton et al., 1991; Vermeij, 1996).

Global ocean temperatures have increased on average 0.1 degree Celsius between 1961 and 2003 (range 0.0–0.5 degree Celsius, depending on depth and location), with temperatures expected to climb another 0.5–2.5 degrees Celsius by 2100 (Levitus et al., 2005; IPCC, 2007). Physiological processes are temperature-dependent; animals and plants restrict growth and reproduction under suboptimal conditions, and under extreme conditions may die (fig. 16-1). Intertidal species are adapted to temperature fluctuation and so might be expected to be resilient to warming, but some, like the snail *Tegula funebralis,* already exist at their upper thermal limit and cannot adapt to warming beyond the range of temperatures they usually experience (reviewed by Tomanek, 2008). Shallow subtidal congeners, by contrast, are more adaptable to warming, despite a narrower background level of temperature variation. However, some species that experience little to no variation in temperature lose their ability to acclimate to changing thermal conditions. Therefore, both species with very narrow thermal tolerances and those with extremely broad thermal tolerances live close to their thermal limits. The adaptive capacity of these organisms to respond to climate change may be most limited (Tomanek, 2008).

Anthropogenic increases in atmospheric carbon dioxide have made the oceans steadily more acidic, with an average drop in ocean pH of 0.1 units (IPCC, 2007; Guinotte and Fabry, 2008). The effects of this acidification are only just beginning to be studied, but some organisms may be particularly affected. Basic cellular processes like protein synthesis are compromised when extracellular pH balance is disrupted; lower invertebrates and especially small-bodied species and life-forms (e.g., larvae, Dupont et al., 2008) are less able to maintain this acid-base balance (Pörtner, 2008). Calcifying organisms (e.g., plankton-like coccolithophores and pteropods, deep-sea corals, echinoderms, and mollusks) are also likely to be affected, as vast areas gradually become undersaturated in aragonite and calcite, the building blocks of shells and skeletons. Species living at high latitudes and those that produce shells of magnesium-calcite or aragonite will be most susceptible to acidification (fig. 16-1). However, adaptive capacity of most species to changing pH is unknown.

Interactions between increasing greenhouse gases and ozone depletion may lead to increased ultraviolet radiation in many parts of the globe (Hartmann et al., 2000). Increased radiation damages cells, which increases stress on organisms, in turn making organisms more susceptible to other stresses, both climate-related and otherwise. Shallow-water planktonic species and those with planktonic life stages may be most susceptible to ultraviolet radiation (Lesser and Barry, 2003) (fig. 16-1). Furthermore, these environmental factors are not changing in isolation. Rising temperatures, more frequent hypoxia events, and increased acidification, for example, are likely to co-occur and will undoubtedly interact in complex ways within organisms' physiologies (Pörtner, 2008).

Ecophysiological needs set the context for many of the other drivers of marine extinction risk that will be described below. A species' basic physiological needs for survival, growth, and reproduction set a minimum bar for escaping population extinction in a particular place. However, depending on a species' motility, population attributes (e.g., growth rate), and habitat characteristics and other features, it may be able to escape some or all direct ecophysiological effects of climate change simply by moving or growing out of inhospitable conditions.

### *Individual and Population-Level Attributes*

Many basic traits of individuals and populations can lead to enhanced extinction risk (for two influential lists see Ehrlenfeld, 1970 and Terborgh, 1974). At the individual level, large body size, high trophic level, habitat or trophic specialization, and limited dispersal/mobility are consistently associated with extinction risk (McKinney, 1997); we discuss several of these in more detail below. These traits can influence either an individual's mortality probability or the number of individuals, and therefore, population abundance and vulnerability (McKinney, 1997). Populations close to their minimum viable population size are at greater risk of extinction because even small disturbances can result in catastrophic losses of individuals.

Climate change likely interacts with population traits in ways similar to other disturbances. Population traits associated with extinction probability in both modern and fossil data sets include low mean population size, limited range, low density, and population fluctuations, while low intrinsic growth rate (r), seasonal aggregations, and low genetic variation have also been linked to modern extinctions (McKinney, 1997). Building on this empirical foundation, the IUCN criteria for determining vulnerability include recent population reduction and continued decline, small population size, extreme fluctuation in population size or range, small range size (either extent of occurrence or area of occupancy), range fragmentation, shrinking range or habitat, and sparse or small subpopulations. Below we address aspects of species' ranges. Small population size is particularly problematic for some species, which suffer negative growth rates due to Allee effects. For example, endangered white abalone (*Haliotis sorenseni*), the first marine invertebrate species listed under the US Endangered Species Act, was reduced by 99.9 percent over 30 years, primarily by overharvest (NOAA, 2001). Like many marine species, it reproduces via external fertilization of spawned gametes. Fertilization success requires population densities of at least 0.15 per square meter (1.6 per square foot); current densities of 0.0002 per square meter (0.002 per square foot) are far below this threshold (Babcock and Keesing, 1999; Hobday and Tegner, 2000).

It remains unclear how climate variables will interact with individual traits like body size to influence extinction risk (or conversely, adaptive capacity). For example, smaller organisms and life stages are less able to buffer changes in extracellular pH and so are more susceptible to ocean acidification (Pörtner, 2008). Large-bodied bivalve species were more likely to shift their geographical ranges in response to Pleistocene climate change, though the process behind this, whether local extinction or expansion, remains unclear (Roy et al., 2001). On the Pacific coast of South America selective extinction among epifaunal, short-ranged, small-bodied bivalves during a late Neogene mass extinction event may have been due to anoxia, cooling, and/or destruction of protected bays (Rivadeneira and Marquet, 2007). However, other paleontological studies have shown no size selectivity of mass extinction events (reviewed by McKinney, 1997). For many marine fishes, large body size leads to enhanced extinction risk (reviewed by Reynolds et al., 2005), although this is likely because larger bodied species are more heavily exploited and/or because body size is correlated with late age of maturity, slow growth rates, and long life span, all of which are also associated with demographic vulnerability to strong and rapid disturbances (Pimm et al., 1988; Dulvy et al., 2003). Although overharvest puts large-bodied marine fish species at risk, habitat degradation tends to threaten small-bodied marine and freshwater fishes differentially (Olden et al., 2007).Teasing out the unique contribution of body size to climate change vulnerability remains a research challenge.

To the extent that climate changes lead to increased variability in ocean conditions (e.g., increased storm frequency or severity; increased extreme temperature events), they may increase the probability of catastrophic losses from small and/or isolated populations. In addition, as extreme events become more frequent, populations with low intrinsic growth rates may not be resilient enough to recover from one disturbance before the next hits. Finally, individual and/or population growth will allow some populations to track hospitable habitat conditions even in a changing ocean climate, but those with slow growth rates may not be able to adjust their distributions quickly enough (see further discussion below).

### *Range Size and Population Structure*

A species' spatial extent (i.e., its range) and the distribution of populations within that range can have important consequences for its vulnerability to extinction (Gaston and Fuller, 2009). Species with larger ranges should be more resilient to disturbances and local extirpations, as they are likely to exist in locations beyond the disturbance. This resilience is further influenced by the distribution of populations within that range. Distributional patterns can vary dramatically (Sagarin and Gaines, 2002a, b); species that are evenly distributed across the range are likely most resilient, while highly patchy populations are likely most vulnerable because loss of a few patches can disrupt connectivity among populations and increase chances of extinction of the entire metapopulation (IUCN, 2001; Gaston and Fuller, 2009).

Where climate change results in reduced range size or increased isolation, extinction risk will increase for small-range or patchily distributed species. Furthermore, species most abundant at range edges may be highly susceptible to climate change if shifting ranges squeeze these populations into smaller areas. The terrestrial literature is replete with such examples (Parmesan and Yohe, 2003; Parmesan, 2006). Marine examples are less well documented, although recent analogous work shows that assemblages of demersal fishes have been moving into deeper (colder) waters as sea temperatures have increased in the North Sea over the past 25 years, constricting their available habitat (Dulvy et al., 2008).

### *Habitat Needs*

Reliance on specific habitats can enhance a species' extinction risk, particularly where habitats have limited distribution (Carlton et al., 1991; Vermeij, 1996; McKinney, 1997), because any reduction in available habitat will likely reduce the abundance of species relying on the habitat and limit the potential for adaptive range shifts. Changing ocean conditions are likely to shift the distribution of many marine habitats, but perhaps most drastically, sea ice, which is being lost at an alarming rate in the Arctic (DeWeaver, 2007). Polar bears, walrus, ringed seals, and other arctic animals that depend on sea ice for shelter and as a hunting ground are finding it increasingly difficult to sustain themselves through the Arctic summer as sea ice breakup continues to shift earlier into spring (Learmonth et al., 2006; chapter 8, this volume).

For other species, especially those that depend upon coastal habitats, sea-level rise could lead to a significant loss in habitat, particularly in places where landward migration of habitat is limited by coastal development and shoreline hardening. In the United States, major estuaries may lose from 20 to 70 percent of important intertidal habitat for migratory shorebirds within the next century due to sea-level rise (Galbraith et al., 2002). Coastal habitat losses will likely reduce population sizes and potentially increase extinction risk, in particular for species that require coastal ecosystems such as mangroves and estuaries as nursery habitats. Extinction risk to migrating species such as salmon and eels, which depend on both oceanic and riverine habitats, may increase significantly with climate change as habitat conditions are disrupted in both systems by rising temperatures and sea level (Mueter et al., 2002; Battin et al., 2007).

For pelagic species, suitable habitats with hospitable ocean conditions may develop in new locations. As long as these new locations are accessible via dispersal of adults or young, these species are unlikely to be habitat-limited. However, these habitats must be present at the appropriate time as well as climatically appropriate. For example, many pelagic species depend on productive locations (e.g., upwelling zones, fronts, etc.) or times of the year for feeding. Changes in ocean conditions are already starting to alter the phenology of planktonic communities, and differences in species' responses to climate change have resulted in mismatches in marine trophodynamics (Edwards and Richardson, 2004). Documented changes in marine plankton communities are much larger in magnitude than phenological changes on land and are likely to affect the distributions of other dependent species (e.g., marine mammals, Learmonth et al., 2006).

### *Species Interactions and Food Web Dynamics*

Specialized species interactions may make species more vulnerable to extinction (McKinney, 1997; chapter 18, this volume). For example, specialized trophic relationships can lead species to follow others into endangerment or extinction, as with the eelgrass limpet, *Lottia alveus,* which went extinct in the Atlantic ocean basin in the early 1930s after the eelgrass, *Zostera marina,* upon which it was specialized to live and feed, succumbed to a wasting disease (Carlton et al., 1991). Eelgrass populations in the entire range of the limpet were wiped out, though *Zostera* survived in refuge populations in brackish waters where the apparently stenohaline limpet could not follow it. As sea-level rise and changing ocean climate affect the distribution of many coastal and biogenic habitats, additional cases like this one may arise, where species with very specific habitat requirements or specialized relationships with other species cannot follow the path their hosts take through a changing ocean. Furthermore, species with varied tolerances for environmental variation and different adaptive capacities will result in reshuffled ecological communities as climate changes.

This same kind of disruption is affecting predator-prey relationships and entire food webs, especially in pelagic systems. Phytoplankton productivity is tightly coupled to ocean conditions, with blooms when nutrients and light are plentiful, followed by die-offs. Many ocean food webs have evolved to take advantage of phytoplankton blooms and the zooplankton production that follows, and climate change is already demonstrably changing these food webs in some places (Grebmeier et al., 2006). This production is often highly seasonal, and annual migrations of species such as whales are timed to take advantage of it (Learmonth et al., 2006). In addition, many marine species have a planktonic larval form; feeding larvae depend on planktonic production to grow and recruit successfully into juvenile and adult populations. Changes in ocean phenology may leave some species without prey to eat (Edwards and Richardson, 2004). Other species may benefit as they are released from predation by such changes. Yet other species may be exposed to novel predators as more mobile species move into new habitats.

In other places, such as upwelling zones, steadier levels of high productivity are fed by cold, nutrient-rich water being upwelled from the deep. Coastal upwelling may intensify under climate change (Bakun, 1990), although El Niño events, which can result in reduced upwelling and deepening of the thermocline in the eastern Pacific, may become more severe (Hansen et al., 2006). These combined effects remain poorly understood, but could have major effects on ocean productivity and upwelling-dependent species such as kelp in upwelling regions (Harley et al., 2006). Recent climate-related changes in upwelling and productivity in the California Current have led to precipitous declines in zooplankton productivity and related declines in the breeding success of seabirds, including vulnerable species such as marbled murrelets (Becker et al., 2007).

### *Dispersal Abilities*

To survive, species must be able to reach hospitable conditions where they can feed, grow, reproduce, and survive over the long term. In general, species capable of longer-range dispersal should be better equipped to move as climate change alters the distribution of suitable habitat. This dispersal may be accomplished at the adult, juvenile, or larval stage. In particular, many marine species have a planktonic larval phase, leading to the potential for long distance dispersal in many marine populations, although even species with long larval periods may be retained close to natal populations (e.g., Cowen et al., 2000; Armsworth et al., 2001; Swearer et al., 2003). Additionally, successful range expansion will depend on recruitment of sufficient numbers to new areas to establish sustainable populations. Given the variability inherent in long distance larval dispersal, this may be a difficult hurdle to overcome.

Predicting extinction risk from larval dispersal distances is difficult. Data on the distribution, abundance, and population structure dating back to the 1970s in England and Scotland showed that the warm water intertidal topshells, *Gibbula umbilicalis* and *Osilinus lineatus,* both of which have very short larval periods (about 3 days), have expanded their northern range limits by about 80 kilometers (50 miles), extending into the colder waters of the English Channel and northeast Atlantic in response to recent warming (Mieszkowska et al., 2006; Mieszkowska et al., 2007). In contrast, warm water barnacles (*Chthamalus* spp.), which have much longer larval periods of several weeks, have not been as successful in extending their ranges. Short distance dispersal may overcome Allee effects and allow for establishment of new subpopulations; however, short distance dispersers may remain vulnerable to local catastrophic events. This suggests that the interaction among dispersal mode, pelagic larval duration, and adaptation may be more complicated than "longer is better."

### *Cumulative Impacts, Synergistic Interactions, and Nonlinearities*

Climate change is not the only threat to species, and so extinction risk will also be a function of the presence and intensity of other stresses (e.g., overharvest, pollution, invasive species) affecting the factors described above, and the potential interaction of these threats with climate change (fig. 16-1). Climate change consistently emerged as the top threat to marine systems at global and regional scales (Halpern et al., 2008; Halpern et al., 2009; Selkoe et al., 2009), but the cumulative impact of all threats is much greater than climate threats alone. In particular, overharvest of species is a dominant driver of extinction risk for many if not most marine species to date (Baillie et al., 2004; Kappel, 2005). The cumulative impacts of nonclimate threats can increase habitat fragmentation and decrease population size, increasing extinction probability directly as well as indirectly by exacerbating the threats caused by climate change. Stressor overlap can also lead to synergistic impacts that are greater than the sum of the parts, although the nature and extent of these synergies remain difficult to predict (Crain et al., 2008; Darling and Cote, 2008). For example, increasing carbon dioxide concentration in the oceans, combined with warming, may lead to the development of vast "dead zones" where aerobic respiration is no longer possible by most creatures (Brewer and Peltzer, 2009). Such synergies are implicit in the methodology used by IUCN to classify species as threatened—in most cases species are listed because several criteria are exceeded.

An additional challenge to predicting extinction risk for species is that most factors described above respond in nonlinear ways to changing climate stress (Pörtner, 2008; Koch et al., 2009). When such nonlinearities involve thresholds, where small changes in driver variables can lead to large changes in response variables, as with many of the ecophysiological traits, small changes in climate can lead to sudden and potentially unexpected and dramatic changes in species abundance and extinction probability. In some cases these thresholds are known and understood (e.g., temperature tolerance limits), but in many cases they are not, and we still know relatively little about whether or how such thresholds interact with the suite of other factors discussed here.

## A Framework for Predicting Marine Extinctions

Our discussion above and the schematic in figure 16-1 suggest no simple rules of thumb or a single model for predicting vulnerability to extinction for all marine species. Few factors directly affect population sizes of species, especially for the climate-related threats, and the indirect pathways of impact vary from relatively simple to quite complex. Two lines of evidence, however, can help shed some light on this issue. First, some geographies and ecosystems are currently much more heavily affected than others, and species in these areas are therefore likely at much greater risk of extinction. Across all threat types, intertidal and shallow subtidal ecosystems, such as coral reefs, mangroves, and seagrass beds as well as continental shelves, are the most affected systems globally (Halpern et al., 2008), in large part because human populations are concentrated in coastal areas and heavily use and impact these nearshore ecosystems. Species that use these ecosystems, partially or exclusively, are likely more prone to extinction synergies with climate change than other species. Hotspots of human impact on the oceans, such as the North Sea and East and South China Seas, regardless of ecosystem type used by species, are also likely to have high extinction probability for species endemic to or highly abundant in those regions.

Second, we can build on the IUCN approach to classifying extinction probability to more directly account for the variables addressed here, and in turn try to isolate the role of climate change. IUCN criteria focus on reproductive capacity and population size, distribution, and trends, which are the ultimate drivers of extinction probability. Above we explored the many proximate causes of declining population trends that can provide helpful "early warning" signs and in some cases easier-to-measure metrics of extinction probability. We conducted a simple accounting of these variables for a small subset of marine species already listed by IUCN (but not necessarily at risk from climate change; table 16-1). This accounting approach ignores

TABLE 16-1. Extinction vulnerability factors affected by climate change for select IUCN Red List species.

| | *Mediterranean Monk Seal* | *Hawaiian Monk Seal* | *Vaquita* | *Galápagos Fur Seal* |
|---|---|---|---|---|
| Scientific Name | *Monachus monachus* | *Monachus schauinslandi* | *Phocoena sinus* | *Arctocephalus galapagoensis* |
| Taxon | Mammal | Mammal | Mammal | Mammal |
| IUCN Rank | | | | |
| VU/EN/CR | CR | CR | CR | EN |
| Harvested currently Y/N | N | N | N | N |
| Ecophysiology | | | | |
| Water temperature | | | | |
| Narrow or very broad vs. medium | medium | medium | medium | medium |
| Cold/warm | warm | warm | warm | mixed |
| Salinity | | | | |
| Estuarine/marine | marine | marine | marine | marine |
| Turbidity | | | | |
| Photosynthetic/heterotrophic | heterotrophic | heterotrophic | heterotrophic | heterotrophic |
| pH | | | | |
| Mg-calcite/aragonite/calcite | calcite | calcite | calcite | calcite |
| Poor/good extracellular pH balance | good | good | good | good |
| UV | | | | |
| Shallow/deep | shallow | shallow | shallow | shallow |
| Habitat dependence | | | | |
| Critical habitat specialists | N | Y (beach) | N | N |
| Sea ice | N | N | N | N |
| Nursery | Y | Y | N | Y |
| Breeding aggregation | Y | Y | N | N |
| Migratory | N | N | N | N |
| Diad- or anadromous Y/N | N | N | N | N |
| Interspp dependencies | | | | |
| Obligate/strong/weak | weak | weak | weak | weak |
| Dispersal | | | | |
| Adult | | | | |
| Sessile/sed/mobile/highly mobile | mobile | mobile | mobile | mobile |
| Larval | | | | |
| No/short/long | No | No | No | No |
| Individual attributes | | | | |
| Body size (m) | 2.5 | 2.25 | 1.5 | 1.4 |
| Trophic level | predator | predator | predator | predator |
| Population attributes | | | | |
| Pop size | 400 | 591 | 570 | 17,500 |
| Age at maturity (yrs.) | 4 | 7.5 | 3–6 | 5 |
| Life span (yrs.) | 20 | 15 | 20 | 20 |
| Selection (r/K) | K | K | K | K |
| Range | | | | |
| Size | | | | |
| IUCN bins | medium | very large | medium | very large |
| Subpopulation structure | | | | |
| Frag/non-frag (# subpopulations) | frag | frag | non-frag | non-frag |
| % range occupied | small | 10 | N/A | 90 |
| Max subpop size | 130 | 75 | N/A | N/A |

## Table 16-1. Continued.

| | Fin Whale | Steller Sea Lion | Marine Otter | Northern Fur Seal |
|---|---|---|---|---|
| Scientific Name | *Balaenoptera physalus* | *Eumetopias jubatus* | *Lontra felina* | *Callorhinus ursinus* |
| Taxon | Mammal | Mammal | Mammal | Mammal |
| IUCN Rank | | | | |
| VU/EN/CR | EN | EN | EN | VU |
| Harvested currently Y/N | N | Y | Y | Y |
| Ecophysiology | | | | |
| Water temperature | | | | |
| Narrow or very broad vs. medium | very broad | medium | medium | medium |
| Cold/warm | mixed | cold | cold | cold |
| Salinity | | | | |
| Estuarine/marine | marine | marine | marine | marine |
| Turbidity | | | | |
| Photosynthetic/heterotrophic | heterotrophic | heterotrophic | heterotrophic | heterotrophic |
| pH | | | | |
| Mg-calcite/aragonite/calcite | calcite | calcite | calcite | calcite |
| Poor/good extracellular pH balance | good | good | good | good |
| UV | | | | |
| Shallow/deep | shallow | shallow | shallow | shallow |
| Habitat dependence | | | | |
| Critical habitat specialists | N | N | N | N |
| Sea ice | N | N | N | N |
| Nursery | Y | Y | N | Y |
| Breeding aggregation | Y | Y | N | Y |
| Migratory | Y | N | N | Y |
| Diad- or anadromous Y/N | N | N | N | N |
| Interspp dependencies | | | | |
| Obligate/strong/weak | weak | weak | weak | weak |
| Dispersal | | | | |
| Adult | | | | |
| Sessile/sed/mobile/highly mobile | highly | mobile | mobile | highly |
| Larval | | | | |
| No/short/long | No | No | No | No |
| Individual attributes | | | | |
| Body size (m) | 26 | 2.9 | 0.9 | 1.75 |
| Trophic level | krill eater | predator | invert pred | predator |
| Population attributes | | | | |
| Pop size | 85,000 | >20,000 | 1,000 | 1,000,000 |
| Age at maturity (yrs.) | 8 | 4.5 | 2 | 4 |
| Life span (yrs.) | 26 | 10 | | 25 |
| Selection (r/K) | K | K | K | K |
| Range | | | | |
| Size | | | | |
| IUCN bins | global | very large | large | ocean |
| Subpopulation structure | | | | |
| Frag/non-frag (# subpopulations) | frag | frag (2) | frag | frag |
| % range occupied | | | 10 | small |
| Max subpop size | 18,000 | >10,000 | 300 | 680,000 |

TABLE 16-1. Continued.

| | *Hooded Seal* | *Black Abalone* | *Pinto Abalone* | *Green Turtle* |
|---|---|---|---|---|
| Scientific Name | *Cystophora cristata* | *Haliotis cracherodii* | *Haliotis kamtschatkana* | *Chelonia mydas* |
| Taxon | Mammal | Mollusk | Mollusk | Reptile |
| IUCN Rank | | | | |
| VU/EN/CR | VU | CR | EN | EN |
| Harvested currently Y/N | Y | Y | Y | Y |
| Ecophysiology | | | | |
| Water temperature | | | | |
| Narrow or very broad vs. medium | narrow | medium | narrow | very broad |
| Cold/warm | cold | mixed | cold | warm |
| Salinity | | | | |
| Estuarine/marine | marine | marine | marine | marine |
| Turbidity | | | | |
| Photosynthetic/heterotrophic | heterotrophic | heterotrophic | heterotrophic | heterotrophic |
| pH | | | | |
| Mg-calcite/aragonite/calcite | calcite | aragonite | aragonite | calcite |
| Poor/good extracellular pH balance | good | poor | poor | good |
| UV | | | | |
| Shallow/deep | shallow | shallow | shallow | shallow |
| Habitat dependence | | | | |
| Critical habitat specialists | Y (ice) | N | N | Y (beach) |
| Sea ice | Y | N | N | N |
| Nursery | Y | N | N | Y |
| Breeding aggregation | N | N | N | N |
| Migratory | N | N | N | Y |
| Diad- or anadromous Y/N | N | N | N | N |
| Interspp dependencies | | | | |
| Obligate/strong/weak | weak | strong | weak | weak |
| Dispersal | | | | |
| Adult | | | | |
| Sessile/sed/mobile/highly mobile | highly | sessile | sessile | highly |
| Larval | | | | |
| No/short/long | No | short | short | No |
| Individual attributes | | | | |
| Body size (m) | 2.3 | 0.2 | 0.16 | 1.5 |
| Trophic level | predator | herbivore | herbivore | herbivore |
| Population attributes | | | | |
| Pop size | 600,000 | | | 90,000 |
| Age at maturity (yrs.) | 5 | 4 | 4 | 33 |
| Life span (yrs.) | 27 | 50 | 15 | 50 |
| Selection (r/K) | K | r | r | K |
| Range | | | | |
| Size | | | | |
| IUCN bins | very large | large | large | global |
| Subpopulation structure | | | | |
| Frag/non-frag (# subpopulations) | frag (3) | frag | non-frag | frag |
| % range occupied | | | | |
| Max subpop size | 125,000 | | | 28,000 |

TABLE 16-1. Continued.

| | Waved Albatross | Northern Rockhopper Penguin | Ganges Shark |
|---|---|---|---|
| Scientific Name | *Phoebastria irrorata* | *Eudyptes moseleyi* | *Glyphis gageticus* |
| Taxon | Bird | Bird | Shark/Ray |
| IUCN Rank | | | |
| VU/EN/CR | CR | EN | CR |
| Harvested currently Y/N | N | Y | N |
| Ecophysiology | | | |
| Water temperature | | | |
| Narrow or very broad vs. medium | N/A | narrow | narrow |
| Cold/warm | N/A | cold | warm |
| Salinity | | | |
| Estuarine/marine | marine | marine | estuarine |
| Turbidity | | | |
| Photosynthetic/heterotrophic | heterotrophic | heterotrophic | heterotrophic |
| pH | | | |
| Mg-calcite/aragonite/calcite | N/A | N/A | calcite |
| Poor/good extracellular pH balance | N/A | good | good |
| UV | | | |
| Shallow/deep | shallow | shallow | shallow |
| Habitat dependence | | | |
| Critical habitat specialists | N | N | Y |
| Sea ice | N | N | N |
| Nursery | Y | N | N |
| Breeding aggregation | Y | Y | N |
| Migratory | N | N | N |
| Diad- or anadromous Y/N | N | N | N |
| Interspp dependencies | | | |
| Obligate/strong/weak | weak | weak | strong |
| Dispersal | | | |
| Adult | | | |
| Sessile/sed/mobile/highly mobile | highly | highly | mobile |
| Larval | | | |
| No/short/long | No | No | No |
| Individual attributes | | | |
| Body size (m) | 0.9 | 0.55 | 2 |
| Trophic level | predator | predator | predator |
| Population attributes | | | |
| Pop size | 35,000 | 120,000 | |
| Age at maturity (yrs.) | 5 | 4 | |
| Life span (yrs.) | 45 | 30 | |
| Selection (r/K) | K | K | K |
| Range | | | |
| Size | | | |
| IUCN bins | large | large | very small |
| Subpopulation structure | | | |
| Frag/non-frag (# subpopulations) | non-frag | frag (4) | non-frag |
| % range occupied | | | |
| Max subpop size | N/A | 45,000 | N/A |

## Table 16-1. Continued.

| | *Barndoor Skate* | *Bocaccio Rockfish* | *Shortnose Sturgeon* |
|---|---|---|---|
| Scientific Name | *Dipturus laevis* | *Sebastes paucispinus* | *Acipenser brevirostrum* |
| Taxon | Shark/Ray | Bony Fish | Bony Fish |
| IUCN Rank | | | |
| VU/EN/CR | EN | CR | VU |
| Harvested currently Y/N | Y | Y | Y |
| Ecophysiology | | | |
| Water temperature | | | |
| Narrow or very broad vs. medium | narrow | medium | medium |
| Cold/warm | cold | mixed | mixed |
| Salinity | | | |
| Estuarine/marine | marine | marine | estuarine |
| Turbidity | | | |
| Photosynthetic/heterotrophic | heterotrophic | heterotrophic | heterotrophic |
| pH | | | |
| Mg-calcite/aragonite/calcite | calcite | calcite | calcite |
| Poor/good extracellular pH balance | good | good | good |
| UV | | | |
| Shallow/deep | deep | deep | shallow |
| Habitat dependence | | | |
| Critical habitat specialists | N | N | N |
| Sea ice | N | N | N |
| Nursery | N | N | Y |
| Breeding aggregation | N | N | Y |
| Migratory | N | N | Y |
| Diad- or anadromous Y/N | N | N | Y |
| Interspp dependencies | | | |
| Obligate/strong/weak | weak | weak | weak |
| Dispersal | | | |
| Adult | | | |
| Sessile/sed/mobile/highly mobile | mobile | sedentary | mobile |
| Larval | | | |
| No/short/long | No | long | short |
| Individual attributes | | | |
| Body size (m) | 1.5 | 0.9 | 1.4 |
| Trophic level | predator | predator | predator |
| Population attributes | | | |
| Pop size | | | |
| Age at maturity (yrs.) | 8 | 5 | 4–12 |
| Life span (yrs.) | 16 | 50 | 67 |
| Selection (r/K) | K | r | r |
| Range | | | |
| Size | | | |
| IUCN bins | very large | very large | very large |
| Subpopulation structure | | | |
| Frag/non-frag (# subpopulations) | frag | frag | frag |
| % range occupied | | | |
| Max subpop size | | | 35,000 |

synergies among factors and assumes equal importance of each, but provides a mechanism for better assessing species' vulnerabilities to climate change. Our efforts to fill in this table produced several salient take-home lessons: (i) for many species, some key attributes are unknown, (ii) different sets of factors will likely be more important for different taxonomic groups—for example, mammals are particularly sensitive to habitat factors, while mollusks may be more sensitive to ecophysiological factors, and (iii) ecophysiology and dispersal factors are not well considered by IUCN for non-coral species but will likely play a key role in extinction vulnerability, especially in combination with other factors. Lesson (i) highlights key challenges for such assessments, lesson (ii) provides some methods for quickly predicting vulnerability for groups of species, and lesson (iii) emphasizes the need to consider a broader suite of factors when assessing extinction vulnerability if the true impact of climate change is to be fairly assessed.

The IUCN Red List presently classifies 363 marine species as threatened by climate change (i.e., climate change is a threat to the species and its status is vulnerable or worse). A full 78 percent of these are associated with coral reefs, with only eighty species from other ecosystems threatened by climate change. We suspect this is a significant underestimate.

## Conclusions

Predictions of species extinction on land have relied on bioclimatic models that evaluate species' environmental tolerances and the changing landscape of available habitat (chapters 2 and 4, this volume). Such approaches are not likely to prove as effective in the ocean because major data gaps remain. We know very little about the distribution of physical and climatic conditions that define habitats (e.g., much of the ocean floor has not been mapped) or the true home range size and traits of most species. In addition, detailed predictions of exactly how ocean conditions are likely to change are still nascent. The picture is further complicated by the number and variety of extinction drivers described above and the potentially complex ways they interact. Calcareous species, those dependent on ice habitat, and those whose life cycles are intimately connected to estuarine and freshwater habitats that will be affected by sea-level rise and changing precipitation are clearly susceptible. The Census of Marine Life (http://www.coml.org)

has documented 335,163 marine taxa and 178,214 valid marine species to date (Appeltans et al., 2010). Species vulnerable to ocean acidification—scleractinian corals (1,997 known valid species) and shelled mollusks (40,533 known valid species)—represent approximately 24 percent of known marine species (Appeltans et al., 2010). Although certainly not all corals and mollusks will go extinct due to acidification, many additional marine species will be affected by acidification, warming, and other impacts of anthropogenic climate change. It is impossible at this stage to give a definitive estimate of the number of species at risk, but it is likely to be greater than 5 percent, and perhaps much greater. Five percent would equate to about 9,000 species, considering only those that have already been described by science, and a much higher number if all species (described and undescribed) are included. It is particularly worrisome that many foundation species—salt marsh plants, kelp, oysters, and of course, corals—are likely very sensitive to climate change; the loss of these species could lead to dramatic cascading losses of species dependent on the habitat created by these foundation species, although the nature and extent of such cascading effects remains largely unknown.

## Acknowledgments

Thanks to Josh Lawler for conversations that guided our development of this framework for assessing climate impacts, and Mary O'Connor for comments on an early draft. Support was provided by the National Center for Ecological Analysis and Synthesis at the University of California-Santa Barbara and by the Packard Foundation.

### REFERENCES

Appeltans, W., P. Bouchet, G. A. Boxshall, K. Fauchald, D. P. Gordon, B. W. Hoeksema, G. C. B. Poore, et al., eds. 2010. "The World Register of Marine Species." Accessed on May 21, 2010. Available online at http://www.marinespecies.org

Armsworth, P. R., M. K. James, and L. Bode. 2001. "When to press on or turn back: Dispersal strategies for reef fish larvae." *American Naturalist* 157: 434–450.

Babcock, R., and J. Keesing. 1999. "Fertilization biology of the abalone *Haliotis laevigata*: Laboratory and field studies." *Canadian Journal of Fisheries and Aquatic Sciences* 56: 1668–1678.

Baillie, J. E. M., C. Hilton-Taylor, and S. N. Stuart. 2004. *2004 IUCN Red List of Threatened Species. A Global Species Assessment*. Gland, Switzerland: IUCN.

Bakun, A. 1990. "Global climate change and intensification of coastal ocean upwelling." *Science* 247: 198–201.

Battin, J., M. W. Wiley, M. H. Ruckelshaus, R. N. Palmer, E. Korb, K. K. Bartz, and H. Imaki. 2007. "Projected impacts of climate change on salmon habitat restoration." *Proceedings of the National Academy of Sciences, USA* 104: 6720–6725.

Becker, B. H., M. Z. Peery, and S. R. Beissinger. 2007. "Ocean climate and prey availability affect the trophic level and reproductive success of the marbled murrelet, an endangered seabird." *Marine Ecology Progress Series* 329: 267–279.

Brewer, P. G., and E. T. Peltzer. 2009. "Limits to marine life." *Science* 324: 347–348.

Brook, B. W., N. S. Sodhi, and C. J. A. Bradshaw. 2008. "Synergies among extinction drivers under global change." *Trends in Ecology & Evolution* 23: 453–460.

Carlton, J. T., G. J. Vermeij, D. R. Lindberg, D. A. Carlton, and E. C. Dubley. 1991. "The first historical extinction of a marine invertebrate in an ocean basin: The demise of the eelgrass limpet *Lottia alveus*." Woods Hole, MA: Marine Biological Laboratory.

Carlton, J. T., J. B. Geller, M. L. Reaka-Kudla, and E. A. Norse. 1999. "Historical extinctions in the sea." *Annual Review of Ecology and Systematics* 30: 515–538.

Clarke, A. 1993. "Temperature and extinction in the sea: A physiologist's view." *Paleobiology* 19: 499–518.

Cowen, R. K., K. M. M. Lwiza, S. Sponaugle, C. B. Paris, and D. B. Olson. 2000. "Connectivity of marine populations: Open or closed?" *Science* 287: 857–859.

Crain, C. M., K. Kroeker, and B. S. Halpern. 2008. "Interactive and cumulative effects of multiple human stressors in marine systems." *Ecology Letters* 11: 1304–1315.

Darling, E. S., and I. M. Cote. 2008. "Quantifying the evidence for ecological synergies." *Ecology Letters* 11: 1278–1286.

DeWeaver, E. 2007. "Uncertainty in climate model projections of Arctic sea ice decline: An evaluation relevant to polar bears." US Dept. of the Interior, US Geological Survey.

Dulvy, N. K., Y. Sadovy, and J. D. Reynolds. 2003. "Extinction vulnerability in marine populations." *Fish and Fisheries* 4: 25–64.

Dulvy, N. K., S. I. Rogers, S. Jennings, V. Stelzenmuller, S. R. Dye, and H. R. Skjoldal. 2008. "Climate change and deepening of the North Sea fish assemblage: A biotic indicator of warming seas." *Journal of Applied Ecology* 45: 1029–1039.

Dupont, S., J. Havenhand, W. Thorndyke, L. Peck, and M. Thorndyke. 2008. "Near-future level of $CO_2$-driven ocean acidification radically affects larval survival and development in the brittlestar *Ophiothrix fragilis*." *Marine Ecology Progress Series* 373: 285–294.

Edwards, M., and A. J. Richardson. 2004. "Impact of climate change on marine pelagic phenology and trophic mismatch." *Nature* 430: 881–884.

Ehrlenfeld, D. W. 1970. *Biological Conservation*. New York: Holt, Rinehart & Winston.

Galbraith, H., R. Jones, R. Park, J. Clough, S. Herrod-Julius, B. Harrington, and G. Page. 2002. "Global climate change and sea level rise: Potential losses of intertidal habitat for shorebirds." *Waterbirds* 25: 173–183.

Gaston, K. J., and R. A. Fuller. 2009. "The sizes of species' geographic ranges." *Journal of Applied Ecology* 46: 1–9.

Grebmeier, J. M., J. E. Overland, S. E. Moore, E. V. Farley, E. C. Carmack, L. W. Cooper, K. E. Frey, et al. 2006. "A major ecosystem shift in the northern Bering Sea." *Science* 311: 1461–1464.

Guinotte, J. M., and V. J. Fabry. 2008. "Ocean acidification and its potential effects on marine ecosystems." In *Year in Ecology and Conservation Biology 2008*, Annals of New York Academy of Sciences, 320–342. Oxford, UK: Blackwell Publishing.

Halpern, B. S., S. Walbridge, K. A. Selkoe, C. V. Kappel, F. Micheli, C. D'Agrosa, J. Bruno, et al. 2008. "A global map of human impact on marine ecosystems." *Science* 319: 948–952.

Halpern, B. S., C. V. Kappel, K. A. Selkoe, F. Micheli, C. Ebert, C. Kontgis, C. M. Crain, et al. 2009. "Mapping cumulative human impacts to California Current marine ecosystems." *Conservation Letters* 2: 138–148.

Hansen, J., M. Sato, R. Ruedy, K. Lo, D. W. Lea, and M. Medina-Elizade. 2006. "Global temperature change." *Proceedings of the National Academy of Sciences, USA* 103: 14288–14293.

Harley, C. D. G., A. R. Hughes, K. M. Hultgren, B. G. Miner, C. J. B. Sorte, C. S. Thornber, L. F. Rodriguez, L. Tomanek, and S. L. Williams. 2006. "The impacts of climate change in coastal marine systems." *Ecology Letters* 9: 228–241.

Hartmann, D. L., J. M. Wallace, V. Limpasuvan, D. W. J. Thompson, and J. R. Holton. 2000. "Can ozone depletion and global warming interact to produce rapid climate change?" *Proceedings of the National Academy of Sciences, USA* 97: 1412–1417.

Hobday, A. J., and M. J. Tegner. 2000. *Status review of white abalone* (Haliotis sorenseni) *throughout its range in California and Mexico*. US Dept. of Commerce, National Oceanic and Atmospheric Administration, National Marine Fisheries Service, Southwest Region Office.

IPCC. 2007. *Synthesis Report*. Geneva, Switzerland: IPCC, 56 pp.

IUCN. 2001. *IUCN Red List categories and criteria*. Version 3-1. Gland, Switzerland: IUCN—The World Conservation Union.

Justic, D., N. N. Rabalais, and R. E. Turner. 1996. "Effects of climate change on hypoxia in coastal waters: A doubled $CO_2$ scenario for the northern Gulf of Mexico." *Limnology and Oceanography* 41: 992–1003.

Kappel, C. V. 2005. "Losing pieces of the puzzle: Threats to marine, estuarine, and diadromous species." *Frontiers in Ecology and the Environment* 3: 275–282.

Koch, E. W., E. B. Barbier, B. R. Silliman, D. J. Reed, G. M. E. Perillo, S. D. Hacker, E. F. Granek, et al. 2009. "Non-linearity in ecosystem services: Temporal and spatial variability in coastal protection." *Frontiers in Ecology and the Environment* 7: 29–37.

Learmonth, J. A., C. D. MacLeod, and M. B. Santos. 2006. "Potential effects of climate change on marine mammals." *Oceanography and Marine Biology: An Annual Review* 44: 431–464.

Lesser, M. P., and T. M. Barry. 2003. "Survivorship, development, and DNA damage in echinoderm embryos and larvae exposed to ultraviolet radiation" (290–400 nm). *Journal of Experimental Marine Biology and Ecology* 292: 75–91.

Levitus, S., J. Antonov, and T. Boyer. 2005. "Warming of the world ocean, 1955–2003." *Geophysical Research Letters* 32: L0260.

McKinney, M. L. 1997. "Extinction vulnerability and selectivity: Combining ecological and paleontological views." *Annual Review of Ecology and Systematics* 28: 495–516.

Mieszkowska, N., M. A. Kendall, S. J. Hawkins, R. Leaper, P. Williamson, N. J. Hardman-Mountford, and A. J. Southward. 2006. "Changes in the range of some common rocky shore species in Britain—A response to climate change?" *Hydrobiologia* 555: 241–251.

Mieszkowska, N., S. J. Hawkins, M. T. Burrows, and M. A. Kendall. 2007. "Long-term changes in the geographic distribution and population structures of *Osilinus lineatus* (Gastropoda: Trochidae) in Britain and Ireland." *Journal of the Marine Biological Association of the United Kingdom* 87: 537–545.

Mueter, F. J., R. M. Peterman, and B. J. Pyper. 2002. "Opposite effects of ocean temperature on survival rates of 120 stocks of Pacific salmon (*Oncorhynchus* spp.) in northern and southern areas." *Canadian Journal of Fisheries and Aquatic Sciences* 59: 456–463.

NOAA (National Oceanic and Atmospheric Administration). 2001. "Endangered and threatened species: Endangered status for white abalone." *Federal Register* 66: 29046–29055.

Olden, J. D., Z. S. Hogan, and M. J. Vander Zanden. 2007. "Small fish, big fish, red fish, blue fish: Size-biased extinction risk of the world's freshwater and marine fishes." *Global Ecology and Biogeography* 16: 694–701.

Parmesan, C. 2006. "Ecological and evolutionary responses to recent climate change." *Annual Review of Ecology, Evolution and Systematics* 37: 637–669.

Parmesan, C., and G. Yohe. 2003. "A globally coherent fingerprint of climate change impacts across natural systems." *Nature* 421: 37–42.

Pimm, S. L., H. L. Jones, and J. Diamond. 1988. "On the risk of extinction." *American Naturalist* 132: 757–785.

Pörtner, H. O. 2008. "Ecosystem effects of ocean acidification in times of ocean warming: A physiologist's view." *Marine Ecology Progress Series* 373: 203–217.

Reynolds, J. D., N. K. Dulvy, N. B. Goodwin, and J. A. Hutchings. 2005. "Biology of extinction risk in marine fishes." *Proceedings of the Royal Society B: Biological Sciences* 272: 2337–2344.

Rivadeneira, M. M., and P. A. Marquet. 2007. "Selective extinction of late Neogene bivalves on the temperate Pacific coast of South America." *Paleobiology* 33: 455–468.

Roy, K., D. Jablonski, and J. W. Valentine. 2001. "Climate change, species range limits and body size in marine bivalves." *Ecology Letters* 4: 366–370.

Sagarin, R. D., and S. D. Gaines. 2002a. "The 'abundant centre' distribution: To what extent is it a biogeographical rule?" *Ecology Letters* 5: 137–147.

Sagarin, R. D., and S. D. Gaines. 2002b. "Geographical abundance distributions of coastal invertebrates: Using one-dimensional ranges to test biogeographic hypotheses." *Journal of Biogeography* 29: 985–997.

Scavia, D., J. C. Field, D. F. Boesch, R. W. Buddemeier, V. Burkett, D. R. Cayan, M. Fogarty, et al. 2002. "Climate change impacts on US coastal and marine ecosystems." *Estuaries and Coasts* 25: 149–164.

Selkoe, K. A., B. S. Halpern, C. M. Ebert, E. C. Franklin, E. R. Selig, K. S. Casey, J. Bruno, and R. J. Toonen. 2009. "A map of human impacts to a 'pristine' coral reef ecosystem, the Papahanaumokuakea Marine National Monument." *Coral Reefs* doi:10.1007/s00338-00009-00490-z.

Swearer, S. E., G. E. Forrester, M. A. Steele, A. J. Brooks, and D. W. Lea. 2003. "Spatio-temporal and interspecific variation in otolith trace-elemental fingerprints in a temperate estuarine fish assemblage." *Estuarine Coastal and Shelf Science* 56: 1111–1123.

Terborgh, J. 1974. "Preservation of natural diversity: The problem of extinction-prone species." *Bioscience* 24: 715–722.

Tomanek, L. 2008. "The importance of physiological limits in determining biogeographical range shifts due to global climate change: The heat–shock response." *Physiological and Biochemical Zoology* 81: 709–717.

Vermeij, G. J. 1996. "Marine biological diversity: Muricid gastropods as a case study." In *Evolutionary Paleobiology,* edited by D. Joablonski, D. H. Erwin, and J. H. Lipps, 355–375. Chicago: University of Chicago Press.

## Chapter 17

# *Climate Change and Freshwater Fauna Extinction Risk*

N. LeRoy Poff, Julian D. Olden, and David L. Strayer

Fresh waters—rivers, streams, lakes, ponds, wetlands—cover less than 1 percent of the earth's surface, yet their biodiversity is unrivaled. Fully 10 percent of all known animal species and a third of all vertebrate species, including about 40 percent of the world's fishes, live in fresh waters. Other well represented groups include insects, crustaceans, mites, and mollusks (table 17-1). Further, an estimated 20,000–200,000 freshwater animal species (mostly invertebrates, including those cryptic species inhabiting ground waters) have yet to be described (Strayer, 2006). Despite this rich diversity, extinction risk of freshwater species has been largely overlooked (Strayer and Dudgeon, 2010).

The freshwater fauna contain a disproportionate number of imperiled species (Master et al., 2000; Strayer and Dudgeon, 2010). The International Union for Conservation of Nature Red List (IUCN, 2007) identifies 2,832 freshwater species as extinct, critically endangered, endangered, or vulnerable, but this list is biased by including only the largest-bodied and best-studied taxa (e.g., fish, mollusks, odonates) from well surveyed regions. The few very approximate estimates of actual imperilment for all freshwater fauna (Wilcove and Master, 2005; Strayer, 2006) suggest that perhaps 10,000–20,000 species are presently at serious risk of extinction, even in the absence of rapid climate change. Simple extrapolations of recent trends, combined with current extent of species imperilment, suggest future extinction rates for

TABLE 17-1. Summary of the known freshwater fauna of the world, excluding wholly parasitic forms.

| *Phylum* | *Approximate number of described species* | *Number of IUCN-listed species* |
|---|---|---|
| Porifera (sponges) | 219 | 0 |
| Cnidaria (hydras, jellyfish) | 30 | 0 |
| Platyhelminthes (flatworms) | 1,300 | 1 |
| Nemertea | 22 | 0 |
| Gastrotricha | 300 | 0 |
| Micrognathozoa | 1 | 0 |
| Rotifera | 2,000 | 0 |
| Nematoda (roundworms) | 1,900 | 0 |
| Nematomorpha (horsehair worms) | 330 | 0 |
| Annelida (oligochaetes, leeches, polychaetes) | 1,800 | 0 |
| Mollusca (snails, mussels, clams) | 5,000 | 497 |
| Bryozoa (moss animalcules) | 90 | 0 |
| Entoprocta | 2 | 0 |
| Crustacea | 12,000 | 254 |
| Chelicerata (mites) | 6,000 | 0 |
| Tardigrada (water bears) | 60 | 0 |
| Uniramia (insects) | 75,000 | 241 |
| Chordata (vertebrates) | 18,000 | 2,064 |

Compiled from Strayer (2006) and Balian et al. (2008). "IUCN-listed species" include species listed by IUCN (2007) in the following categories: "extinct," "extinct in the wild," "critically endangered," "endangered," and "vulnerable."

freshwater fish, crayfish, mussels, gastropods, and amphibians to be about an order of magnitude greater than the projected extinction rates for land and marine birds, reptiles, and mammals (Ricciardi and Rasmussen, 1999).

The high extinction risk of freshwater species is, seemingly paradoxically, explained in large part by the very feature that has created such diverse systems: habitat isolation. Freshwater environments are embedded in a terrestrial landscape that strongly restricts the movement or dispersal of aquatic organisms (e.g., amphibians, fishes, large crustaceans, mollusks), including those with aerial adults (insects) or resting stages (small crustaceans). Even where habitats are connected hydrologically in stream and river networks, individuals are often limited in their ability to move upstream and downstream because natural biophysical features pose barriers (e.g., Pringle, 2003). For example, large rivers generally act as movement barriers for species inhabiting headwater streams. In the face of rapid climate change, freshwater species will be challenged to move freely through river corridors to find more favorable habitats. This challenge is greatly exacerbated by human disruption of hydrologic connectivity through extensive construction of dams. Species that are unable to move will be challenged to either tolerate changing local environmental conditions or possibly genetically adapt to them (Poff et al., 2002; Allan et al., 2005).

The current global freshwater biodiversity crisis (Dudgeon et al., 2006; Strayer and Dudgeon, 2010) stems from many types of human activity: severe alteration of natural runoff patterns, fragmentation of river corridors by dams, increased addition of sediment and nutrients from poor land use practices, and introduction and spread of harmful nonnative species. Climate change is expected to intensify the threats to freshwater fauna, although it will be difficult to identify the increased risk specifically attributable to climate change versus other forms of anthropogenic global change (Sala et al., 2000). The challenge of doing so is further complicated by recognition of another key feature of freshwater ecosystems: species vulnerability has a regional context. Geographic variation in climate change will induce region-specific thermal and hydrologic deviations from recent historical climatic conditions. This, combined with regional variation in species richness, in species sensitivities to climate deviations and in habitat fragmentation, requires regional-scale analysis of vulnerability. For example, tropical species have evolved under more constant thermal regimes compared to temperate species, and thus they may be more sensitive to a unit increase in temperature (e.g., Deutsch et al., 2008; Tewksbury et al., 2008).

Beyond the modeling challenges of regional-scale projections of climate change and species vulnerability, there is added uncertainty associated with human responses to climate change. Freshwater systems are already heavily impacted by human activities (Malmqvist and Rundle, 2002; Allan, 2004; Palmer et al., 2008; Vörösmarty et al., 2010), and the manner by which human societies respond or adapt to climate change through water development and management will bear strongly and directly on species extinction risk (Poff, 2009; Strayer and Dudgeon, 2010). Strategic, proactive management actions by humans will increasingly be recognized to be of paramount importance to sustaining freshwater diversity in a rapidly changing world.

## Inability of Current Modeling Frameworks to Predict Global Freshwater Extinction Risk

Developing models to forecast extinction risk for individual freshwater species or groups of species is still in its infancy and suffers from significant knowledge gaps. To date, modeling approaches fail to capture the fundamental aspects that define freshwater vulnerability: highly dispersal-limited species living in linear, dendritic landscapes characterized by numerous natural barriers and human-caused fragmentation. Moreover, existing models are based on the (often implicit) assumption that the current distributions and abundances of aquatic species are in equilibrium with some largely static set of environmental conditions, whether in space or time. In this view, species themselves are typically assumed to show no variation across spatial or temporal gradients and to occupy a fixed niche space that cannot dynamically respond or adapt to changing environmental conditions.

Aside from simple extrapolation of current trends (e.g., Riccardi and Rasmussen, 1999), two general types of models have been used to estimate extinction risk in fresh waters. First, species-area modeling is based on the universal statistical relationship between area sampled and number of species observed. A regression on many paired observations of area (catchment size, lake area, or river discharge, lake volume) and species tallied yields a relation that is often applied at broad geographic extents across multiple basins—for example, Australia (Poff et al., 2001) or globally (Xenopolous et al., 2005). This relation has been used to predict the numbers of species that would become vulnerable to extinction if specific flow reductions were to occur under future climatic conditions or human activities (e.g., Xenopoulos et al., 2005; Xenopolous and Lodge, 2006).

These models, although appealingly simple, suffer from important limitations stemming from underlying equilibrial assumptions and scale-dependence (see Botkin et al., 2007; McGarvey and Hughes, 2008). For example, the underlying assumption of an "equilibrium" between area and fish richness is violated by the fact that in many rivers the total number of species has increased greatly in recent history due to the spread and establishment of nonnative species, generally without concurrent extinction of native species (Leprieur et al., 2008). Further, these relationships typically are based on static environmental measures (catchment size, mean annual discharge) that fail to capture seasonal dynamics that may act as bottlenecks on species richness. For example, intermittent Australian streams have much lower fish richness than do perennial streams of similar catchment size, probably due to high variation in runoff including stream drying that creates extinction-colonization cycles (Poff et al., 2001; see also Angermeier and Schlosser, 1989). Similarly, rivers of similar catchment area may support fewer top vertebrate predator species if seasonal drying is part of the river's natural flow regime (Sabo et al., 2010).

A second modeling approach uses climate change predictions to evaluate how species ranges will shift with climate-induced modifications to the environment. So-called bioclimatic models presume that the geographic range of a species can be effectively defined in terms of a small number of climatic variables, most prominently temperature and precipitation. These models have been used widely to predict the spread of invasive species in the terrestrial and aquatic realm, and they are now being used to project how native species ranges may shift with climate change (see Jeschke and Strayer, 2008 for a thorough review).

Bioclimatic models are subject to criticism of limited biological realism. For example, they assume that biotic interactions are unimportant in setting a species' range and that species have fixed niches not subject to change under novel selection regimes, despite some recent evidence to the contrary (Broennimann et al., 2007; Pearman et al., 2008). More subtly, when applied to native species, these models predict many "unoccupied" locales where environmental conditions are suitable but the species does not naturally occur, because species are unable to move or disperse to these favorable habitats. Failure to incorporate species ability to move along and across river networks of varying connectivity will continue to limit the value of these models to predict extinction risk for native freshwater species (Jeschke and Strayer, 2008).

## Developing a Framework for Modeling the Threat to Freshwater Diversity Posed by Climate Change

To credibly estimate global freshwater extinction risk requires an approach that explicitly incorporates the key features of freshwater ecosystems and their particular sources of vulnerability. Adopting the terminology of Turner et al. (2003), we can identify vulnerability to extinction as a function of three components: extrinsic exposure to climate change, intrinsic sensitivity of species to altered environmental

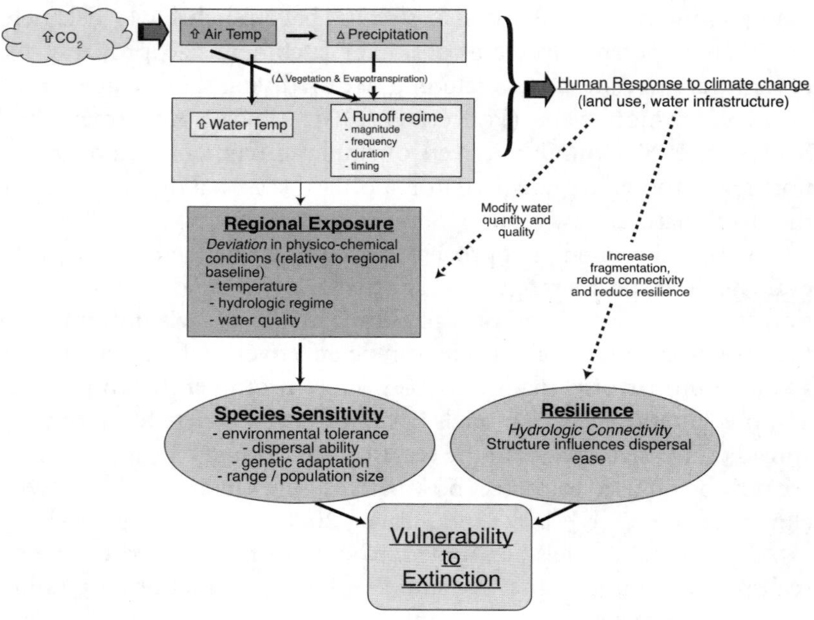


FIGURE 17-1. General conceptual model showing how freshwater species extinction vulnerability is a function of exposure to regional climate change, intrinsic species sensitivity, and landscape resilience. Vulnerability is potentially exacerbated or ameliorated by human response to climate change.

conditions, and resilience, or a species' ability to cope with the change (cf. Dawson et al., 2011). This vulnerability can be ameliorated or exacerbated by the human response to climate change. Figure 17-1 captures the general conceptual model that summarizes the major components of this framework, which we discuss more fully below.

## *Extrinsic Exposure to Climate Change: Importance of Regional Context*

Patterns of precipitation and temperature, which act as environmental drivers for freshwater ecosystems, vary substantially from region to region, reflecting geographic variation in recent historical climatic patterns. Geographic projections of changes in air temperature and precipitation also bear a strong regional signature, as shown in figure 17-2 for the continental United States. Streamflow is an integrated response to temperature and precipitation, and it likewise shows a

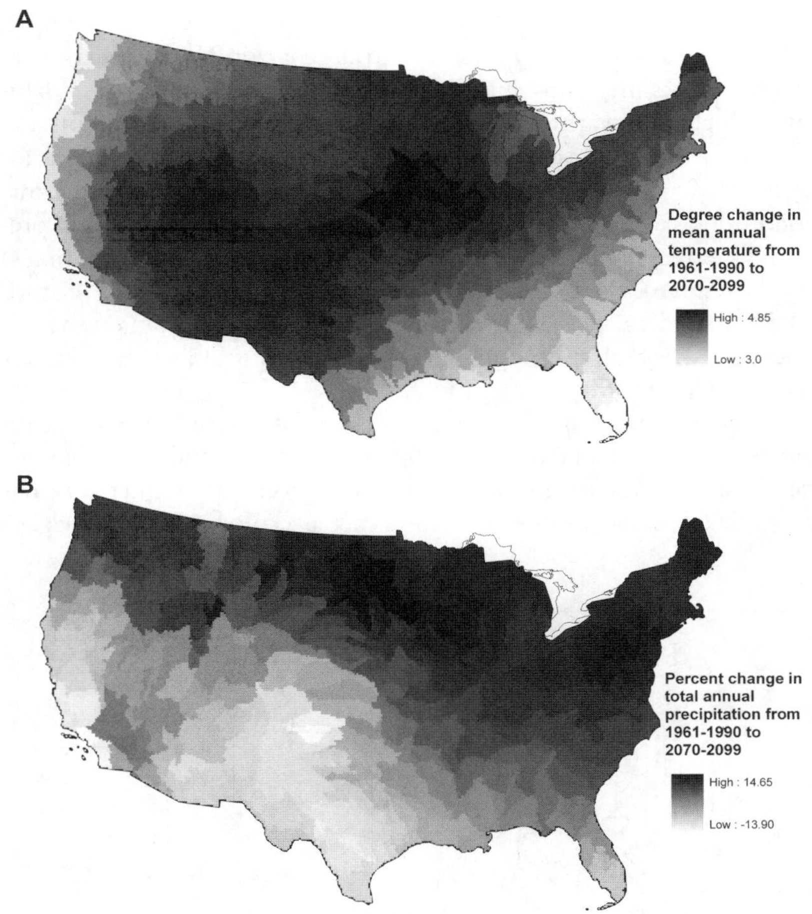


FIGURE 17-2. Projected changes in mean annual temperature (A) and precipitation (B) for the United States based on a multimodel ensemble of sixteen global climate models (median value of the first model run, Girvetz et al., 2009) under greenhouse emission scenario A2 (IPCC, 2007) for a comparison of 1961–1990 versus 2070–2099.

strong regional signature in projected deviations from contemporary conditions (fig. 17-3). Clearly, the *extrinsic exposure* of species to climate change has a strong regional context.

Water temperature regulates organismal bioenergetics and population size and ultimately sets the geographic limits to a species' range (Magnuson et al., 1979; Vannote and Sweeney, 1980; Caissie, 2006). Modeled future changes in water temperature suggest extensive local extirpations and potential range shifts for many species. For example, coldwater fish species (e.g., salmon, trout, char) in North America are projected to suffer substantial reductions in thermally suitable habitat with climate warming (see Mohseni et al., 2003; also Eaton and Scheller, 1996; Rahel et al., 1996; Wegner et al., 2011). For aquatic insects, whose development and maturation depend on annual thermal conditions, Sweeney et al. (1992) estimated a northward shift in preferred thermal regimes for some species of 422 kilometers (262 miles) for a 4-degree Celsius increase in air temperature. These kinds of studies do not address species extinction risk per se, but they suggest that extinction risk is high for species having limited range, low thermal tolerance, and limited dispersal ability.

Changes in precipitation will translate into altered hydrologic regimes in streams and rivers and inflows to wetlands and lakes. Temporal variation in the magnitude, timing, duration, and frequency of hydrologic extremes (floods, low flows) act as ecological bottlenecks to

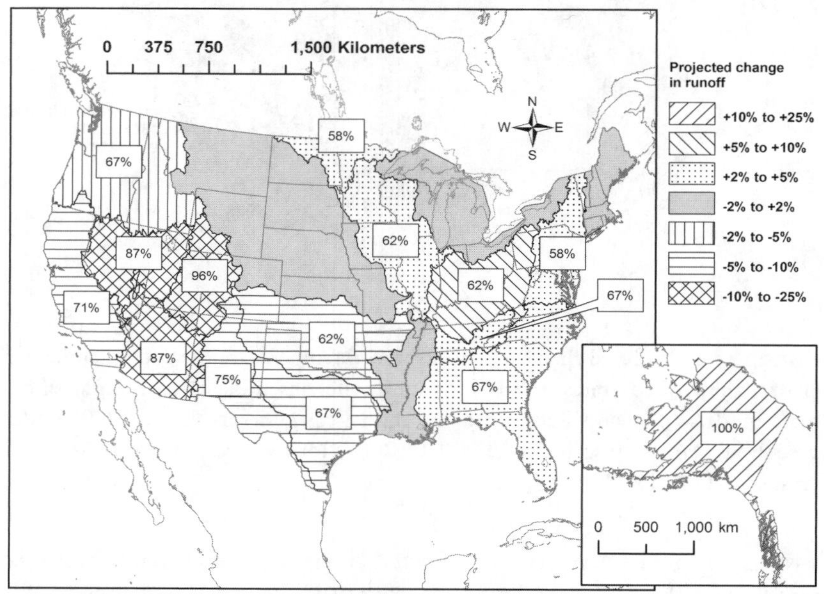


FIGURE 17-3. Median changes in runoff interpolated to US Geological Survey water resources regions from Milly et al. (2005) from twenty-four pairs of global climate model simulations. The numbers on the map represent percentage agreement among models for projected changes in runoff. Reprinted from Lettenmaier et al. (2009).

exert strong selective force on organisms and shape species adaptations and abundance (Poff et al., 1997; Lytle and Poff, 2004). Ecologically distinctive types of flow regimes vary geographically with prevailing climatic and geologic controls on runoff. Examples in the United States include predictable spring flood snowmelt-dominated systems in the montane West, stable groundwater-fed systems in limestone or glaciated geologies, and flashy, unpredictable rainfall-dominated systems that can be perennial or seasonally intermittent in humid or desert regions (Poff, 1996). The amount and timing of runoff can also directly influence the extent to which spatially separated habitats are connected to one another (Pringle, 2003). A change in precipitation regime can modify connectivity within river networks, for example, by shifting perennial streams (or stream segments) to intermittent ones (Seager et al., 2007) that fragment habitat and disconnect populations of aquatic organisms.

### *Intrinsic Sensitivity of Freshwater Fauna: Species Traits That Influence Vulnerability to Climate-Induced Extinction*

Species differ widely in their sensitivities to environmental change based on intrinsic traits that determine their ability to accommodate local environmental change. Unfortunately, the sensitivities of most freshwater species to environmental conditions are poorly known. It is, however, possible to group species broadly according to key biological attributes (traits) that reflect relative sensitivity to altered thermal and hydrologic regimes. Given the tremendous diversity of the freshwater fauna, a species-specific focus is infeasible for estimating global extinction risk, so a traits-based approach for taxonomic groups is appealing. We are now just beginning to understand how species traits relate to different stages of the extinction process (rarity, extirpation, and extinction) and how these traits respond to individually and interactively different sources of environmental change (see Olden et al., 2008). In principle, many sensitivity traits could be selected; here we identify four that are particularly important in assessing vulnerability and can be attributed to many taxonomic groups.

#### ENVIRONMENTAL TOLERANCE

A species' tolerance for a wide range of environmental conditions, or to environmental extremes, is likely to be a key predictor of its sensitivity to rapid climate change. For example, in the temperate zone, cold-adapted species appear to be more sensitive to warming than warm-adapted species (see Morgan et al., 2001). Thermal tolerance attributes have been described for various temperate freshwater groups, such as fish (Eaton and Scheller, 1996; Frimpong and Angermeier, 2009), aquatic insects (Poff et al., 2006; Yuan, 2006), and groundwater crustaceans (Issartel et al., 2005). Likewise, species inhabiting hydrologically stable habitats are more sensitive to imposed hydrologic variation than are species living in hydrologically variable environments (Poff and Ward, 1989). Species tolerant of more variable (disturbed) habitats have traits suited for resisting and recovering from disturbance (rapid reproduction, high mobility, trophic generalization, etc.; see Townsend, 1989).

Tolerance may be correlated to other sensitivity traits. For example, broadly tolerant species often occur over a broader geographic extent, and because they can accommodate greater variance in environmental conditions, they would be expected to have greater opportunities to disperse to more favorable environments as local environmental conditions change. By contrast, species with narrow tolerances are often localized in their distribution (small range size), although they can, in principle, be widely distributed habitat specialists. Two examples are some groundwater species that occur in relatively narrow thermal limits in geographically disjunct aquifers (Issartel et al., 2005), and aquatic insects with isolated populations restricted to small streams in montane alpine zones (Finn et al., 2006).

### DISPERSAL ABILITY

Another key trait that will influence a species' extinction risk is its ability to move (disperse) through river networks or across the terrestrial landscape to locate suitable habitats when local conditions become unfavorable. Freshwater species possess either a solely aquatic lifestyle (all life stages) or an amphibiotic one (including a terrestrial phase, characteristic of most amphibians and aquatic insects). Although we have limited knowledge of dispersal ability for most species (Bohonak and Jenkins, 2003), some generalizations are possible.

For example, species that are strong fliers as adults (e.g., insects such as odonates, some beetles, and some stoneflies) or are passively dispersed (some small crustaceans and mollusks) readily cross drainage divides within a continent. Strong aerial dispersers will be less sensitive to rapid environmental change than weak dispersers, because they can selectively colonize most favorable habitats across broad geographic extents (e.g., Bonada et al., 2007).

By contrast, large unionoid mussels, groundwater crustaceans, and large-bodied aquatic species (e.g., fish, crayfish) largely depend on connected waterways of suitable quality to move among drainages and extend their range. Where there are barriers, natural or human-caused, ability to move will be restricted.

### GENETIC ADAPTATION

Accommodation of freshwater species to new environmental conditions could occur via either phenotypic adjustment or selection for heritable genetic variation within a species (Ghalambor et al., 2007). In general, our understanding of the genetic basis for freshwater species adaptation to changing environmental conditions is poor (Hughes et al., 2009), and there is limited firm evidence for genetic adaptation to specific environmental regimes (Lytle and Poff, 2004; Lytle et al., 2008). Many examples exist of interpopulation differences in character states that could be either phenotypic or genetically based. For example, the generation time for some aquatic insect species varies with local thermal regime (Ward and Stanford, 1982). Hassall et al. (2007) found that odonate species have advanced their timing of emergence coincident with climate warming over the twentieth century, although it is not clear if this is a phenotypic or genotypic response. For fish, Blanck and Lamouroux (2007) examined trait variation for wide-ranging European fish species and found large among-population variation in growth rate, mortality rate, and length of breeding season, but not in fecundity and body size. It is possible that geographically variable traits offer some opportunity for genetic response to climate change, and more efforts to document such trait variability could advance our ability to project species extinction risk (see discussion in Ghalambor et al., 2007).

### RANGE SIZE AND/OR POPULATION SIZE

The range size of a species is a potential indicator of its vulnerability to rapid environmental change. In general, species with smaller geographic ranges are expected to be more vulnerable to global extinction under rapid environmental change, due to limited ability to withstand stochastic environmental and demographic fluctuations (Lawton and May, 1995), as has been shown for fish (Angermeier, 1995). Such species are often characterized by small population size, which further increases extinction risk.

Our knowledge of most freshwater species' ranges is incomplete (Heino et al., 2009), although some taxa such as fish are reasonably well described (e.g., Lee et al., 1980; Leprieur et al., 2008). Information gleaned from published sources is summarized in figure 17-4, which shows the distribution for selected taxa in North America based on the number of US states and Canadian provinces (unadjusted for areal extent or latitudinal range) in which they are known to occur. In some poorly dispersing groups (like prosobranch snails, crayfish, capniid stoneflies), the majority of species are restricted to just one or two states (and typically with a restricted range within a state). Groups with stronger dispersal ability generally have broader geographic ranges; these include odonates, fish, pulmonate snails, and small sphaeriid bivalves (which can be passively dispersed), but some species with small ranges are also included. Species with large ranges might be presumed to be at lower risk, especially where a large latitudinal range is encompassed, but these broadly distributed species dominate just a few groups.

For a few taxa, such as freshwater fish, a relatively good understanding of sensitivity to environmental conditions allows some assessment of risk associated with contemporary human activities and with future climate change. Figure 17-5, which illustrates three such indicators for native fish species in the United States, makes clear that species sensitivity varies regionally. "At risk" species are of most concern in the Southeast and Southwest; thermally sensitive cool- and cold-water species occur mostly in the colder climates of the montane West and northern tier of states; and small-bodied species (indicative of small home-range size and/or relatively poor dispersal ability) are most prevalent in the most southerly states. These maps, when considered with regional exposure to climate change (figs. 17-2 and 17-3) and with landscape resilience, suggest significant regional variation in vulnerabilities of fish faunas to climate change.

### *Freshwater Resilience: Habitat Connectivity and Human Fragmentation of Riverine Corridors*

Although resilience could be considered solely an attribute of the species (e.g., genetic adaptation), here we consider resilience more

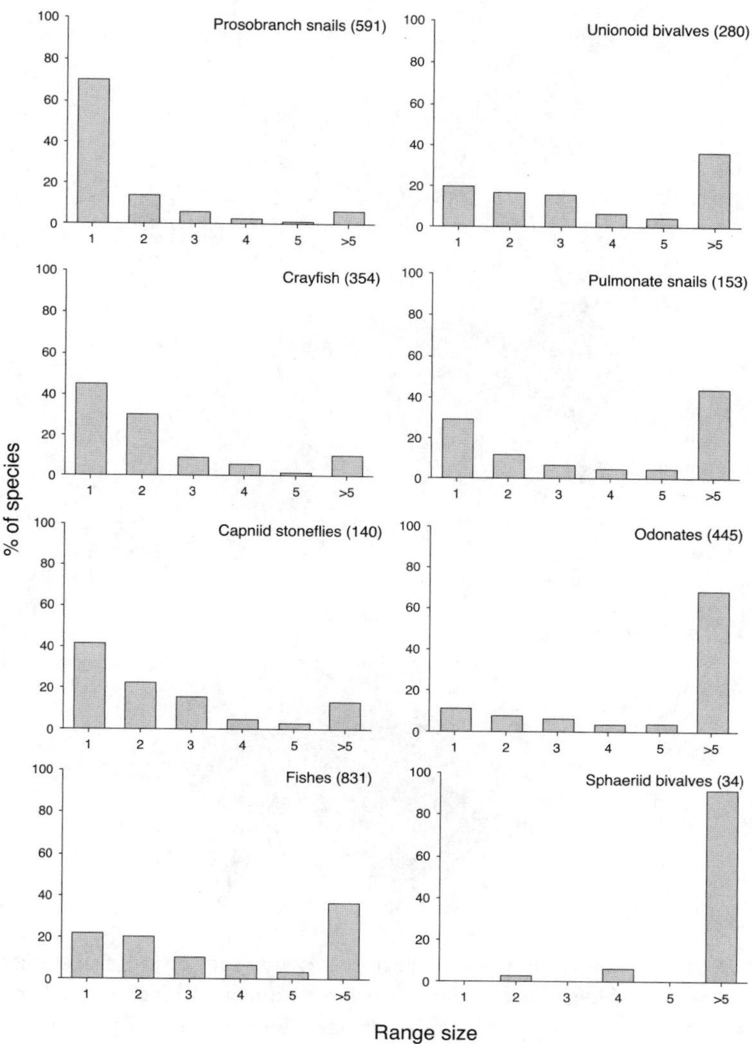


FIGURE 17-4. Range sizes (expressed as the number of US states and Canadian provinces where the species is known to occur) for native species belonging to selected groups of North American freshwater animals. The number of species in each taxonomic group for which data were available is given in parentheses. Because of widespread extirpations of unionoid mussel populations, the data for this group exclude states and provinces from which the species has already been extirpated. Data for fish include diadromous as well as purely freshwater species. Because of incomplete distributional information, data for capniid stoneflies may underestimate range sizes. Data from NatureServe (2009) and Mackie (2007).

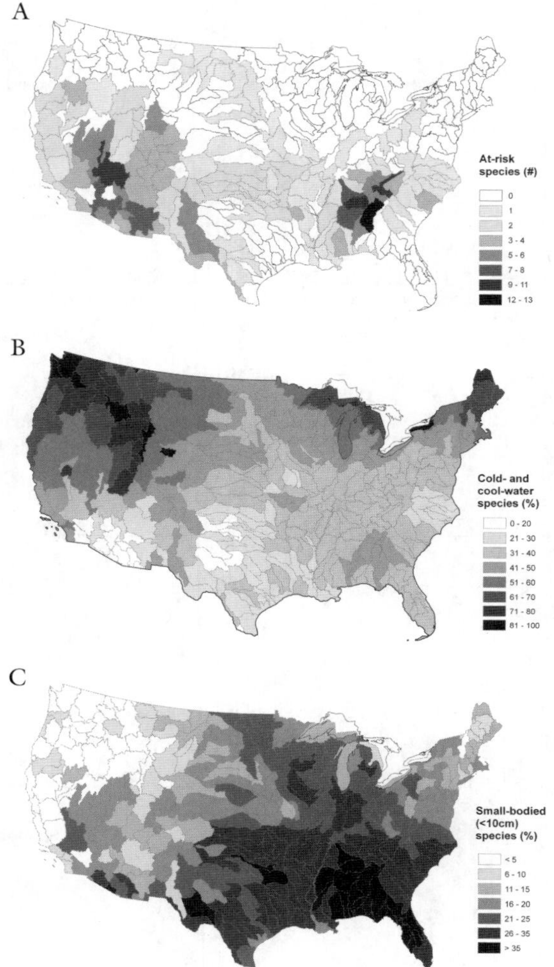


FIGURE 17-5. Maps of three measures of sensitivity of freshwater fish species for sixth-level Hydrologic Unit Codes across the continental United States. Sensitivity was characterized as (A) the number of highly imperiled and imperiled fish species according to the conservation ranking of NatureServe (G1 and G2 categories), (B) the percentage of native species exhibiting a temperature preference for cold or cool waters (less than 26 degrees Celsius), and (C) the percentage of native species with a maximum body length of less than 10 centimeters (3.9 inches) total length (termed small-bodied species). We assume that increasing values of these three metrics reflect more vulnerable catchments to the impacts of climate change due to greater numbers of at-risk species, more species sensitivity to future warming trends (thermal guild), and more restricted ability to disperse to more favorable habitats (body size).

broadly as a feature of the landscape, namely the connectivity of aquatic habitats that can allow species the opportunity to move to more hospitable habitats as local conditions change. The insular nature of freshwater systems and the associated strong dispersal limitation of many freshwater species implies that many of these species are vulnerable to extinction risk from rapidly changing environmental conditions. Therefore, the connectivity among suitable habitats is a critical element of species resilience to climate change.

The degree of natural connectivity of aquatic habitat in an area can be represented in various ways, one of which is the density of stream channels in a region. More surface water indicates not only total aquatic habitat, but also reduced overland distance among similar habitats. This can be especially important for species living in small, headwater streams and having limited ability to disperse overland (Finn et al., 2007; Hughes et al., 2009). Drainage density thus represents a natural source of resilience for freshwater species, and it varies greatly at regional scales within the United States, largely reflecting patterns in annual precipitation along an arid-humid gradient (fig. 17-6A).

Basin orientation and altitudinal range also influence species resilience to climate change. For example, basins with little topographic relief and with an east-west orientation, such as in the US Great Plains, provide minimal opportunity for aquatic species to escape increasing temperatures (Matthews and Zimmerman, 1990). Similarly, cold-adapted species in alpine streams may have limited thermal refuge as temperatures warm (Poff et al., 2002; Hering et al., 2009). By contrast, north-south–oriented basins provide greater resilience. As an example, during Pleistocene glaciation fishes in North America moved southward ahead of advancing ice sheets and recolonized in the North as the glaciers retreated, but in Europe the Alps offered no southern refuge and many fish species became extinct (Mahon, 1984; Oberdorff et al., 1997).

Humans have dramatically increased the insular nature of freshwater habitats in the last one hundred years by erecting tens of thousands of large dams and millions of small structures globally. These structures not only fragment river drainages and interrupt upstream-downstream connectivity (Jackson et al., 2001; Nilsson et al., 2005), but they also homogenize regional variation in climate-driven streamflow dynamics (Poff et al., 2007). In the United States the fragmentation of rivers has been dramatic, as illustrated in figure 17-6B. Populations of many freshwater species that were potentially connected via

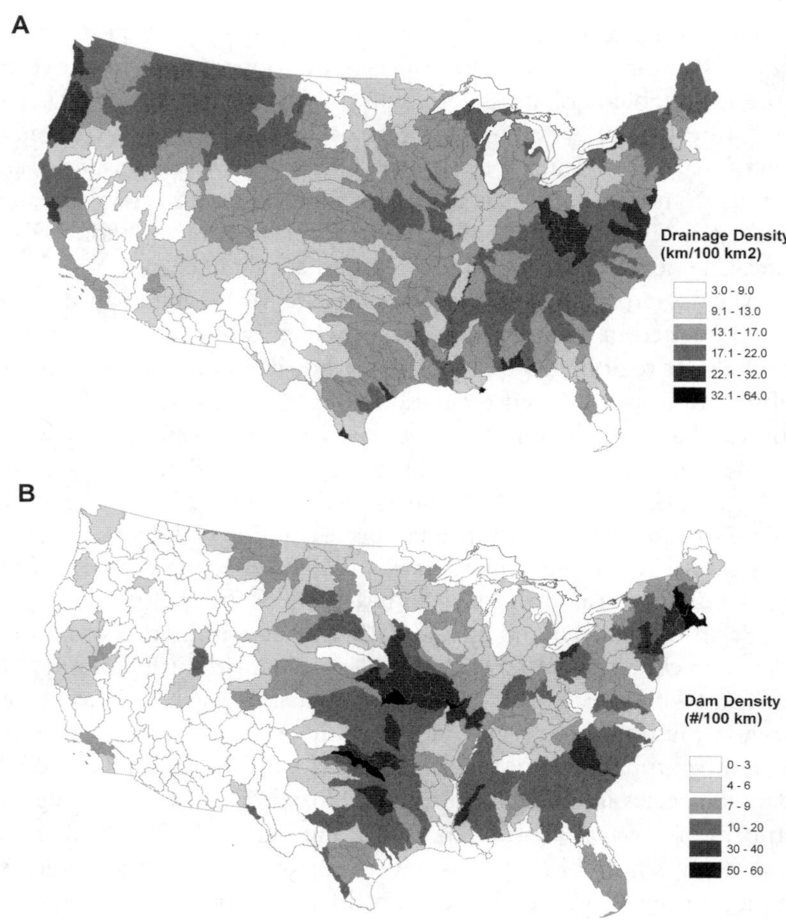


FIGURE 17-6. Maps showing natural (A) and anthropogenic (B) fragmentation at the sixth-level HUC for the United States. Dam density is for dams greater than 2 meters (6.5 feet) in height based on the National Inventory of Dams (USACE, 2007). See text for discussion.

dispersal in recent historical times now occur as isolated subpopulations, which have an increased risk of local extirpation (Winston et al., 1991). The effects of human-caused fragmentation of freshwater systems on dispersal rates of species that move solely through river channels (e.g., fish) have scarcely been studied. But simple metapopulation models suggest that the freshwater biota might be subject to a very large extinction debt from this fragmentation (Strayer, 2008), and there is some evidence that highly fragmented populations have high extinction rates (Fagan et al., 2002).

## Vulnerability of Freshwater Species to Regional Climate Change

Exposure to climate change, combined with species sensitivity and landscape resilience, generate vulnerability to extinction. As stated earlier, we have many knowledge gaps about species-level sensitivities, although some major freshwater faunal groups can be categorically separated from one another to make qualitative distinctions in extinction vulnerability. Table 17-2 presents a general summary of relative vulnerability of particular faunal groups, showing how possession of particular species traits makes these groups differentially vulnerable to rapid climate change.

Figure 17-7 presents a general conceptualization of species vulnerability in terms of exposure, sensitivity, and resilience. Species tolerance to varying environmental conditions can be envisioned as ranging from broad (eurytopic species) or narrow (stenotopic species), and species dispersal (movement) ability from strong to weak. Strongly dispersing species are able to move across the terrestrial matrix, whereas weak dispersers comprise both species with limited overland movement (e.g., amphibians, some insects) and those solely reliant on movement through water (e.g., fish and many mollusks and crustaceans). For the purposes of this illustration, other aspects of sensitivity can be considered to co-vary with these two dimensions. For example, potential for genetic adaptation can be viewed as functionally similar to environmental tolerance, so adaptable species would show a similar response as eurytopic species. Likewise, species with small ranges or population sizes would be expected to show a similar response as weak dispersers.

Tolerant (eurytopic) species (fig. 17-7A) show relatively low vulnerability along the exposure axis, because they are able to tolerate changing conditions. With increasing exposure, tolerances can be exceeded, but the ability to move to colonize suitable new habitats will reflect innate dispersal ability coupled with the extent of landscape fragmentation. So, for example, weak dispersers in highly fragmented landscapes would be more vulnerable than strong overland dispersers, which would show relatively low vulnerability.

TABLE 17-2. Traits of freshwater species relevant to predicting species extinction risk from climate change.

| *Trait* | *High-risk Taxa* | *Low-risk Taxa* |
|---|---|---|
| Small natural range size | Groundwater species, many decapods (crayfish), unionoid mussels, prosobranch snails, benthic fishes | Many species in glaciated regions |
| Reductions in range size from past human activities | Habitat specialists | Habitat and trophic generalists |
| Low environmental tolerance | Groundwater taxa | Species in temporally variable waters |
| High sensitivity to hydrologic change | Species restricted to highly stable or seasonally predictable waters | Species of intermittent waters |
| Poor cross-basin dispersal ability | Groundwater specialists, non-diadromous fishes, unionoid bivalves, most decapods, insects without aerial or amphibious stages | Strong fliers (odonates, some caddisflies, stoneflies), diadromous species, amphibious species, species that are easily passively dispersed in active (small pulmonate snails) or resting (ectoprocts, cladocerans) stages |
| Ranges in areas of current or future water stress for humans | Species with small ranges in highly affected regions (cf. first category) | Wide-ranging species |

For intolerant (stenotopic) species (fig. 17-7B), vulnerability increases relatively rapidly in response to exposure to changing environmental conditions compared to eurytopes. Strong overland dispersers are less vulnerable than weak dispersers across all levels of landscape fragmentation. Weakly dispersing stenotopes will have the greatest overall extinction risk (e.g., the groundwater fauna in table 17-2), particularly if they also have a small geographic range.

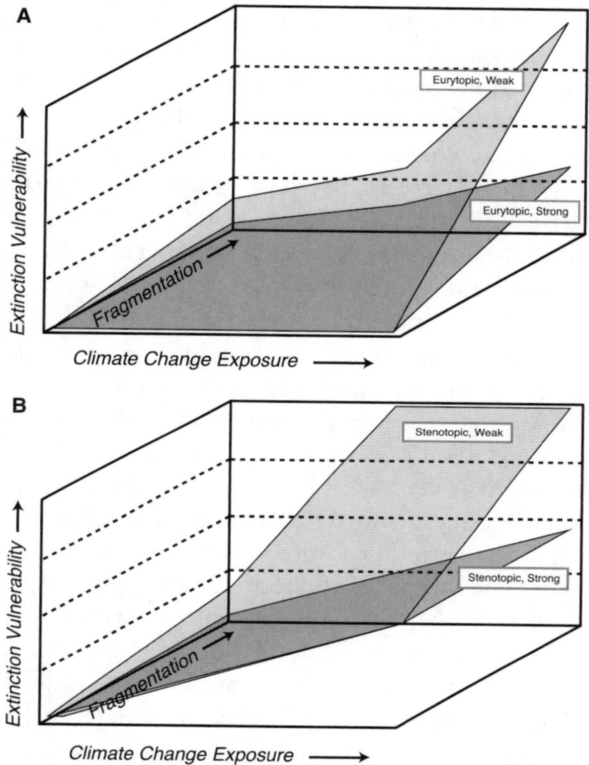


FIGURE 17-7. Extinction vulnerability response surfaces for freshwater species having different intrinsic sensitivities as a function of climate change exposure and habitat fragmentation (inverse of resilience). Sensitivity is defined in terms of two traits: environmental tolerance (broadly tolerant or eurytopic species, and weakly tolerant or stenotopic species) and dispersal ability (species that are strong overland dispersers, and species that are weak overland dispersers and/or require hydrologic connectivity to move). Extinction vulnerability is represented separately for eurytopic (upper panel) and stenotopic (lower panel) species.

## Conclusions

In conclusion, a precise estimate of global extinction rates for freshwater species is clearly beyond our grasp at the moment. Given the regional context of extrinsic exposure and landscape resilience, modeling efforts to estimate extinction need to be conducted by region and then aggregated to larger national and global scales. We have identified a general framework with specific elements that need to be better understood and captured to ensure credible modeling estimates. Despite the limitations of current models to robustly forecast extinction risk for freshwater species, we are highly confident that freshwater extinction rates due to climate change in the twenty-first century will equal or exceed that of most terrestrial systems given the exceptionally high background (nonclimate) rates (Riccardi and Rasmussen, 1999). With perhaps tens of thousands of freshwater species currently at risk of extinction (Wilcove and Master, 2005; Strayer, 2006) on a global scale, we can reasonably expect many thousand species to be lost as temperatures warm and precipitation regimes become more variable. These at-risk species belong to particularly vulnerable groups, as characterized by a variety of biological attributes (sensitivity traits) and narrow habitat requirements (table 17-2). However, as we have emphasized, the ultimate vulnerability of even these most sensitive species will depend on resolving some key questions: What is the geographic range of the species? What is the projected exposure to climate change across the geographic range? What degree of connectivity is there among suitable (or transitional) habitats that will allow movement during climate change? Additional challenges are posed by considering how climate-induced changes in species interactions (Sala et al., 2000; Van der Putten and Visser, 2010) and in the spread of invasive species (Rahel and Olden, 2008) will further exacerbate extinction risk for freshwater species.

To better characterize species vulnerability, we argue that a traits-based approach holds much promise because we cannot hope to assemble adequate information for every species (e.g., Angermeier, 1995; O'Grady et al., 2004; Olden et al., 2007, 2008). Species sensitivity is likely to be related to particular traits, which can be broadly ascribed across many freshwater taxa (table 17-2). Lumping species according to trait sensitivities can allow rough estimates of extinction vulnerability to be made, once these are placed in context of region-specific temperature and hydrologic change and landscape fragmentation.

The challenge in this approach is to identify traits that can be linked directly (mechanistically) to climatically driven environmental changes, that can be measured, and that can be generally applied at regional to global scales. This will require advances in current approaches to quantifying trait characteristics of species, as has recently been undertaken by the International Union for Conservation of Nature for some well studied groups, including birds, amphibians, and reef-building corals (see http://www.sciencedaily.com/releases/2008/10/081013142545.htm). Traits are generally attributed to whole species, but ideally we need to account for intraspecific variation in traits if we hope to predict extinction risk at regional scales. In a hierarchical framework we can begin to identify some traits that can be assigned as "mean state" values for species. We might, for example, be able to make coarse assessments at regional scales based on body size, relative dispersal ability, and possibly thermal tolerance (see Angermeier, 1995; Poff et al., 2006; Blanck and Lamouroux, 2007; Olden et al., 2007).

More information on dispersal ability of species is sorely needed. This information can inform our understanding of how many and which species are currently dispersal-limited (as opposed to being limited by local environmental conditions or biotic interactions), thereby allowing us to know when it is appropriate to apply existing modeling approaches (e.g., bioclimatic models) or whether we need to explicitly include dispersal. Knowledge of dispersal limitation can also guide management across modern fragmented landscapes and inform the need for potential assisted dispersal (McLachlan et al., 2007; Hoegh-Guldberg et al., 2008; Olden et al., 2010) of freshwater species.

Finally, a very real—and easily overlooked—driver of future freshwater extinctions will be the human response to climate change. Global analyses of future "water stress" indicate that certain regions of the world are likely to suffer disproportionately in the future as expanding populations confront anticipated reductions in runoff from climate change (Palmer et al., 2008). Freshwater ecosystems are already highly stressed (Dudgeon et al., 2006) and fragmented by water conveyance and storage infrastucture, and human management of this infrastructure is a key to future freshwater sustainability, including species persistence. Such changes may also promote the range expansion and impact of invasive species, thus further enhancing extinction risk (Rahel and Olden, 2008). We should anticipate engineering responses to climate change or prevent the most destructive of these responses. Incorporating ecosystem needs for freshwater in water resources planning to enhance water quality and connectivity will augment system resilience, and this, combined with enhanced flexibility in management of water infrastructure, can reduce extinction risk, as can establishment of conservation reserves that aim to maintain habitat connectivity and natural variation in environmental processes (Poff, 2009; Strayer and Dudgeon, 2010; Poff and Richter, 2011).

## REFERENCES

Allan, J. D. 2004. "Landscapes and riverscapes: The influence of land use on stream ecosystems." *Annual Review of Ecology, Evolution and Systematics* 35: 257–284.

Allan, J. D., M. A. Palmer, and N. L. Poff. 2005. "Climate change and freshwater ecosystems." In *Climate Change and Biodiversity,* edited by T. E. Lovejoy and L. Hannah, 274–290. New Haven, CT: Yale University Press.

Angermeier, P. L. 1995. "Ecological attributes of extinction-prone species: Loss of freshwater fishes of Virginia." *Conservation Biology* 9: 143–158.

Angermeier, P. L., and I. J Schlosser. 1989. "Species-area relationships for stream fishes." *Ecology* 70: 1450–1462.

Balian, E. V., C. Lévêque, H. Segers, and K. Martens, eds. 2008. "Freshwater animal diversity assessment." *Hydrobiologia* 595: 1–637.

Blanck, A., and N. Lamouroux. 2007. "Large-scale intraspecific variation in life-history traits of European freshwater fish." *Journal of Biogeography* 34: 862–875.

Bohonak, A. J., and D. G. Jenkins. 2003. "Ecological and evolutionary significance of dispersal by freshwater invertebrates." *Ecology Letters* 6: 783–796.

Bonada, N., S. Dolédec, and B. Statzner. 2007. "Taxonomic and biological trait differences of stream macroinvertebrate communities between Mediterranean and temperate regions: Implications for future climatic scenarios." *Global Change Biology* 13: 1658–1671, doi:10.1111/j.1365-2486.2007.01375.x

Botkin, D. B., H. Saxe, M. G. Araujo, R. Betts, R. H. W. Bradshaw, T. Cedhagen, P. Chesson, et al. 2007. "Forecasting the effects of global warming on biodiversity." *BioScience* 57: 227–236.

Broennimann, O., U. A. Treier, H. Müller-Schärer, W. Thuiller, A. T. Peterson, and A. Guisan. 2007. "Evidence of climatic niche shift during biological invasion." *Ecology Letters* 10: 701–709.

Caissie, D. 2006. "The thermal regime of rivers: A review." *Freshwater Biology* 51: 1389–1406.

Dawson, T. P., S. T. Jackson, J. I. House, I. C. Prentice, and G. M. Mace. 2011. "Beyond predictions: Biodiversity conservation in a changing climate." *Science* 332: 53–58.

Deutsch, C. A., J. J. Tewksbury, R. B. Huey, K. S. Sheldon, C. K. Ghalambor, D. C. Haak, and P. R. Martin. 2008. "Impacts of climate warming on terrestrial ectotherms across latitude." *Proceedings of the National Academy of Sciences, USA* 105: 6668–6672.

Dudgeon, D., A. H. Arthington, M. O. Gessner, Z. I. Kawabata, D. J. Knowler, C. Lévêque, R. J. Naiman, et al. 2006. "Freshwater biodiversity: Importance, threats, status and conservation challenges." *Biological Review* 81: 163–182.

Eaton, J. G., and R. M. Scheller. 1996. "Effects of climate warming on fish thermal habitat in streams of the United States." *Limnology & Oceanography* 41: 1109–1115.

Fagan, W. F., P. J. Unmack, C. Burgess, and W. L. Minckley. 2002. "Rarity, fragmentation, and extinction risk in desert fishes." *Ecology* 83: 3250–3256.

Finn, D. L., D. M. Theobald, W. C. Black IV, and N. L. Poff. 2006. "Spatial population genetic structure and limited dispersal in a Rocky Mountain alpine stream insect." *Molecular Ecology* 15: 3553–3566.

Finn, D. S., M. Blouin, and D. A. Lytle. 2007. "Population genetic structure reveals terrestrial affinities for a headwater stream insect." *Freshwater Biology* 52: 1881–1897.

Frimpong, E. A., and P. L. Angermeier. 2009. "Fishtraits: A database of ecological and life-history traits of freshwater fishes of the United States." *Fisheries* 34: 487–495.

Ghalambor, C. K., J. K. McKay, S. P. Carroll, and D. N. Reznick. 2007. "Adaptive versus non-adaptive phenotypic plasticity and the potential for contemporary adaptation in new environments." *Functional Ecology* 21: 394–407.

Girvetz, E. H., C. Zganjar, G. T. Raber, E. P. Maurer, P. Kareiva, and J. J. Lawler. 2009. "Applied climate-change analysis: The Climate Wizard Tool." *PLoS ONE* 4 (12): e8320. doi:10.1371/journal.pone.0008320.

Hassall, C., D. J. Thompson, G. C. French, and I. F. Harvey. 2007. "Historical changes in the phenology of British Odonata are related to climate." *Global Change Biology* 13: 933–941.

Heino, J., R. Virkkala, and H. Toivonen. 2009. "Climate change and freshwater biodiversity: Detected patterns, future trends and adaptations in northern regions." *Biological Reviews* 84: 39–54.

Hering, D., A. Schmidt-Kloiber, J. Murphy, S. Lücke, C. Zamora-Muñoz, M. López-Rodríguez, T. Huber, and W. Graf. 2009. "Potential impact of climate change on aquatic insects: A sensitivity analysis for European caddisflies (Trichoptera) based on distribution patterns and ecological preferences." *Aquatic Sciences—Research Across Boundaries* 71: 3–14.

Hoegh-Guldberg, O., L. Hughes, S. McIntyre, D. B. Lindenmayer, C. Parmesan, H. P. Possingham, and C. D. Thomas. 2008. "Assisted colonization and rapid climate change." *Science* 321: 345–346.

Hughes, J. M., D. J. Schmidt, and D. S. Finn. 2009. "Genes in streams: Using DNA to understand the movement of freshwater fauna and their riverine habitat." *BioScience* 59: 573–583, doi:10.1525/bio.2009.59.7.8

Issartel, J., F. Hervant, Y. Voituron, D. Renault, and P. Vernon. 2005. "Behavioural, ventilatory and respiratory responses of epigean and hypogean crustaceans to different temperatures." *Comparative Biochemistry and Physiology* 141: 1–7.

IPCC. "IPCC Fourth Assessment Report: Climate Change 2007." Accessed 17

June 2011. Available at http://www.ipcc.ch/publications_and_data/ar4/syr/en/contents.html

IUCN. 2007. "2007 IUCN Red List of threatened species." Accessed 27 January 2009. Available at http://www.iucnredlist.org

Jackson, R. B., S. R. Carpenter, C. N. Dahm, D. M. McKnight, R. J. Naiman, S. L. Postel, and S. W. Running. 2001. "Water in a changing world." *Ecological Applications* 11: 1027–1045.

Jeschke, J. M., and D. L. Strayer. 2008. "Usefulness of bioclimatic models for studying climate change and invasive species." *Annals of the New York Academy of Sciences* (The Year in Ecology and Conservation Biology) 1134: 1–24.

Lawton, J. H., and R. M. May. 1995. *Extinction Rates*. Oxford, UK: Oxford University Press.

Lee, D. S., C. R. Gilbert, C. H. Hocutt, R. E. Jenkins, D. E. McAllister, and J. R. Stauffer Jr. 1980. *Atlas of North American Freshwater Fishes*. Raleigh, NC: North Carolina State Museum of Natural History.

Leprieur, F., O. Beauchard, S. Blanchet, T. Oberdorff, and S. Brosse. 2008. "Fish invasions in the world's river systems: When natural processes are blurred by human activities." *PLoS Biology* 6 (2): e28. doi:10.1371/journal.pbio.0060028.

Lettenmaier, D. P., D. Major, N. L. Poff, and S. Running. 2008. "Water Resources." 121–150, In *The Effects of Climate Change on Agriculture, Land Resources, Water Resources, and Biodiversity in the United States. A Report by the U.S. Climate Change Science Program and the Subcommittee on Global Change Research, Synthesis and Assessment Product 4.4*. M. Walsh (managing editor), P. Backlund, A. Janetos, and D. Schimel (convening lead authors). U.S. Department of Agriculture, Washington D.C. Accessed 17 June 2011. Available at http://www.climatescience.gov/Library/sap/sap4-3/final-report/default.htm

Lytle, D. A., and N. L. Poff. 2004. "Adaptation to natural flow regimes." *Trends in Ecology & Evolution* 19: 94–100.

Lytle, D. A., M. T. Bogan, and D. S. Finn. 2008. "Evolution of aquatic insect behaviours across a gradient of disturbance predictability." *Proceedings of the Royal Society of London B* 275: 453–462.

Mackie, G. L. 2007. "Biology of corbiculid and sphaeriid clams of North America." *Ohio Biological Survey Bulletin,* New Series 15 (3): ix + 436 pp.

Magnuson, J. J., L. B. Crowder, and P. A. Medvick. 1979. "Temperature as an ecological resource." *American Zoologist* 18: 331–343.

Mahon, R. 1984. "Divergent structure in fish taxocenes of north temperate streams." *Canadian Journal of Fisheries and Aquatic Sciences* 41: 330–350.

Malmqvist, B., and S. Rundle. 2002. "Threats to the running water ecosystems of the world." *Environmental Conservation* 29: 134–153.

Master, L. L., B. A. Stein, L. S. Kutner, and G. A. Hammerson. 2000. "Vanishing assets: Conservation status of U.S. species." In *Precious Heritage: The Status of*

*Biodiversity in the United States,* edited by B. A. Stein, L. S. Kutner, and J. S. Adams, 93–118. New York: Oxford University Press.

Matthews, W. J., and E. G. Zimmerman. 1990. "Potential effects of global warming on native fishes of the southern Great Plains and the Southwest." *Fisheries* 15: 26–32.

McGarvey, D. J., and R. M. Hughes. 2008. "Longitudinal zonation of Pacific Northwest (U.S.A.) fish assemblages and the species-discharge relationship." *Copeia* 2: 311–321.

McLachlan, J. S., J. J. Hellmann, and M. W. Schwartz. 2007. "A framework for debate of assisted migration in an era of climate change." *Conservation Biology* 21: 297–302.

Milly, P. C. D., K. A. Dunne, and A. V. Vecchia. 2005. "Global pattern of trends in streamflow and water availability in a changing climate." *Nature* 438: 347–350. doi:10.1038/nature04312.

Mohseni, O., H. G. Stefan, and J. G. Eaton. 2003. "Global warming and potential changes in fish habitat in U.S. streams." *Climatic Change* 59: 389–409.

Morgan, I. J., D. G. McDonald, and C. M. Wood. 2001. "The cost of living for freshwater fish in a warmer, more polluted world." *Global Change Biology* 7: 345–355.

NatureServe. 2009. *NatureServe Explorer: An online encyclopedia of life [web application].* Version 7.1. NatureServe, Arlington, VA. Accessed 16–17 February 2009. Available at http://www.natureserve.org/explorer

Nilsson, C., C. A. Reidy, M. Dynesius, and C. Revenga. 2005. "Fragmentation and flow regulation of the world's large river systems." *Science* 308: 405–408.

Oberdorff, T., B. Hugueny, and J.-F. Guégan. 1997. "Is there an influence of historical events on contemporary fish species richness in rivers? Comparisons between Western Europe and North America." *Journal of Biogeography* 24: 461–467.

O'Grady, J. J., D. H. Reed, B. W. Brook, and R. Frankham. 2004. "What are the best correlates of predicted extinction risk?" *Biological Conservation* 118: 513–520.

Olden, J. D., Z. S. Hogan, and M. J. Vander Zanden. 2007. "Small fish, big fish, red fish, blue fish: Size-biased extinction risk of the world's freshwater and marine fishes." *Global Ecology and Biogeography* 16: 694–701.

Olden, J. D., N. L. Poff, and K. Bestgen. 2008. "Trait synergisms and the rarity, extirpation, and extinction risk of endemic fishes in the Colorado River Basin." *Ecology* 89: 847–856.

Olden, J. D., M. J. Kennard, J. J. Lawler, and N. L. Poff. "Diving into murky waters: Challenges and opportunities for implementing managed relocation to mitigate the threat of climate change to freshwater biodiversity." *Conservation Biology*. (forthcoming).

Palmer, M. A., C. A. Reidy Liermann, C. Nilsson, M. Flörke, J. Alcamo, P. S. Land, and N. Bond. 2008. "Climate change and the world's river basins:

Anticipating management options." *Frontiers in Ecology and Environment* 6: 81–89.

Pearman, P. B., A. Guisan, O. Broennimann, and C. F. Randin. 2008. "Niche dynamics in space and time." *Trends in Ecology & Evolution* 23: 149–158.

Poff, N. L. 1996. "A hydrogeography of unregulated streams in the United States and an examination of scale-dependence in some hydrological descriptors." *Freshwater Biology* 36: 71–91.

Poff, N. L. 2009. "Managing for variation to sustain freshwater ecosystems." *Journal of Water Resources Planning and Management* 135: 1–4.

Poff, N. L., and B. D. Richter. "Aquatic ecosystem sustainability in 2050." In *Environment and Water Resources in 2050: A Vision and Path Forward.* American Society of Civil Engineers special publication. (forthcoming).

Poff, N. L., and J. V. Ward. 1989. "Implications of streamflow variability and predictability for lotic community structure: A regional analysis of streamflow patterns." *Canadian Journal of Fisheries and Aquatic Sciences* 46: 1805–1818.

Poff, N. L., J. D. Allan, M. B. Bain, J. R. Karr, K. L. Prestegaard, B. Richter, R. Sparks, et al. 1997. "The natural flow regime: A new paradigm for riverine conservation and restoration." *BioScience* 47: 769–784.

Poff, N. L., P. L. Angermeier, S. D. Cooper, P. S. Lake , K. D. Fausch, K. O. Winemiller, L. A. K. Mertes, et al. 2001. "Fish diversity in streams and rivers." In *Scenarios of Future Biodiversity,* edited by F. S. Chapin, O. E. Sala, and R. Huber-Sannwald, 315–349. New York: Springer-Verlag.

Poff, N. L., M. Brinson, and J. B. Day. 2002. "Freshwater and Coastal Ecosystems and Global Climate Change: A Review of Projected Impacts for the United States." Arlington, VA: Pew Center on Global Climate Change. Accessed 17 June 2011. Available at http://www.pewclimate.org/global-warming-in-depth/all_reports/aquatic_ecosystems

Poff, N. L., J. D. Olden, N. K. M. Vieira, D. S. Finn, M. P. Simmons, and B. C. Kondratieff. 2006. "Functional trait niches of North American lotic insects: Trait-based ecological applications in light of phylogenetic relationships." *Journal of the North American Benthological Society* 25: 730–755.

Poff, N. L., J. D. Olden, D. Merritt, and D. Pepin. 2007. "Homogenization of regional river dynamics by dams and global biodiversity implications." *Proceedings of the National Academy of Sciences, USA* 104: 5732–5737.

Pringle, C. M. 2003. "What is hydrologic connectivity and why is it ecologically important?" *Hydrological Processes* 17: 2685–2689.

Rahel, F. J., and J. D. Olden. 2008. "Assessing the effects of climate change on aquatic invasive species." *Conservation Biology* 22: 521–533.

Rahel, F. J., C. J. Keleher, and J. L. Anderson. 1996. "Habitat loss and population fragmentation for coldwater fishes in the Rocky Mountain Region in response to climate warming." *Limnology & Oceanography* 41: 1116–1123.

Ricciardi, A., and J. B. Rasmussen. 1999. "Extinction rates of North American freshwater fauna." *Conservation Biology* 13: 220–222.

Sabo, J. L., J. C. Finlay, T. Kennedy, and D. M. Post. 2010. "The role of discharge variation in scaling of drainage area and food chain length in rivers." *Science* 330: 965–967.

Sala, O. E., F. S. Chapin, J. J. Armesto, E. Berlow, J. Bloomfield, R. Dirzo, E. Huber-Sanwald, et al. 2000. "Biodiversity—global biodiversity scenarios for the year 2010." *Science* 287: 1770–1774.

Seager, R., M. F. Ting, I. Held, Y. Kushnir, J. Lu, G. Vecchi, H. P. Huang, et al. 2007. "Model projections of an imminent transition to a more arid climate in southwestern North America." *Science* 316: 1181–1184.

Strayer, D. L. 2006. "Challenges for freshwater invertebrate conservation." *Journal of the North American Benthological Society* 25: 271–287.

Strayer, D. L. 2008. Freshwater Mussel Ecology: A Multifactor Approach to Distribution and Abundance. Berkeley: University of California Press.

Strayer, D. L., and D. Dudgeon. 2010. "Meeting the challenges of freshwater biodiversity conservation." *Journal of the North American Benthological Society* 29: 344–358.

Sweeney, B. W., J. K. Jackson, J. D. Newbold, and D. H. Funk. 1992. "Climate change and the life histories and biogeography of aquatic insects in eastern North America." In *Global Climate Change and Freshwater Ecosystems,* edited by P. Firth and S. G. Fisher, 143–176. New York: Springer-Verlag.

Tewksbury, J. J., R. B. Huey, and C. A. Deutsch. 2008. "Putting the heat on tropical animals." *Science* 320: 1296–1297.

Townsend, C. R. 1989. "The patch dynamics concept of stream community ecology." *Journal of the North American Benthological Society* 8: 36–50.

Turner, B. L. II, R. E. Kasperson, P. A. Matson, J. J. McCarthy, R. W. Corell, L. Christensen, N. Eckley, et al. 2003. "A framework for vulnerability analysis in sustainability science." *Proceedings of the National Academy of Sciences, USA* 100: 8074–8079.

USACE (United States Army Corps of Engineers). 2007. *National Inventory of Dams,* Federal Emergency Management Agency, Washington, D.C. Accessed 17 June 2011. Available at http://nid.usace.army.mil

Van der Putten, M. M., and M. E. Visser. 2010. "Predicting species distribution and abundance responses to climate change: Why it is essential to include biotic interactions across trophic levels." *Philosophical Transactions of the Royal Society B: Biological Sciences* 365: 2025–2034.

Vannote, R. L., and B. W. Sweeney. 1980. "Geographic analysis of thermal equilibria: A conceptual model for evaluating the effect of natural and modified thermal regimes on aquatic insect communities." *The American Naturalist* 115: 667.

Vörösmarty, C. J., P. B. McIntyre, M. O. Gessner, D. Dudgeon, A. Prusevich, P. Green, S. Glidden, et al. 2010. "Global threats to human water security and river biodiversity." *Nature* 467: 555–561. doi:10.1038/nature09440.

Ward, J. V., and J. A. Stanford. 1982. "Thermal responses in the evolutionary ecology of aquatic insects." *Annual Review of Entomology* 27: 97–117.

Wenger, S. J., D. J. Isaak, C. H. Luce, H. M. Neville, K. D. Fausch, J. B. Dunham, D. C. Dauwalter, M. K. Young, M. M. Elsner, B. E. Rieman, A. F. Hamlet, and J. E. Williams. 2011. "Flow regime, temperature, and biotic interactions drive differential declines of trout species under climate change." *Proceedings of the National Academy of Sciences, USA* 108: 14175–14180.

Wilcove, D. S., and L. L. Master. 2005. "How many endangered species are there in the United States?" *Frontiers in Ecology and the Environment* 3: 414–420.

Winston, M. R., C. M. Taylor, and J. Pigg. 1991. "Upstream extirpation of four minnow species due to damming of a prairie stream." *Transactions of the American Fisheries Society* 120: 98–105.

Xenopoulos, M. A., and D. M. Lodge. 2006. "Going with the flow: Using species-discharge relationships to forecast losses in fish biodiversity." *Ecology* 87: 1907–1914.

Xenopolous, M. A., D. M. Lodge, J. Alcamo, M. Märker, K. Schulze, and D. P. Van Vuuren. 2005. "Scenarios of freshwater fish extinctions from climate change and water withdrawal." *Global Change Biology* 11: 1557–1564.

Yuan, L. L. 2006. "Estimation and application of macroinvertebrate tolerance values." U.S. Environmental Protection Agency, National Center for Environmental Assessment, Washington, D.C. Report EPA/600/P-04/116F.

# Chapter 18

## *Climate Change Impacts on Species Interactions: Assessing the Threat of Cascading Extinctions*

LESLEY HUGHES

Complex networks of interacting species play important roles in the maintenance of biodiversity, the stability of food webs, and the ecosystem services that communities provide. Predicting the myriad of impacts that climate change will have on ecological interactions is a complex task. Species will respond individualistically to climatic and atmospheric changes. The geographic ranges and/or temporal coincidence of species that currently interact may therefore progressively move apart, while species that do not presently co-occur may do so in the future. Novel species combinations will result, and many present-day relationships between species may become increasingly decoupled. Changes in species interactions have enormous potential to alter community structure and composition, and these impacts may be even greater than the direct effects of a changed climate.

This chapter addresses the circumstances by which changes in the nature or intensity of species interactions could lead to extinction of one or more of the interacting partners. Such predictions are necessarily highly speculative and are not yet captured in most species or community models used to assess extinction risk. Recent developments in several families of models, however, present promising avenues of research.

## Which Interactions Will Make Species Most Vulnerable?

Each time a species undergoes a change in geographic range, phenology, or population size, cascading impacts to other species (e.g., pollinators, competitors, predators) may be expected. Indeed, these indirect effects may have greater impacts on many species than the direct effects of changes in atmospheric carbon dioxide, temperature, sea level, or rainfall. Progressive decoupling of present-day interactions between species will result in cascading changes in trophic interactions, food web structure, and ecosystem processes (Traill et al., 2010). Species with specialist requirements for particular habitats, mutualists, or hosts will be intrinsically more vulnerable to rapid change than those with more generalist habits.

We can distinguish between two broad categories of extinction resulting from changed species interactions—those due to (i) increasing negative interactions with other species (in the case of parasites, predators, or competitors), and to (ii) declining positive interactions (in the case of a mutualist, host, or prey). In the latter case, the most extreme outcome is *coextinction* (or secondary extinction) that will occur if a species goes extinct because another species on which it is dependent is extinguished (Stork and Lyal, 1993). Where this triggers extinctions among multiple interacting species, *extinction cascades* or chains of extinctions (sensu Diamond, 1989) may result.

## Mechanisms of Change

Within each category of potential extinction, we can also identify four broad mechanisms by which changed interactions will occur: (i) phenology-mediated, (ii) spatially mediated, (iii) population-mediated, and (iv) changes mediated by diet quality. These mechanisms are not mutually exclusive.

### *Phenology-Mediated Changes*

Studies of phenology provide some of the strongest evidence for climate change impacts on species (Hughes, 2000; Rosenzweig et al., 2008; Thackeray et al., 2010), and these changes are already resulting in altered synchrony of life cycles between species (match/mismatch hypothesis, sensu Cushing, 1990; fig. 18-1). Such disruption may be particularly important for species whose successful reproduction relies on a narrow window of seasonal food resources (e.g., Both et al., 2009). Disruption in synchrony may be especially critical for migratory animals; migration is initiated by cues at wintering rather than breeding sites. A growing disjunction between phenology in overwintering areas and that in summer breeding grounds has been documented for several bird species (e.g., Inouye et al., 2000; Both and Visser, 2001).

## *Spatially Mediated Changes*

As climate zones shift, geographic ranges are also shifting, especially in those species most capable of dispersal, such as birds, butterflies, and some pelagic marine organisms. Any decrease or fragmentation in range size may also reduce the opportunity for interaction with other species, assuming that such changes will not occur in the same direction and at the same rate among the participants.

## *Population-Mediated Changes*

Individualistic species' responses to climate change mean that there will be winners and losers, and therefore impacts on population sizes. However, our basic lack of knowledge about whether most food webs are regulated by top-down or bottom-up processes limits our ability to predict impacts on populations of most species.

Analyses of population trends associated with large-scale meteorological phenomena do offer some insights. Variation in the North Atlantic Oscillation, for example, mediates the influence of density-dependence (Forchhammer et al., 2002), predator populations (Wilmers et al., 2006), and synchrony in predator-prey dynamics (Post and Forchhammer et al., 2004). Within species, populations that fluctuate dramatically may be more vulnerable to extinction because the effective size of these populations is lower (Heino, 1998). Changes in population sizes depend not only on the type of interactions with other species in a food web, but also on the strength of those interactions and how disturbances such as extreme climatic events affect interaction strength (e.g., Harmon et al., 2009).

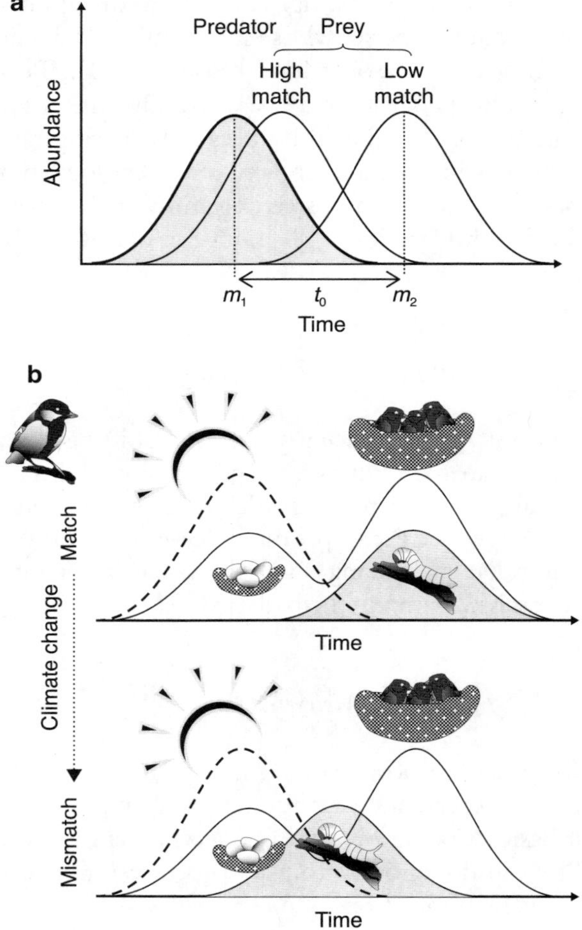


FIGURE 18-1. The match/mismatch hypothesis. (A) Interaction between two trophic levels explained by this hypothesis. A high match is represented by temporal overlap of the predator and its prey. An increase in time lag ($t_0$) between two population peaks ($m_1$, $m_2$: mean peak time for populations 1 and 2, respectively) leads to a low match. Adapted from Cushing (1990). (B) Example of mismatch due to climate change. The environmental cues (dashed line), triggering onset of egg laying, change in asynchrony to the environmental conditions prevailing when chicks are reared and when birds' energetic demands are highest, as shown for the great tit (Visser et al., 1998) from Durant et al. (2007).

### *Changes Mediated by Diet Quality*

Plants grown at elevated carbon dioxide are generally less nutritious due to increased carbon:nitrogen ratios and reduced digestibility. This fundamental shift in the quality of the base of all food chains will have cascading impacts on all species, although most immediately to herbivores (see below).

## Which Interactions Are the Most at Risk?

Although climate change may affect virtually all interactions among species, some impacts can be identified as being more likely to cause major disruptions to populations and therefore more likely to lead to species extinctions.

### *Plant-Herbivore Interactions*

Plants and their herbivores represent almost 50 percent of described species on Earth (Strong et al., 1984), and the interactions among them affect many aspects of communities and ecosystems (Coley, 1998). Further, it has been estimated that 22–47 percent of plant species globally are threatened with extinction (Pitman and Jorgensen, 2002). Because many plants have at least one species-specific insect herbivore (Strong et al., 1984; Janzen, 2003), the potential for co-extinction of many host plant–specific invertebrates is extremely high. The loss of tropical butterfly species from Singapore, for example, has been attributed to the loss of specific larval host plants (Koh et al., 2004a).

Plant-herbivore interactions will be affected directly by climate impacts on the participants, and indirectly, via nutritional impacts of enhanced carbon dioxide on the host plants. Herbivores feeding on plants grown at elevated carbon dioxide often develop more slowly, and have reduced body size and fecundity as adults, despite increased consumption rates. Elevated carbon dioxide can also affect competition among herbivorous insects (Stacey and Fellowes, 2002) and their susceptibility to natural enemies (Stiling et al., 2002).

Notwithstanding the experimentally demonstrated impacts of carbon dioxide on invertebrate herbivores, it seems unlikely that this particular driver will, by itself, lead to extinctions. Carbon dioxide is increasing steadily, rather than abruptly, exerting an ongoing selection pressure on herbivores to adapt to the altered food supply. Some vertebrate herbivores, however, may be more vulnerable, partly due to their reduced capacity for rapid evolutionary adaptation (e.g., Kanowski, 2001).

Although carbon dioxide impacts may not be catastrophic for most herbivores, the potential for plant-herbivore interactions to be affected by changes in distributions of the partners is high. Many herbivores, especially flying insects, are considerably more mobile than their hosts and thus have greater capacity to respond to shifting climate zones by altering their geographic range. Colonization of new hosts is likely for the more generalist species, potentially leading to reductions in herbivore pressure on the original hosts and increased pressure for the new partners.

Decoupling of interactions via changes in the timing of life cycles is also likely. Herbivorous insects that currently emerge synchronously with the bud burst of their host plants may starve if they hatch too early, or be forced to eat tougher, less digestible leaves if they emerge too late; either situation could result in a reduction in plant damage, as well as risks to the herbivores. Warmer springs disrupt the otherwise tight synchrony between bud burst in the oak *Quercus robur* and the winter moth *Operophtera brumata* egg hatch (Visser and Holleman, 2001). As springs become progressively warmer, the winter moth eggs have been hatching up to three weeks before bud burst, leading to higher mortality and lower reproductive success, with impacts on birds that prey on the moth larvae (Visser et al., 2004). Similar trophic asynchronies have also been recorded in aquatic systems; rising sea temperatures are affecting the recruitment of bivalves such as *Macoma balthica* (Philippart et al., 2003) by advancing their spawning and thus creating a mismatch with the light-dependent phytoplankton on which they feed.

### *Interactions among Hosts, Pathogens, Parasites, and Parasitoids*

Forty percent of known species are parasitic, and recent studies of food webs suggest that about 75 percent of trophic links involve a parasite species (Dobson et al., 2008). Parasite diversity may be an underappreciated component of many ecosystems; numerical estimates of parasite diversity in coral reefs, for example, indicate that parasites could total more than ten times the number of fish species (Justine, 2007). Virtually all aspects of existing host-parasite relationships will be affected by climate change, including the dynamics and seasonality of disease transmission, probability of host-switching, virulence of infections, host susceptibility, parasite and host survival rates, immunocompetence, size and location of contact zones, and the quality and quantity of vector-breeding sites (e.g., Patz et al., 2000; Kovats et al., 2001; Wilmers et al., 2006; Dobson et al., 2008; Acevedo-Whitehouse and Duffus, 2009).

Climate change has been implicated in the phenomenon of emerging infectious diseases that often occur when parasite species begin infecting hosts with which they have no previous history of association (Brooks and Hoberg, 2000; Gould and Higgs, 2008). New diseases appear to be emerging at unusually high rates, particularly in marine environments where disease incidence will interact with, and exacerbate, the impacts of other stresses such as ocean acidification (fig. 18-2; Harvell et al., 2002; Brandt and McManus, 2009; but see Smith et al., 2006; Lafferty, 2009).

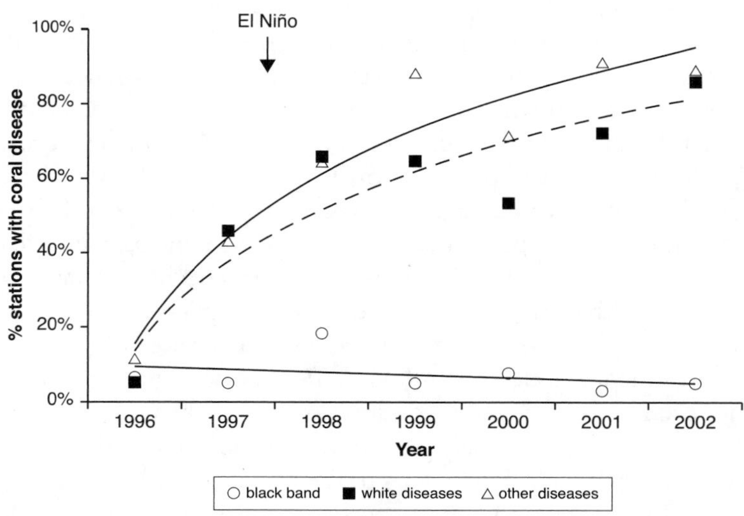


FIGURE 18-2. The percentage of the Florida Keys Coral Reef Monitoring Project stations where disease increased significantly between 1996 and 2002, with 1997–1998 El Niño noted. The data come from 105 randomly chosen stations that were censused annually by uniform survey methods (Lafferty et al., 2004).

Because parasites, parasitoids, and pathogens regulate both vertebrate and invertebrate populations (e.g., Stireman et al., 2005; Wilmers et al., 2006), affect concentrations of environmental pollutants (e.g., Sures, 2003), and stabilize food webs (Dobson et al., 2008), subsequent impacts on communities and ecosystem function are expected as a result of changes to host-parasite relationships.

Although parasites and disease have been frequently documented as causing extirpation of populations, the role of disease in global species extinctions is poorly understood. Smith et al. (2006) estimated that of the more than 830 species extinctions over the last 500 years documented by the International Union for Conservation of Nature, only 3.7 percent have been attributed in part to infectious disease. However, it is interesting to note that impacts of climate change on the disease chytridiomycosis have been implicated in perhaps the first documentation of a climate change–related extinction in amphibians (Pounds et al., 2006; chapter 6, this volume).

What types of parasites or hosts are likely to suffer increased extinction risk? On the parasite side, species with complex life cycles (those that need multiple host species) may be less successful at dispersing geographically than those with direct life cycles (Poulin and Morand, 2004; Koh et al., 2004b). Highly host-specific parasites will also suffer disproportionately as host populations fluctuate because they will be more sensitive than generalists to variation in host emergence time or development rate and may miss narrow windows of host vulnerability. Inefficiently transmitted species will most likely be lost first because they will be less likely to persist at low host densities. Although a parasite species that uses a range of host species will not necessarily go extinct if one of its host species declines to extinction, the parasite will decline as each potential host species is lost or itself declines in range and abundance (Altizer et al., 2007). Additionally, given the evidence of minimum thresholds of host density below which parasites cannot sustain recruitment (Anderson and May, 1986), many parasites will go extinct even before their host disappears. Arguably, the least endangered parasites will be sexually transmitted parasites and pathogens that are transmitted by infected females to their offspring. Although highly host-specific, these pathogens can persist in smaller populations (Dobson et al., 2008).

On the host side, infectious disease can drive populations temporarily or permanently to low numbers or densities, predisposing them to extinction by other forces (de Castro and Bolker, 2005; Gerber et al., 2005). Infectious diseases that are host-specific and density-dependent are unlikely to be the sole cause of species extinction because they typically die out when the host population falls below a threshold density (Anderson and May, 1986). Infectious diseases that are frequency-dependent and/or use reservoir hosts are more likely to induce extinction in their hosts (de Castro and Bolker, 2005).

One of the few comprehensive studies to investigate the potential impact of climate change on a host-parasite relationship predicts a parasite-induced collapse of populations of the marine amphipod *Corophium volutator*, an abundant species on the mudflats of the Danish Wadden Sea (Mouritsen et al., 2005; Poulin and Mouritsen, 2006). A simulation model parameterized with field and experimental data indicated that a warming of less than 4 degrees Celsius will cause population crashes of the amphipod due to mortality induced by microphallid trematodes to which it is host.

### *Predator-Prey Interactions*

The consumption of one species by another drives the flow of energy in ecosystems. The population dynamics of predators can regulate prey populations, and vice versa. In particular, increases or decreases in populations of top predators can have strong cascading impacts through food webs; marine systems appear particularly sensitive to such top-down regulation (e.g., Guidetti, 2007; Baum and Worm, 2009). Increasing asynchrony and intensity of predator-prey interactions associated with recent climate trends have been found in a wide variety of taxa, including frogs and newts (Beebee, 1995), Antarctic sea birds (van Franeker et al., 2001), phytoplankton and herbivorous zooplankton (Winder and Schindler, 2004), wolves (Wilmers et al., 2006), Arctic lemmings and their predators (Gilg et al., 2009), birds, and nest predators (Martin, 2007).

Recent work suggests that warming may directly affect predator-prey interactions in some invertebrate species by decreasing handling time, increasing predator attack rates, and decreasing the ratio of per capita feeding rate to metabolic rate (Vucic-Pestic et al., 2010). As a result, predator-prey oscillations may be dampened in some interactions, thus stabilizing their dynamics. However, warming has also been shown to reduce ingestion efficiency, potentially resulting in an increased risk of starvation (Rall et al., 2010).

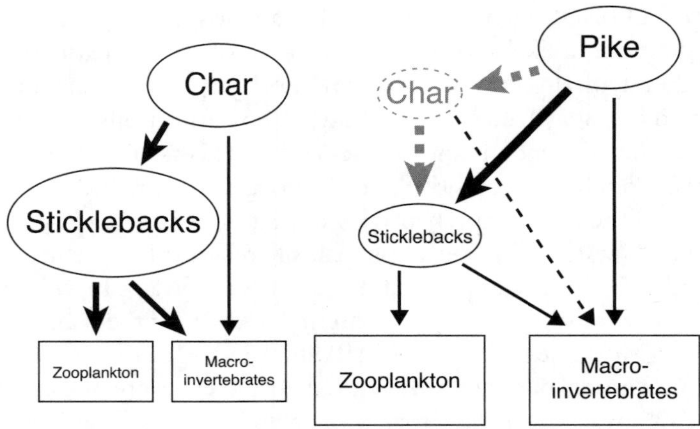


FIGURE 18-3. Schematic food web for Arctic char and ninespined stickleback pikes (left) and predicted changes in food web structure (right) in the presence of pike. Gray hatched figures and consumption arrows represent transient food web structure during pike establishment (Byström et al., 2007).

Invasions by top predators can have strong cascading effects in ecosystems. Climate change–related changes in abundance of a top predator, the northern pike (*Esox lucius*), have been implicated in the local extirpation of the Arctic char (*Salvelinus alpinus*) in a subarctic lake ecosystem (fig. 18-3, Byström et al., 2007). This replacement apparently took place quickly (in less than 4 years) following invasion by the pike, a more warm-adapted species.

Predators may also mediate the effects of climate fluctuations and changes in climatic extremes. On Isle Royale, Michigan, and in Yellowstone National Park, the grey wolf (*Canis lupus*) buffers the effects of large-scale climate phenomena on prey populations (Wilmers et al., 2006) and on the dynamics of the scavenger community for which they provide carrion (Wilmers and Getz, 2005).

### *Pollination*

Changes in temperature, rainfall, and carbon dioxide directly influence the level of resources available to pollinators (e.g., Bond, 1995; Tylianakis et al., 2008; Mitchell et al., 2009). Many plants are responding to warming by flowering earlier (e.g., Rosenzweig et al., 2008), and the response appears linear in many cases (e.g., Menzel et al., 2006). Climatic variables will also directly affect pollinator distribution and abundance (Mitchell et al., 2009) and shifts in the composition of pollinator assemblages may affect the quality of the pollinator service provided (Bond, 1995; Mitchell et al., 2009). Linear relationships are also found for the effect of warming on pollinator phenology (e.g., Menzel et al., 2006). Future phenological mismatches between plants and pollinators are likely, resulting in changes to diet breadth of the pollinators (Hegland et al., 2009).

Extrapolating from phenological shifts observed over the past century, Memmott et al. (2007) predicted that phenologies of plants and pollinators will be advanced on average 1–3 weeks by the end of the twenty-first century. As a result, Memmot et al. estimated 17–50 percent of pollinator species in a prairie-forest transition zone in Illinois may suffer some level of disruption in food supply. This analysis indicated that although the impacts on specialized pollinators were predicted to be greatest, even generalist pollinators could also experience considerable reductions in resources.

The important question here is whether any of these changes will lead to complete reproductive failure in plants or to complete loss of food resources for a pollinator. In general, the impacts of mismatches between plants and their pollinators would be expected to be more severe on the pollinators because of their nutritional dependence on plant resources. Pollinators such as hummingbirds might be especially vulnerable due to their high metabolic requirements for nectar (Hegland et al., 2009). Further, the generally shorter life spans of insect pollinators compared to plants will make them more sensitive to climatic variability (Hegland et al., 2009).

Factors influencing which plant species may be more at risk include plant breeding system, length of flowering period, pollination mode, and the degree of reproductive dependence on seeds (Bond, 1995). Most plants are pollinated by several species, often of widely diverse taxonomic origin, and pollination by a single species seems rare, although figs, *Yucca,* and orchids are notable exceptions (Bond, 1995; Waser and Ollerton, 2006; Swarts and Dixon, 2009). Most pollinator systems therefore have a degree of built-in redundancy and are thus less vulnerable to future disruption (Hegland et al., 2009). In highly seasonal climates, such as those in alpine areas and at high latitudes, inclement weather often disrupts pollinating insects, and most plants seem well insured against pollinator failures (Bond, 1995).

Wind-pollination, self-pollination, and asexual propagation are common, and insect-pollinated flowers are often visited by many species (Bond, 1995). By contrast, tropical floras may contain species at higher risk. Lowland tropical forests, in particular, have unusually low levels of self-pollination, very high levels of dioecy, and more specialized pollinator relationships (Bond, 1995; Bawa and Dayanandan, 1998).

Cascading impacts of reduced pollinator success may also be expected in some regions. The most severe El Niño–associated drought on record in Borneo (1997–1998) caused a substantial break in the production of fig inflorescences and led to the local extinction of pollinating fig wasps (Harrison, 2000). Predictions of more frequent and severe El Niño–associated drought in the lowland dipterocarp forests of southeast Asia (IPCC, 2001) are therefore of concern because figs represent an important food resource for a wide variety of animal taxa (Terbough, 1986; Harrison, 2000).

In many cases, disruption of pollinator relationships may not lead directly to extinctions, but may nonetheless contribute to population declines, rendering species more vulnerable to extinction from other stresses. Introduction of the brown tree snake (*Boiga irregularis*) to Guam resulted in the extinction of several native bird pollinators. Cascading impacts to the plant species these birds pollinated, including lower seed set and recruitment, were subsequently recorded (Mortensen et al., 2008).

### *Dispersal*

As for plant-pollinator systems, the majority of plant-disperser relationships are fairly general, with the seeds of most plants being dispersed by several animal species, and most dispersers relying on several plant species for resources. Such diffuse interactions will presumably buffer most populations against decoupled relationships. Breakdowns in dispersal mutualisms seem rare, but when they occur they can be devastating. Fruit production on Barro Colorado Island in Panama, for example, tends to be high during El Niño events and uses much of the plants' reserves. When a mild dry season follows, fruit production is very low and has been known to lead to the death of many frugivorous vertebrates (Wright et al., 1999). Predictions that El Niño events in the future may become more frequent and/or more intense (IPCC, 2001) are therefore a cause for concern in these systems, as for obligate plant-pollinator relationships.

### *Competition*

Although climate change will undoubtedly affect competitive interactions among species (e.g., Poloczanska et al., 2008), identifying situations where such changes will lead to species extirpation is even more difficult than for the other types of interactions because the role of competition in determining the distribution and abundance of species, relative to other factors, is poorly understood. Carbon dioxide enrichment, for example, may promote woody plant invasion of grasslands through its effect on competitive interactions between grass and tree seedlings (Bloor et al., 2008), but extrapolating to species extinctions is tenuous at best. Similarly, vines and scramblers in Amazonian rain forest may be relatively more advantaged than the woody species on which they grow (e.g., Phillips et al., 2002), leading to changes in the structure and function of many plant communities, but not necessarily to global extinctions. Perhaps one of the most likely roles that competition may play in species extinction is on coral reefs, where competition for space is intense. Faster growing species are expected to take advantage of new colonization sites as sea levels rise. Coupled with the multiple stresses of coral bleaching due to high ocean temperatures, and the impacts of ocean acidification (chapter 16, this volume), such competitive effects may be critical.

## Moving from Anecdote to Quantification

The assessment of extinction risk under climate change as a result of altered biotic interactions is at a fledgling stage. However, recent advances in three distinctive families of models offer promising, though somewhat disparate, avenues of research.

### *Species Distribution Models*

Most attempts to forecast climate change impacts on biodiversity have relied on bioclimatic envelope modeling, whereby present-day distributions of species are combined with environmental variables to project potential distributions of species under future climate scenarios (chapter 4, this volume). One of the most frequently identified shortcomings in these models is that biotic interactions are not included (e.g., Davis et al., 1998). Recent analyses have shown that inclusion of additional predictor variables representing the presence-absence of known competitors can significantly increase the predictive power of models (reviewed in Guisan and Thuiller, 2005). A handful of studies have attempted to overcome the limitation of species distribution models in terms of lack of biotic interactions either by explicitly including the distribution of an interacting species as a predictive variable (Araújo and Luoto, 2007), or by examining the overlap of the current and predicted geographic ranges of the interacting species (Schweiger et al., 2008). Preston et al. (2008), for example, showed that projections of range change of two endangered species, the Quino checkerspot butterfly (*Euphydryas editha quino*) and the California gnatcatcher (*Polioptila californica californica*), were reduced by 68–100 percent when availability of habitat was included in the models, compared to using climate variables only.

Perhaps the most exciting future prospect in this area of research is the development of "hybrid models" that couple "classic" species distribution modelling with mechanistic models that incorporate demographic, genetic, and other processes (Araújo et al. 2006). For example, Keith et al. (2008) combined a time series of habitat suitability models with spatially explicit stochastic population models to explore factors that influence the viability of plant species populations under climate change scenarios in the South African fynbos.

## *Coextinction Models*

Recent development of models to estimate the probability of coextinction also holds promise. Koh et al. (2004b) proposed two methods to estimate loss of dependent species as increasing numbers of hosts were extinguished. They estimated that at least two hundred species extinctions have already occurred via this mechanism and that a further 6,300 affiliate species are "co-endangered"—that is, likely to go extinct if their currently endangered hosts become extinct (fig. 18-4). Extending these models further, Dunn et al. (2009) estimated that the number of potential extinctions of parasite species could be an order of magnitude greater than that of hosts; extinction of species of five North American carnivores was estimated to result in the coextinction of fifty-six parasite species.

Estimates of potential coextinction in helminth parasites illustrate the magnitude of the potential impact of future change. Dobson et al.

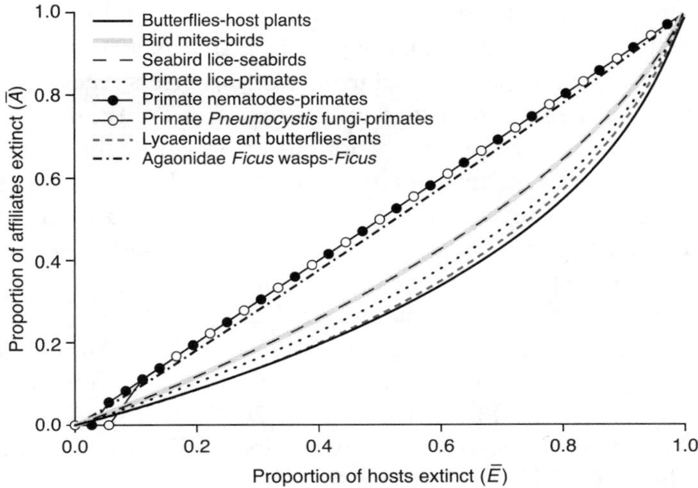


FIGURE 18-4. Proportion of affiliate species expected to go extinct through co-extinction for a given proportion of host extinction in eight affiliate-host systems (Koh et al., 2004b).

(2008) estimated that there are between 75,000 and 300,000 helminth species parasitizing vertebrates, and of these, 3–5 percent are threatened with extinction in the next 50–100 years. Avian extinctions are projected to be a major driver of parasite extinctions because the bulk of helminth diversity occurs in birds. Rare and specific tropical parasites were predicted to be lost rapidly as tropical bird species decline. However, common parasite species that can use a range of host species in the temperate zone may be significantly buffered against extinction (Stork and Lyal, 1993; Bush and Kennedy, 1994). This suggests that the relationship between loss of host species and loss of parasite species will tend to be concave (Koh et al., 2004b). At best, the relationship may be sigmoidal in shape, with the point of inflexion determined by the relative proportion of species that are host-specific (Dobson et al., 2008).

### *Food Web and Community Models*

Structural comparisons of the networks formed by interacting species are beginning to offer useful insights into the sensitivity of these networks to perturbations such as climate change (Tylianakis et al., 2008; Allesina et al., 2009; Dupont et al., 2009). Networks of mutualists such as plants and their pollinators or dispersers tend to show a nested structure that is robust to species loss. In contrast, networks of species with antagonistic interactions (predator-prey, host-parasite) appear to be more compartmentalized, with each compartment including a group of strongly interacting plants and animals, but few interactions among different compartments (Tylianakis et al., 2008). These types of networks may be more sensitive to disruption if particular species, especially top predators, are lost.

Topological analysis of natural food webs (e.g., Dunne et al., 2004; Memmot et al., 2007) and local stability analysis of model food webs (e.g., Pimm, 1980; Jonsson et al., 2010) also show that connectance of an ecological community and its trophic complexity affect its response to species loss. These studies have been extended using methods of permanence analysis (Ebenman et al., 2004) to investigate the effects of both direct and indirect interactions on the response of communities to species loss. Although not specifically directed at assessing the impacts of climate change, these studies nonetheless provide useful insights as to which communities may be most vulnerable in the future. Eklöf and Ebenman (2006), for example, showed that complex communities are, on average, more resistant to species loss than simple communities; the number of secondary extinctions decreases with increasing connectance (fig. 18-5).

## Conclusions

More than 30 years ago, Janzen (1974) raised the specter of the "most insidious sort of extinction, the extinction of ecological interactions." Predicting the direct effects of future changes in atmospheric carbon dioxide and climate for even single, well studied species is a challenging task. Predicting future impacts on the multitudinous interactions among species is more difficult by orders of magnitude. For some relationships, current knowledge about cues that trigger components of the life cycle, or factors that determine range margins, mean that decoupling of present-day relationships can be predicted with some confidence. Similarly, a lack of dramatic impacts on generalized, diffuse relationships, such as those between most plants and their pollinators, may also be expected. However, anticipating the effects of novel spe-

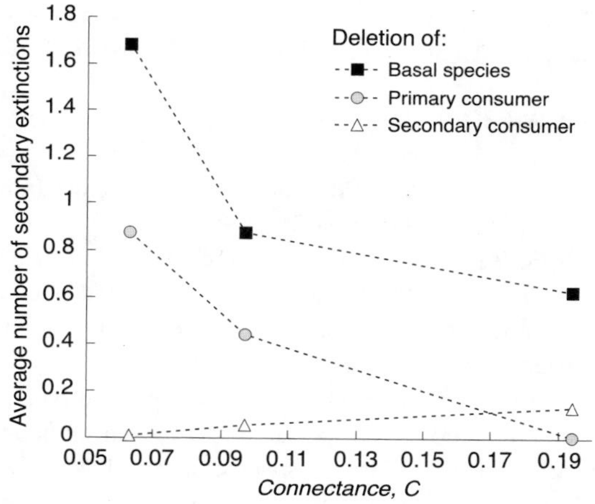


FIGURE 18-5. Average number of secondary extinctions as a function of web connectance and trophic level of deleted species. Results for rectangular web without omnivory. Each data point is an average calculated from 4,000 deletion events (Eklöf and Ebenman, 2006).

cies combinations is largely speculative and likely to remain so until far greater research emphasis is given to measuring impacts on multispecies communities.

## REFERENCES


Acevedo-Whitehouse, K., and Duffus, A. L. J. 2009. "Effects of environmental change on wildlife health." *Philosophical Transactions of Royal Society B-Biological Sciences* 364: 3429–3438.

Allesina, S., Bodini, A., and Pascual, M. 2009. "Functional links and robustness in food webs." *Philosophical Transactions of the Royal Society B-Biological Sciences* 364: 1701–1709.

Altizer, S., Nunn, C. L., and Lindenfors, P. 2007. "Do threatened hosts have fewer parasites? A comparative study in primates." *Journal of Animal Ecology* 76: 304–314.

Anderson, R. M., and May, R. M. 1986. "The invasion, persistence and spread of infectious diseases within animal and plant populations." *Philosophical Transactions of the Royal Society of London Series B* 314: 533–570.

Araújo, M. B., and Luoto, M. 2007. "The importance of biotic interactions for modeling species distributions under climate change." *Global Ecology and Biogeography* 16: 743–753.

Araújo, M. B., Thuiller, W., and Pearson, R. G. 2006. "Climate warming and the decline of amphibians and reptiles in Europe." *Journal of Biogeography* 33: 1712–1728.

Baum, J. K., and Worm, B. 2009. "Cascading top-down effects of changing oceanic predator abundances." *Ecology* 78: 699–714.

Bawa, K. S., and Dayanandan, S. 1998. "Global climate change and tropical forest genetic resources." *Climatic Change* 39: 473–485.

Beebee, T. J. C. 1995. "Amphibian breeding and climate." *Nature* 374: 219–220.

Bloor, J. M. G., Barthes, L., and Leadley, P. W. 2008. "Effects of elevated $CO_2$ and N on tree-grass interactions: An experimental test using *Fraxinus excelsior* and *Dactylis glomerata*." *Functional Ecology* 22: 537–546.

Bond, W. J. 1995. "Effects of global change on plant-animal synchrony: Implications for pollination and seed dispersal in Mediterranean habitats." In *Global Change and Mediterranean-Type Ecosystems,* edited by J. M. Moreno and W. C. Oechel, 181–202. New York: Springer-Verlag.

Both, C., and Visser, M. E. 2001. "Adjustment to climate change is constrained by arrival date in a long-distance migrant bird." *Nature* 411: 296–298.

Both, C., van Asch, M., Bijlsma, R. G., van den Burg, A. B., and Visser, M. E. 2009. "Climate change and unequal phenological changes across four trophic levels: Constraints or adaptations?" *Journal of Animal Ecology* 78: 73–83.

Brandt, M. E., and McManus, J. W. 2009. "Disease incidence is related to bleaching extent in reef-building corals." *Ecology* 90: 2859–2867.

Brooks, D. R., and Hoberg, E. P. 2000. "Triage for the biosphere: The need and rationale for taxonomic inventories and phylogenetic studies of parasites." *Comparative Parasitology* 67: 1–25.

Bush, A. O., and Kennedy, C. R. 1994. "Host fragmentation and helminth parasites: Hedging your bets against extinction." *International Journal of Parasitology* 24: 1333–1343.

Byström, P., Karlsson, J., Nilsson, P., van Kooten, T., Ask, J., and Olofsson, F. 2007. "Substitution of top predators: Effects of pike invasion in a subarctic lake." *Freshwater Biology* 52: 1271–1280.

Coley, P. D. 1998. "Possible effects of climate change on plant/herbivore interactions in moist tropical forests." *Climatic Change* 39: 455–472.

Cushing, D. H. 1990. "Plankton production and year-class strength in fish populations: An update of the match/mismatch hypothesis." *Advances in Marine Biology* 26: 249–293.

Davis, A. J., Jenkinson, L. S., Lawton, J. H., Shorrocks, B., and Wood, S. 1998. "Making mistakes when predicting shifts in species range in response to global warming." *Nature* 391: 783–786.

de Castro, F., and Bolker, B. 2005. "Mechanisms of disease-induced extinction." *Ecology Letters* 8: 117–126.

Diamond, J. M. 1989. "Overview of recent extinctions." In *Conservation for the Twenty-First Century,* edited by D. Western, and M. C. Pearl, 37–41. New York: Oxford University Press.

Dobson, A. P., Lafferty, K. D., Kuris, A. M., Hechinger, R. F., and Jetz, W. 2008. "Homage to Linnaeus: How many parasites? How many hosts?" *Proceedings of the National Academy of Sciences* 105: 11482–11489.

Dunn, R. R., Harris, N. C., Colwell, R. K., Koh, L. P., and Sodhi, N. S. 2009. "The sixth mass coextinction: Are most endangered species parasites and mutualists?" *Proceedings of the Royal Society B-Biological Sciences* 276: 3037–3045.

Dunne, J. A., Williams, R. J., and Martinez, N. D. 2004. "Network structure and robustness of marine food webs." *Marine Ecology Progress Series* 273: 291–302.

Dupont, Y. L., Padron, B., Olesen, J. M., and Petanidou, T. 2009. "Spatio-temporal variation in the structure of pollination networks." *Oikos* 118: 1261–1269.

Durant, J. M., Hjermann, D. O., Ottersen, G., and Stenseth, N. C.. 2007. "Climate and the match or mismatch between predator requirements and resource availability." *Climate Research* 33: 271–283.

Ebenman, B., Law, R., and Borvall, C. 2004. "Community viability analysis: The response of ecological communities to species loss." *Ecology* 85: 2591–2600.

Eklöf, A., and Ebenman, B. 2006. "Species loss and secondary extinctions in simple and complex model communities." *Journal of Animal Ecology* 75: 239–246.

Forchhammer, M. C., Post, E., Stenseth, N. C., and Boertmann, D. M. 2002. "Long-term responses in arctic ungulate dynamics to changes in climatic and trophic processes." *Population Ecology* 44: 113–120.

Gerber, L. R., McCallum, H., Lafferty, K. D., Sabo, J. L., and Dobson, A. P. 2005. "Exposing extinction risk analysis to pathogens: Is disease just another form of density dependence." *Ecological Applications* 15: 1402–1414.

Gilg, O., Sittler, B., and Hanski, I. 2009. "Climate change and cyclic predator-prey population dynamics in the high Arctic." *Global Change Biology* 15: 2634–2652.

Gould, E. A., and Higgs, S. 2008. "Impact of climate change and other factors on emerging arbovirus diseases." *Transactions of the Royal Society of Tropical Medicine and Hygiene* 103: 109–121.

Guidetti, P. 2007. "Predator diversity and density affect levels of predation upon strongly interactive species in temperate rocky reefs." *Oecologia* 154: 513–520.

Guisan, A., and Thuiller, W. 2005. "Predicting species distributions: Offering more than simple habitat modeling." *Ecology Letters* 8: 993–1009.

Harmon, J. P., Moran, N. A., and Ives, A. R. 2009. "Species response to environmental change: Impacts of food web interactions and evolution." *Science* 323: 1347–1350.

Harrison, R. D. 2000. "Repercussions of El Niño: Drought causes extinction and the breakdown of mutualism in Borneo." *Proceedings of the Royal Society of London Series B* 267: 911–915.

Harvell, C. D., Mitchell, C. E., Ward, J. R., Altizer, S., Dobson, A. P., Ostfeld,

R. S., and Samuel, M. D. 2002. "Climate warming and disease risks for terrestrial and marine biota." *Science* 296: 2158–2162.

Hegland, S. J., Nielsen, A., Lazaro, A., Bjerknes, A.-L., and Totland, O. 2009. "How does climate warming affect plant-pollinator interactions?" *Ecology Letters* 12: 184–195.

Heino, M. 1998. "Noise colour, synchrony and extinctions in spatially structured populations." *Oikos* 83: 368–375.

Hughes, L. 2000. "Biological consequences of global warming: Is the signal already apparent?" *Trends in Ecology and Evolution* 15: 56–61.

Inouye, D. W., Barr, B., Armitage, K. B., and Inouye, B. D. 2000. "Climate change is affecting altitudinal migrants and hibernating species." *Proceedings of the National Academy of Sciences, USA* 97: 1630–1633.

IPCC. 2001. "Climate Change 2001: The Scientific Basis." Technical summary from Working Group 1, Geneva: Intergovernmental Panel on Climate Change.

Janzen, D. H. 1974. "The deflowering of Central America." *Natural History* 83: 48–53.

Janzen, D. H. 2003. "How polyphagous are Costa Rican dry forest saturniid caterpillars?" In *Arthropods of Tropical Forests. Spatio-Temporal Dynamics and Resource Use in the Canopy,* edited by Y. Basset, V. Novotny, S. E. Miller, and R. L. Kitching, 369–379 Cambridge, UK: Cambridge University Press.

Jonsson, T., Karlsson, P., and Jonsson, A. 2010. "Trophic interactions affect the population dynamics and risk of extinction of basal species in food webs." *Ecological Complexity* 7: 60–68.

Justine, J.-L. 2007. "Parasite biodiversity in a coral reef fish: Twelve species of monogeneans on the gills of the grouper *Epinephelus maculates* (Perciformes: Serransidae) off New Caledonia, with a description of eight new species of *Pseudorhabdosynchus* (Monogenea Diplectanidae)." *Systematic Parasitology* 66: 81–129.

Kanowski, J. 2001. "Effects of elevated $CO_2$ on the foliar chemistry of seedlings of two rainforest trees from north-east Australia: Implications for folivorous marsupials." *Austral Ecology* 26: 165–172.

Keith, D. A., Akçakaya, R., Thuiller, W., Midgley, G. F., Pearson, R. G., Phillips, S. J., Regan, H. M., Araújo, M. B., and Rebelo, T. G. 2008. "Predicting extinction risks under climate change: Coupling stochastic models with dynamic bioclimatic habitat models." *Biology Letters* 4: 560–563.

Koh, L. P., Sodhi, N. S., and Brook, B. W. 2004a. "Co-extinctions of tropical butterflies and their hostplants." *Biotropica* 36: 272–274.

Koh, L. P., Dunn, R. R., Sodhi, N. S., Colwell, R. K., Proctor, H. C., and Smith, V. S. 2004b. "Species coextinctions and the biodiversity crisis." *Science* 305: 1632–1634.

Kovats, R. S., Campbell-Lendrum, D. H., McMichael, A. J., Woodward, A., and Cox, J. S. H. 2001. "Early effects of climate change: Do they include changes in vector-borne disease?" *Proceedings of the Royal Society of London Series B* 356: 1057–1068.

Lafferty, K. D. 2009. "The ecology of climate change and infectious diseases." *Ecology* 90: 888–900.

Martin, T. E. 2007. "Climate correlates of 20 years of trophic changes in a high-elevation riparian system." *Ecology* 88: 367–380.

Memmott, J., Craze, P. G., Waser, N. M., and Price, M. V. 2007. "Global warming and the disruption of plant-pollinator interactions." *Ecology Letters* 10: 710–717.

Menzel, A., Sparks, T. H., Estrella, N., and Roy, D. B. 2006. "Altered geographic and temporal variability in phenology in response to climate change." *Global Ecology and Biogeography* 15: 498–505.

Mitchell, R. J., Flanagan, R. J., Brown, B. J., Waser, N. M., and Karron, J. D. 2009. "New frontiers in competition for pollination." *Annals of Botany* 103: 1403–1413.

Mortensen, H. S., Dupont, Y. L., and Olesen, J. M. 2008. "A snake in paradise: Disturbance of plant reproduction following extirpation of bird flower-visitors on Guam." *Biological Conservation* 141: 2146–2154.

Mouritsen, K. M., Tompkins, D. M., and Poulin, R. 2005. "Climate warming may cause a parasite-induced collapse in coastal amphipod populations." *Oecologia* 146: 476–483.

Patz, J. A., Graczyk, T. K., Geller, N., and Vittor, A. Y. 2000. "Effects of environmental change on emerging parasitic diseases." *International Journal for Parasitology* 30: 1395–1405.

Philippart, C. J. M., van Aken, H. M., Beukema, J. J., Bos, O. G., Cadee, G. C., and Dekker, R. 2003. "Climate-related changes in recruitment of the bivalve *Macoma balthica*." *Limnology and Oceanography* 48: 2171–2185.

Phillips, O. L., Martinez, R. V., Arroyo, L., Baker, T. R., Killeen, T., Lewis, S. L., Malhi, Y., et al. 2002. "Increasing dominance of large lianas in Amazonian forests." *Nature* 418: 770–774.

Pimm, S. L. 1980. "Food web design and the effect of species deletion." *Oikos* 35: 139–149.

Pitman, N. C. A., and Jorgensen, P. M. 2002. "Estimating the size of the world's threatened flora." *Science* 298: 989.

Poloczanska, E. S., Kawkins, S. J., Southward, A. J., and Burrows, M. T. 2008. "Modeling the response of populations of competing species to climate change." *Ecology* 89: 3138–3149.

Post, E., and Forchhammer, M. C. 2004. "Spatial synchrony of local populations has increased in association with the recent Northern Hemisphere climatic trend." *Proceedings of the National Academy of Sciences, USA* 101: 9286–9290.

Poulin, R., and Morand, S. 2004. *Parasite Biodiversity*. Washington, D.C.: Smithsonian Institution Books.

Poulin, R., and Mouritsen, K. N. 2006. "Climate change, parasitism and the structure of intertidal ecosystems." *Journal of Helminthology* 80: 183–191.

Pounds, J. A., Bustamante, M. R., Coloma, L. A., Consuegra, J. A., Fogden, M. P. L., Foster, P. N., La Marca, E., et al. 2006. "Widespread amphibian extinctions from epidemic disease driven by global warming." *Nature* 439: 161–167.

Preston, K. E. L., Rotenberry, J. T., Redak, R. A., and Allen, M. F. 2008. "Habitat shifts of endangered species under altered climate conditions: Importance of biotic interactions." *Global Change Biology* 14: 2501–2515.

Rall, B. C., Vucic-Pestic, O., Ehnes, E. B., Emmerson, M., and Brose, U. 2010. "Temperature, predator-prey interaction strength and population stability." *Global Change Biology* 16: 2145–2157.

Rosenzweig, C., Karoly, D., Vicarelli, M., Neofotis, P., Wu, Q., Casassa, G., Menzel, A., et al. 2008. "Attributing physical and biological impacts to anthropogenic climate change." *Nature* 453: 353–357.

Schweiger, O., Settele, J., Kudrna, O., Klotz, S., and Kühn, I. 2008. "Climate change can cause spatial mismatch of trophically interacting species." *Ecology* 89: 3472–3479.

Smith, K. F., Sax, D. F., and Lafferty, K. D. 2006. "Evidence for the role of infectious disease in species extinction and endangerment." *Conservation Biology* 20: 1349–1357.

Stacey, D. A., and Fellowes, M. D. E. 2002. "Influence of elevated $CO_2$ on interspecific interactions at higher trophic levels." *Global Change Biology* 8: 668–678.

Stiling, P., Cattell, M., Moon, D. C., Rossi, A. M., Hungate, B. A., Hymus, G., and Drake, B. 2002. "Elevated atmospheric $CO_2$ lowers herbivore abundance, but increases leaf abscission rates." *Global Change Biology* 8: 658–667.

Stireman, J. O. III, Dyer, L. A., Janzen, D. H., Singer, M. S., Lill, J. T., Marquis, R. J., Ricklefs, R. E., et al. 2005. "Climatic unpredictability and parasitism of caterpillars: Implications of global warming." *Proceedings of the National Academy of Sciences, USA* 102: 17384–17387.

Stork, N. E., and Lyal, C. H. C. 1993. "Extinction or 'coextinction' rates." *Nature* 366: 307.

Strong, D. R., Lawton, J. H., and Southwood, T. R. E. 1984. *Insects on Plants*. Oxford, UK: Blackwell Scientific Publications.

Sures, B. 2003. "Accumulation of heavy metals by intestinal helminthes in fish: An overview and perspective." *Parasitology* 126: 553–560.

Swarts, N. D., and Dixon, K. W. 2009. "Terrestrial orchid conservation in the age of extinction." *Annals of Botany* 104: 543–556.

Terbough, J. 1986. "Keystone plant resources in the tropical forest." In *Conservation Biology: The Science of Scarcity and Diversity,* edited by M. E. Soulé, 330–344. Sunderland, MA: Sinauer Associates.

Thackeray, S. J., T. H. Sparks, M. Frederiksen, S. Burthe, P. J. Bacon, J. R. Bell, M. S. Botham, et al. 2010. "Trophic level asynchrony in rates of phonological change for marine, freshwater and terrestrial environments." *Global Change Biology* 16: 3304–3313.

Traill, L. W., Lim, M. L. M., Sodhi, N. S., and Bradshaw, C. J. A. 2010. "Mechanisms driving change: Altered species interactions and ecosystem function through global warming." *Journal of Animal Ecology* 79: 937–947.

Tylianakis, J. M., Didham, R. K., Bascompte, J., and Wardle, D. A. 2008. "Global change and species interactions in terrestrial ecosystems." *Ecology Letters* 11: 1351–1363.

van Franeker, J. A., Creuwels, J. C. S., Van Der Veer, W., Cleland, S., and Robertson, G. 2001. Unexpected effects of climate change on the predation of Antarctic petrels. *Antarctic Science* 13: 430–439.

Visser, M. E., Both, C., and Lambrechts, M. M. 2004. "Global climate change leads to mistimed avian reproduction." *Advances in Ecological Research* 35: 89–110.

Visser, M. E., and Holleman, L. J. M. 2001. "Warmer springs disrupt the synchrony of oak and winter moth phenology." *Proceedings of the Royal Society of London Series B-Biological Sciences* 268: 289–294.

Visser, M. E., van Noordwijk, A. J., Tinbergen, J. M., and Lessells, C. M. 1998. "Warmer springs lead to mistimed reproduction in great tits (*Parus major*)." *Proceedings of the Royal Society of London Series B* 265: 1867–1870.

Vucic-Pestic, O., Ehnes, R. B., Ral, B. C., and Brose, U. 2010. "Warming up the system: Higher predator feeding rates but lower energetic efficiencies." *Global Change Biology* 17: 1301–1310. doi:10.1111/j.1365-2486.2010.02329.x

Waser, N. M., and Ollerton, J. 2006. *Plant-Pollinator Interactions from Specialization to Generalization*. Chicago: University of Chicago Press.

Wilmers, C. C., and Getz, W. M. 2005. "Gray wolves as climate change buffers in Yellowstone." *PLoS Biology* 3: 571–576.

Wilmers, C. C., Post, E., Peterson, R. O., and Vucetich, J. A. 2006. "Predator-disease outbreak modulates top-down, bottom up and climatic effects on herbivore population dynamics." *Ecology Letters* 9: 383–389.

Winder, M., and Schindler, D. E. 2004. "Climate change uncouples trophic interactions in an aquatic ecosystem." *Ecology* 85: 2100–2106.

Wright, S. J., Carrasco, C., Calderon, O., and Paton, S. 1999. "The El Niño Southern Oscillation, variable fruit production, and famine in a tropical forest." *Ecology* 80: 1632–1647.

# PART VI

# Conservation Implications

And now for the good news. The good news is that, paradoxically, because we have done so little to prepare for conservation in the face of climate change, there is much room to reduce the number of potential extinctions it may cause. Solutions to the problem must come from both sides—adaptation and mitigation.

Jessica Hellman, Vicky Meretsky, and Jason MacLachlan look at closing the gap in conservation strategies. They examine ways of making protected areas, ecosystem services, and species more robust to the changes in climate that are sure to come. They point out that as change becomes more severe, the adaptive capacity for conservation and nature will be exhausted, and that the longer we wait to adapt the more extreme and more expensive solutions will become. The book concludes in chapter 20 with a look at the overall prospects for reducing extinction risk of species, with particular emphasis on the need to reduce global greenhouse gas emissions and to have binding international agreement on action to reduce human-caused climate change.

The actions outlined in the two chapters of this part are not simple. They are politically complex, and so far have largely eluded cooperative international action. Nonetheless, they are not out of reach—the technical knowledge needed to reduce greenhouse gas emissions is largely available. The actions needed for adaptation of conservation strategies are increasingly clear. What *is* needed is political willpower and concerted global action to avoid unacceptable consequences for people, the ecosystems on which they depend, and nature.

## Chapter 19

# *Strategies for Reducing Extinction Risk under a Changing Climate*

JESSICA J. HELLMANN, VICKY J. MERETSKY, AND JASON S. MCLACHLAN

In a world characterized by increasing mean temperatures and a higher frequency of climatic extremes, species will reorganize their geographic distributions to track changing conditions, evolve new environmental tolerances, or risk local or global extinction. Limitations on rapid evolutionary change for most organisms of conservation concern suggest that range shifts are the most feasible mechanism for evading extinction under climate change, and migration was a common mode of species response following postglacial warming (Davis and Shaw, 2001). Changes in the geographic ranges of species will cause changes in communities, altering biological systems as we know them and affecting ecosystem services and functions. Central to these ecological changes are changes in cultural, economic, and aesthetic values.

In this chapter, we outline five strategies for conservation under climate change—strategies to avoid species-level extinction and the loss of genetic diversity within a species. The concept of biodiversity management under climate change could include a broad range of biodiversity values, such as aesthetics of maintaining an ecosystem type (e.g., forest versus grassland) without concern for individual species identity, or one might be interested in species-level or genotypic-level management not for the sake of conservation but for maximization of an economically valuable good or service. Achieving these goals also will involve aspects of the five strategies presented here. However, we hope that efforts to preserve ecosystems do not undermine actions that target genetic preservation (Chan et al., 2006; Naidoo et al., 2008; Ranganathan et al., 2008).

We do not make recommendations of which strategies should be used because much remains to be studied before this issue can be addressed comprehensively. Instead, we qualitatively compare various strategies to point out complementary and conflicting elements. Table 19-1, for example, describes steps of implementation for each strategy to facilitate comparison and contrast. The main purpose of this chapter is to provide a single place where alternatives are characterized and, at least initially, compared.

## The Need for Conservation Action

We start our review of alternate strategies by considering the need for conservation action that specifically addresses threats due to climate change. The ways climate change might affect species conservation are numerous and mostly poorly understood. Species may change their phenology, with potentially cascading impacts on other species (Walther et al., 2002). The effect of biotic interactions on the abundance of species of interest, including the impact of disease, may change in unanticipated ways as climate changes (Pound et al., 2006; Moorcroft et al., 2006). And there are many other potential nonlinearities as the direct impact of climate on species interacts with other anthropogenic drivers of environmental change, such as habitat fragmentation.

Given the magnitude of impending climate change, however, it is certain that many species will be displaced hundreds of kilometers from their current range (Williams et al., 2007). This has important implications for conservation strategies, which we address directly below. The threat of species displacement is, of course, mitigated by the capacity of species to naturally disperse to safe new habitats. Some species are obviously capable of such translocations, but the extent of natural dispersal is unknown for many species.

Even fecund species with special adaptations for dispersal and large populations may be more dispersal-limited than we might guess. For example, red maple—a widespread tree in eastern North America with broad environmental tolerances, a relatively short time to

Table 19-1. Steps necessary for implementation of five strategies for conserving species, genetic diversity, or species complexes under climate change.

| | *1. Traditional Approach* | *2. Managing Resistance* | *3. Corridors* | *4. Managed Relocation* | *5. Ex situ* |
|---|---|---|---|---|---|
| **Define goal:** | Maximize population size of target species to facilitate natural range change or preserve as long as possible so that alternate strategy(ies) could apply in the future. | | Support passage of specified target species. Also identify which undesirable species might use corridors. | Select target species and time horizon. | |
| **Identify strategy:** | Could include habitat acquisition, modification, or restoration, and/or actions to affect interacting species (not involving major site change; see #2). | Could include prioritizing populations in climate refugia, major site modification (e.g., altering hydrology), use of pesticide or other treatments to affect preda- | Identify available lands and consider necessary width, land cover, and need for restoration. Evaluate necessary rate of spread and if that rate is feasible. | Could include restoration-like establishment with cultivation or widespread seeding with little cultivation. Could involve one-time introduction or re- | Could include captive breeding, breeding for potentially adaptive traits, and/or propagule banking. Choose number of propagules or individuals. Determine strategy to ensure minimal loss of adaptive diversity |

TABLE 19-1. Continued.

| | *1. Traditional Approach* | *2. Managing Resistance* | *3. Corridors* | *4. Managed Relocation* | *5. Ex situ* |
|---|---|---|---|---|---|
| **Identify strategy cont.:** | | tors or competitors, or cultivation (e.g., watering, fertilizing) | | peated implementation over decades. Choose based on colonization parameters (e.g., population growth rate, likelihood of establishment), and desired spatial extent and population size. | (Menges et al., 2004); consider ex situ wild exchange to maintain genetic diversity or need to augment with later collections. |

| | | | | | |
|---|---|---|---|---|---|
| **Identify target region:** find climatically suitable region for future occupancy with suitable growing conditions (e.g., soil type, obligate mutualists). Target region could move over time with changing conditions. | n/a | n/a | Determine where connectivity is needed for target species to reach target region. | If precise target region not known, propagules could be distributed widely. If climate but not growing conditions are suitable, introductions of additional species may be necessary. | Determine ex situ location. If precursor to later managed relocation (MR), identify sites likely to serve as future introduction sites. |
| **Identify source population:** | n/a | n/a | Identify populations least likely to persist under climate change and determine if these genotypes are suitable for target region. | Source(s) must be able to tolerate removal of individuals. For MR or later anticipated MR, source must have proper genotype for target region. | |
| **Collect propagules:** | n/a | n/a | | Determine the frequency of introduction needed. If risk of take is high, captive breeding may be required. Determine number of propagules needed and strategy for capture with minimal collateral effects. | |

TABLE 19-1. Continued.

| | *1. Traditional Approach* | *2. Managing Resistance* | *3. Corridors* | *4. Managed Relocation* | *5. Ex situ* |
|---|---|---|---|---|---|
| **Population establishment:** | n/a | n/a | n/a | Determine method of release at new site(s) (Guerrant et al., 2004; Seddon et al., 2007). Effort per site in site preparation might be scaled with number of sites in which introduction is attempted. Where multiple sites are used and appropriate methods are uncertain, replicate establishment attempts can provide valuable information. | If precursor to MR, consider later population establishment when ex situ conservation is initiated. |

| | | | | |
|---|---|---|---|---|
| **Monitoring and indicators of success and failure:** Evaluate program success and developing risks. Determine indicators/-thresholds for choosing an alternate strategy or augmenting with additional strategies. | Monitor reproduction and recruitment to determine if population is increasing, is stable, or if decrease has been slowed (i.e., buying time). | Monitor movements or spread of individuals to determine if corridor is used. Set benchmarks for adequate passage rates. Monitor corridor to detect unwanted species or disturbances and determine benchmarks for these that will signal a corridor that has become a risk rather than a benefit. | If population too successful, control before becomes invasive. If introduction not successful, supplemental introductions may be needed. | For propagule banking, consider viability testing if gene pool is being protected (Baskin and Baskin, 2004). Continue to monitor source populations for possible in situ conservation and for comparison with ex situ gene pool and individual fitness. |

reproduction, and abundant wind-dispersed seeds—would seem to have a high capacity for spread under future warming. If we assume a doubling of carbon dioxide within 50 years (Intergovernmental Panel on Climate Change scenario A2: IPCC, 2007), figure 19-1 shows that the average rate of spread of red maple is less than 5 kilometers (3 miles). These rates are based on model results from Clark et al. (2003), using measured life history traits and seed dispersal. A high net reproductive rate ($R_0$ = 1,325) was based on recent increases of red maple abundance in eastern forests. Even under the more optimistic IPCC B1 scenario of a doubling of carbon dioxide within 100 years, red maple spreads less than 10 kilometers (6.2 miles). Faster rates may be possible under fortuitous circumstances (Nathan, 2006), but natural dispersal is likely to limit spread. Even when rapid capacity is possible, inherent stochasticity in the underlying population processes make predicting these rates impossible (Clark et al., 2003).

At whatever rate species' ranges shift to accommodate changing climate, dispersal into new territory (and concomitant range retreats elsewhere) has important consequences for genetic diversity. The same

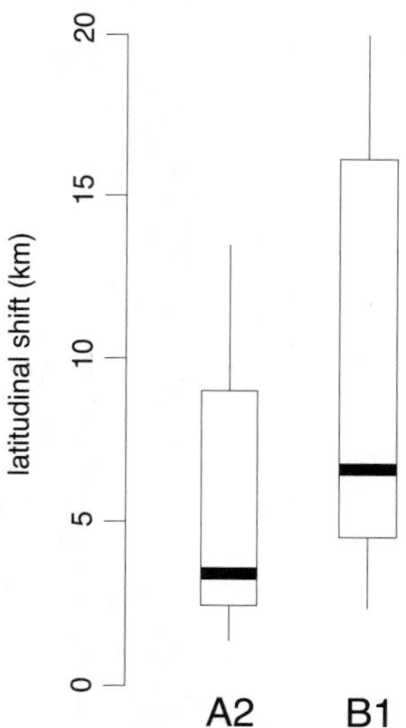


FIGURE 19-1. Total range shifts expected for natural dispersal of red maple by the time of a doubling of atmospheric carbon dioxide under the IPCC A2 and B1 emissions scenarios; based on Clark et al. (2003). Thick line shows the mean expectation of shift. Boxes indicate 50% confidence intervals and whiskers indicate 90% confidence intervals.

long-distance dispersal that might allow species to match the rate of shifting climate envelopes is likely to result in strong founder effects that generally reduce genetic diversity in newly colonized territory (Hewitt, 1996). Furthermore, genotypes in areas of range retreat may be lost (McLaughlin et al., 2002; Hampe and Petit, 2005). The geographic distribution of adaptive (ecotypic) variation in most species is largely unknown, but we have increasingly rich understanding of the distribution of neutral genetic data gathered from the field of phylogeography (e.g., Soltis et al., 2006). We can use such data to estimate how much genetic diversity is threatened by range shifts due to climate change.

Consider how far the southern edge of species' ranges in North America would have to recede northward before major clades in species' phylogenies are put at risk. Using twenty published studies of broadly distributed species spanning a wide range of taxonomic groups with phylogeographic data (references in Soltis et al., 2006; plus Zakharov and Hellmann, 2008), we measured the distance from the southern range limit of each species to the northern range limit of each major clade in that species, using clades described in the original citation. We then calculated the proportion of major clades that would be at risk of extinction under scenarios of a northward retreat ranging from 100 to 1,000 kilometers (62 to 620 miles).

This quick calculation reveals that significant portions of a species's genetic diversity could be lost if local extinctions predominate where local genotypes are concentrated. Many of the species studied would lose genetic diversity if their southern territory shrank by as little as 300 kilometers (186 miles) (fig. 19-2a). Figure 19-2b illustrates the potential for loss of genetic diversity in an example species, *Liriodendron tulipifera*. Sewell et al. (1996) identified three electrophoretic groups in an allozyme analysis of *L. tulipifera,* suggesting that if the southern boundary retreated 180 kilometers (112 miles) (A in fig. 19-2b), the southernmost clade would be at risk of extinction. A retreat of 650 kilometers (404 miles) would put the next most southerly clade at risk (B in fig. 19-2b).

Of course, the extent to which a loss of genetic diversity is of conservation concern depends on its functional significance (Hedrick, 2001; Kohn et al., 2006). In *Liriodendron tulipifera,* for example, the "peninsular" genotype (black circles in fig. 19-2b) is so divergent from other genotypes that it may have originally been a separate isolated species that introgressed with "upland" genotypes (white circles) during Pleistocene range shifts (Parks et al., 1994; Sewell et al., 1996). It

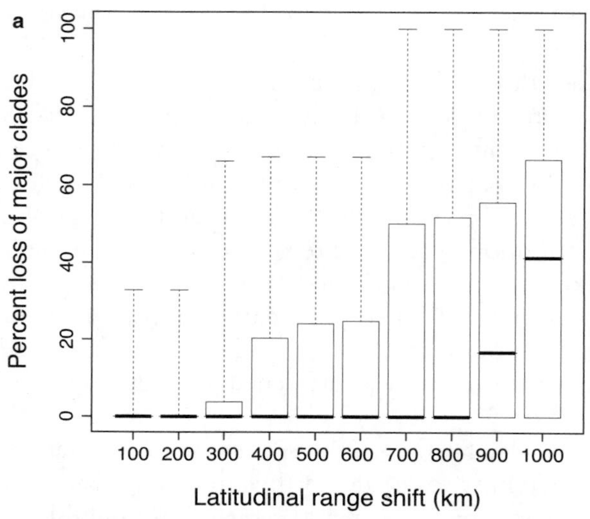


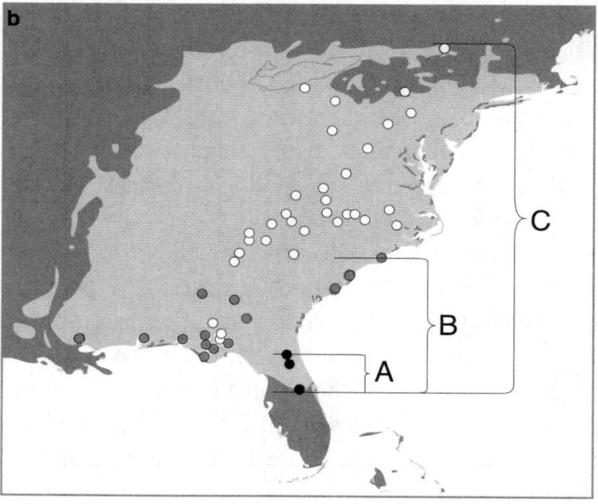


FIGURE 19-2. (a) The loss of genetically distinct clades for twenty species with increasing distance within 1,000 kilometers (620 miles) of a retreating southern range boundary (see text). Thick line indicates the median value; the box is the interquartile range; and whiskers show the range. (b) The distribution of major electrophoretic groups in *Liriodendron tulipifera* (adapted from Sewell et al., 1996). Black circles are the highly divergent peninsular group, gray circles are Sewell et al.'s "coastal intermediate group," and white circles are the common "upland group." The range of *Liriodendron* is depicted in light gray. (A), (B), and (C) are the distances estimated to threaten 33 percent, 67 percent, and 100 percent, respectively, of major clades in this species.

has distinct morphological and ecological features. This variant occurs in a different habitat and with different floristic associations than more northerly variants (Parks et al., 1994). Thus, this may be the sort of functional genetic diversity that we wish to conserve under climate change.

The analyses above suggest that individual taxa, particularly at the level of populations and species, are highly likely to be threatened, even if we have little information about the total percentage of species endangered. Given the interaction of climate change with habitat fragmentation, a high percentage of species may also experience range contraction. As well, genetic diversity is highly likely to be endangered, even in widespread species. Thus, managers should consider the effects of conservation strategies for climate change on genetic as well as species preservation.

## Five Alternate Strategies

Given this background, we now consider five alternate and nonexclusive approaches to conservation under a changing climate with the aim of stimulating discussion and directed research.

### *Established Approaches to Conservation Biology*

Traditional conservation biology encompasses a suite of practices that addresses a wide range of threats other than climate change. In the past, these activities largely considered in situ solutions to in situ problems. In addressing climate change, one would use traditional conservation to reduce other threats so that target individuals can expend more time and energy in coping with climate change. This assumes populations can withstand increased mortality from climate change if mortality from other factors has declined. Common management activities for in situ protection include removal of invasive species, protection from herbivory or predation, and management of disturbance regimes. Land acquisition can enlarge in situ habitat to support larger populations and reduce stochastic risks to small or fragmented populations. Under climate change, larger populations also provide a greater number of colonists to establish new populations naturally through translocation.

There also are established efforts to provide landscape connectivity through land acquisition. These efforts provide stepping stones or continuous corridors to connect larger habitat blocks and enhance effective habitat area (Tewksbury et al., 2002; Bennett, 2003). Such acquisitions also enhance opportunities for species to modify their ranges in response to climate change (see below).

Established approaches are well understood by managers and grow out of existing conservation actions that benefit many species; no special training or public education is needed to extend their use to conservation against climate-change impacts (table 19-1). However, these business-as-usual approaches may not be sufficient to achieve biodiversity conservation objectives (e.g., see McLachlan et al., 2007). More ambitious and more specific actions may be needed to account for the pace of climate change and the needs of particularly vulnerable populations and species.

Managers using established approaches to offset climate-change impacts must add climate-mediated threats to the formal or informal risk assessments they already conduct for managed species. Monitoring will be needed to detect population trends, and thresholds that signal a stressor that has previously been compatible with species' persistence (e.g., competition with an invasive species) should be reduced given potential interactions with climate change. In situ conservation for species that have been successfully conserved historically may need to be renewed or greater protection may need to be afforded to minimize new risk from climate change (Root and Schneider, 2002).

Research needed to guide in situ conservation is generally the same research needed to guide all biodiversity conservation under climate change: improved climate forecasting at local and regional scales and improved understanding of tolerances of species to climate change (including interactions of climate-change tolerance with other threats) (Botkin et al., 2007; Pelini et al., 2009). In particular, information on tolerances of apparently vulnerable species would improve triage and avoid unnecessary efforts aimed at species that appear to be at risk but that can, in fact, tolerate climate change (Dawson et al., 2011).

### *Strategically Managing for Resistance*

For species at the northern or upper extent of their habitat (e.g., mountaintop species, sea-ice–dependent marine mammals), species with habitat too fragmented to permit successful range extension, or species with narrowly defined habitats that are not available outside their present range, range extension is unlikely, leaving aggressive in situ (or ex situ, see below) conservation as the only option(s). In these cases, managers might pursue in situ conservation much more intensively than traditional approaches generally permit (Parmesan, 2006) (table 19-1).

Managing for resistance to climate change could involve aggressive control of competitors and predators or the use of pesticides, herbicides, or exclosures (Millar et al., 2007). These practices are expensive and will likely need to be ongoing if, in fact, they are successful. Protection may require that managers not only enhance the ability of target species to resist climate change but also reduce potential competitor or predator species that are naturally extending their ranges into the management area, and prevent harmful changes to disturbance regimes that are an aspect of changing climate. In some cases, the intensity of such efforts may approach cultivation, essentially creating a wild-animal park or arboretum.

For species that can be protected in relatively small areas, climatic refugia may provide effective options for in situ conservation (Millar et al., 2007). Such refugia often already harbor species that require temperatures cooler than the local norm. Management actions, including manipulating hydrology, vegetation, and disturbance regimes, may enhance the effectiveness of small refugia for long-term conservation.

Intensive in situ conservation runs the continual and increasing risk of creating unsustainable tensions between the active maintenance of constant conditions and the changing climate and species complement of the surrounding landscape (Millar et al., 2007). Management will need to be increasingly heavy-handed and may come into conflict with public perceptions of appropriate land use or appropriate economic priorities. Finally, in situ efforts may fail, even catastrophically (if fire or flood control efforts fail, for example). However, in situ efforts may buy time for alternative strategies to be developed and may buy time for populations to adapt to changing climate.

Conservation of management-dependent endangered species—species that will never truly recover in the sense used in the Endangered Species Act—already employs intensive efforts such as those described above. The number of management-dependent species is likely to grow under climate change, particularly if conservation efforts are limited to in situ approaches. However, if in situ efforts are used in conjunction with the next two approaches, which attempt to extend species' ranges, long-term dependency on intensive efforts may be avoided.

### *Landscape Conservation to Facilitate Natural Range Change*

For relatively mobile species, facilitating natural range change is an obvious management approach (table 19-1). Managing the larger landscape to facilitate natural range change builds from earlier work on corridors (e.g., Bennett, 2003; Beier et al., 2008), and such landscape management could include a variety of types of connectivity. For example, in their discussion of conservation of migration, a problem very similar to geographic range change, Webster et al. (2002) emphasized habitat diversity and created the concept of "migration connectivity."

Landscapes that promote geographic range change under climate change must also have a broad ecological context, including widely spaced stepping stones, tightly focused travel paths, and continuously connected habitats. In some areas, for example, where high-intensity agriculture, industry, or urban development dominate, populations of conservation concern will be restricted to protected areas. Enabling these populations to move in response to climate change will require improving connectivity among protected areas (Williams et al., 2005; Phillips et al., 2008; Hole et al., 2009). In less modified landscapes, however, protected areas may not be the primary focus of connectivity; instead, a variety of public and private lands could support movement through a landscape. Given the time frame of climate change, enabling population movement via these mechanisms will not be a one-time event, but an ongoing process of land use planning for centuries to come.

Landscape management for geographic range change will require a very large spatial scale—a larger scale than the other strategies discussed in this chapter. Management at this scale, involving multiple watersheds, states, provinces, or countries, is rarely performed, likely due to the obvious difficulties (see, for example, Lindenmayer et al., 2010). However, climate change demands new attention to landscape conservation, and it will require unprecedented cooperation between public and private landowners. For example, in the United States, national wildlife refuges have been encouraged to think outside their boundaries in addressing their conservation responsibilities (Fischman, 2003), and several federal agencies have programs that provide modest support for conservation on private lands. But no agency has a legislative mandate or a budget that allows it to plan or to impose regional corridors on the landscape.

Mechanisms that encourage private landowners to participate in landscape conservation initiatives would significantly increase the chances for successful collaboration (Theobald and Hobbs, 2002; Brown and Harris, 2005). Existing public land networks are too sparse, and expansion through acquisition is too expensive to allow corridors to be public-lands-only initiatives. Programs that create easements or contractual conservation on grasslands, wetlands, and forests could be made more responsive to connectivity concerns through changes in the ranking processes used to determine which landowners will receive easement subsidies (Ribaudo et al., 2001; e.g., the US Farm Bill). The use of private lands for climate change mitigation (carbon sequestration) would provide additional funding possibilities for private land contributions to corridors.

In addition to the unprecedented scale of landscape management needed to enable natural range change under climate change, sessile and slowly advancing species are unlikely to be able to advance poleward or upward sufficiently rapidly to balance range loss due to climate change (fig. 19-2a). And of those that can advance quickly, not all are desirable: range expansion of invasive species and disease vectors also may be facilitated.

Presently, research is lacking to allow managers to predict which species will benefit from what kinds of landscape-scale conservation and to what extent.

## *Managed Relocation*

In some cases, we may feel a special responsibility to aid species whose spread is limited, and "managed relocation" (MR) may emerge as an acceptable conservation option. We follow Richardson et al. (2009) in using "managed relocation" over "assisted colonization" or "assisted migration" (Hunter, 2007; Hoegh-Guldberg et al., 2008) for its neutral connotation, its application to both facilitation and prevention of range change, and its full array of steps, from introduction through cultivation and, if necessary, subsequent control or removal. MR is an intervention strategy in which organisms are intentionally moved from areas of historical occupancy to new areas where they are projected to persist under emerging conditions (McLachlan et al., 2007; table 19-1).

The concept of MR is an old idea with a new twist. Humans have regularly moved species intentionally, often with the aim of advancing conservation or maximizing production of an ecosystem service (Griffith et al., 1989). However, MR differs from many of these cases by placing species *outside* their native ranges. The reason for action—potential losses due to climate change—also is a distinguishing characteristic of MR. On its surface, MR is a potentially appealing approach for overcoming fundamental constraints on natural range change, whether inherent or human-caused. However, MR applies only to individual genotypes or species. Groups of species might be moved in concert, but MR is generally not a strategy for community- or ecosystem-level conservation in the way that in situ approaches are, for example. In addition, the difficulty of implementing MR will make it costly, though perhaps less costly than land acquisition for traditional conservation or corridor construction (see table 19-1 and above).

Figure 19-3 illustrates the balance between the potential pros and cons of MR using species endangerment as an example motivation and the risk of introducing novel invasive species as the principal risk. If it achieves its goal, MR would reduce the number of species endangered due to climate change, as illustrated by the shaded area in figure 19-3A (Mueller and Hellmann, 2008). However, MR could further increase risk of invasion, as illustrated by the shaded area in figure 19-3B (ideally, however, species assisted in a migration program should not have features of invasive species). The difference in the shaded values of figures 19-3A and 19-3B captures the relative value of MR, and this value must be positive to warrant its pursuit. Of course, the simple comparison of endangerment (extinction) and invasion risk does not account for interactions such as the possibility that assisted species could cause native species to become endangered. An additional concern with MR is the risk to the original population, particularly if it is imperiled, of removing individuals for establishment elsewhere. A similar comparison of benefits and risk is necessary for MR activities motivated by ecosystem services (e.g., timber production), where benefits avoided are production losses instead of extinctions.

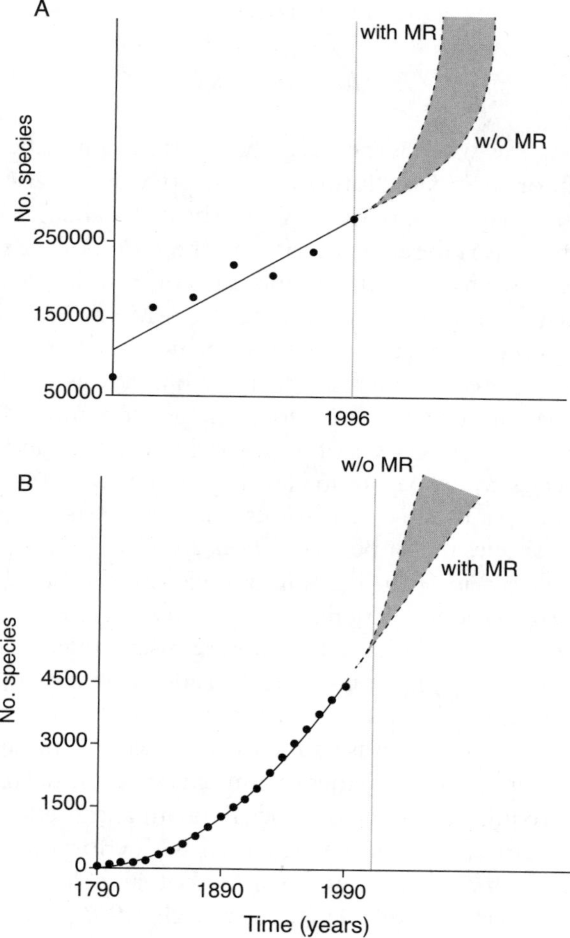


FIGURE 19-3. (A) Schematic illustrating historic and potential future trajectories of the number of endangered species (data from Dawson and Shogren, 2001). Managed relocation is assumed to reduce the number of species endangered by climate change (shaded area). Future projections (dashed lines) are concave-up, reflecting an accelerating rate of climate change under a business-as-usual scenario and assuming that climate change causes species endangerment in a proportional way. (B) Schematic illustrating invasive species (from Office of Technology Assessment, 1993) in the United States without and with MR (see text). Managed relocation is assumed to increase the number of invasive species (shaded area). Data missing for individual taxa were filled with linear interpolation. Future projections are linear to reflect a continuation of historical trends, assuming that the pool of potential invaders for North America remains large.

### *Ex Situ Conservation*

In those instances when a species or population cannot be conserved in its present or anticipated future range, captive propagation or propagule banking may be an option. Where the original range of the species is not its future range, the precise interpretation of "ex situ" is ambiguous; here, we use the term to indicate conservation in specialized facilities such as zoos and botanical gardens (Maunder et al., 2004). The limitations of captive propagation for wildlife conservation have been extensively documented, particularly limitations on the number of species that can be accommodated, the genetic impacts of captive propagation, the disease risks of multispecies facilities, and the failure of some species to propagate in captivity (Conway, 1986; Snyder et al., 1996). For plants, an optimistic estimate suggests that 60 percent of imperiled species might be accommodated (CBD, 2002).

If only a portion of the existing population is to be taken ex situ, careful demographic and genetic analysis is needed to ensure that both in situ and ex situ population fragments are sustainable. Triage for captive propagation often must take into consideration the impacts of the choice of species for the host institution: income and risk for zoos and botanical gardens may vary as a function of choice of species.

Ex situ propagation provides no impetus for eventual repatriation, which may require considerable social and financial effort if land is limiting. Ex situ efforts should be paired with efforts to ensure that repatriation occurs as soon as possible so that genetic impacts are limited and habitat is available (Snyder et al., 1996; Maunder et al., 2004). In some cases, this repatriation might ultimately constitute MR, where individuals are placed outside their historic range.

Even if ex situ conservation efforts can limit genetic impacts, initial propagule selection is still subject to sampling error of the genome, and genetic diversity may still be lost through differential viability (Guerrant and Fiedler, 2004). Furthermore, no propagule bank can guarantee the preservation of its material indefinitely.

## Choosing among Conservation Strategies

Conservation efforts under climate change would rarely be limited to a single approach. Traditional conservation approaches are often already in use at some scale, and may be adequate for many common, mobile, or human-tolerant species. Species well suited or initially limited to in situ conservation should be fairly easy to identify because there will be no other place available or readily available. For others, their present existence in a climate refugium may grant them some initial immunity to climate change or at least buy managers time to plan alternate strategies. In situ conservation may initially be straightforward and inexpensive except where new land acquisition or extensive habitat modification for strategic resistance is needed. Ex situ conservation is best used as a safety net, beginning early enough that collection of captive stock or propagules does not constitute a risk to the wild population. MR may be more widely useful than ex situ conservation, but many questions about MR remain unanswered (Richardson et al., 2009). Corridor establishment requires complex social coordination. Whereas all of these techniques may involve many partners, partnering in corridor establishment is essential at the scale climate change requires, and the choice of partners is fixed, in part, by the landscape. Population monitoring (including monitoring of population size and phenology) and availability of climate models on ecologically relevant scales are the biggest missing links we face in the support system for all approaches to biodiversity conservation. The National Phenology Network (www.usanpn.org) and the National Climate Change and Wildlife Center (nccw.usgs.gov) are working to fill the monitoring gap, but formal trend detection usually requires at least a decade of data.

## Conclusions

Biodiversity conservation under climate change is a staggeringly uncertain task. Adaptive management is the tool most recommended for efficient conservation in the face of uncertainty (Holling and Meffe, 1996), but its implementation to date is incomplete and tentative (e.g., Gunderson and Light, 2006; Jacobson et al., 2006; Walters 2006; but see also, e.g., Bormann et al., 2007; van der Brugge and van Raak, 2007).

Adaptive management requires monitoring of the species and ecosystems that management affects. Monitoring results feed into decision making, resulting in modified goals and actions whose results are in turn monitored, continuing the cycle (Holling, 1978). Few agencies or organizations have an institutional history of or funding for the level of monitoring needed to support activities such as intensive in situ conservation or MR. However, agencies making programmatic changes and practitioners employing the strategies described here to conserve species and communities in the face of climate change are all breaking new ground. Adaptive monitoring (Lindenmayer and Likens, 2009) will be needed to measure progress and to detect threshold conditions. Without such data, chances of success will be seriously diminished.

The flexibility needed to modify management in response to monitoring results is not easy for natural resources agencies that must remain accountable to the public as they continually modify their practices. Federal agencies are required to file National Environmental Policy Act (NEPA) documents, particularly environmental impact statements, that must specify actions in some detail (Mandelker, 2006–2008), and the level of effort required to produce an environmental impact statement presently makes it impractical to update these documents frequently. A change to more numerous, smaller impact statements, and other novel procedures are needed to allow adaptive management and the need to respond to new information.

As increasing numbers of agencies, nongovernmental organizations, volunteers, and researchers explicitly focus on climate change, the need for cross-organization communications also will increase. Formal research to test approaches and detect thresholds will not keep pace with information needs. As practitioners learn from their own experiments, as they encounter unpredictable system behaviors, conversions, collapses, etc., we need information sharing that is faster and less demanding than traditional peer-reviewed publication. Some new venues are trying to meet this need. For example, the Conservation Registry website (www.conservationregistry.org/) allows users to describe, track, and map conservation projects. Similar sites at cses.washington.edu/cig/cases and www.cakex.org provide information on local and state-level adaptation planning and allow users to upload cases to the database. The US Fish and Wildlife Service's *Journal of Fish and Wildlife Management* (www.fws.gov/science/publicationsys.html), an in-house, peer-reviewed journal, provides a publication venue for useful information unsuited for traditional peer review (e.g., due to lesser impact). Blog and wiki sites may offer additional options for informal information sharing among practitioners.

Managers should begin now to acquaint themselves with data requirements and strategies for implementation (table 19-1). Adaptive management provides the best possibility of success if we can find the institutional and societal fortitude to commit to it. The process can be readily communicated to the public to provide context for the likely future errors and the need for flexibility and responsiveness to new data and new approaches as these arise. Many kinds of new partnerships will be needed to protect biodiversity under climate change. We should get started; institutional change at most levels is slower than climate change and we've given climate a head start.

## Acknowledgments

We thank Mark Schwartz and Dov Sax for thoughtful discussion. Jason Dzurisin helped with analyses. The following individuals commented on the manuscript: Lynn Anderson, Jason Dzurisin, Robert Fischman, Travis Marsico, Jessica Mikels-Carrasco, Carsten Nowak, Shannon Pelini, and Teresa Woods.

### REFERENCES

Baskin, C. C., and Baskin, J. M. 2004. "Determining dormancy-breaking and germination requirements from the fewest seeds." In *Ex Situ Plant Conservation,* edited by E. O. Guerrant, Jr., K. Havens, and M. Maunder, 162–179. Washington, D.C.: Island Press.

Beier, P., Majka, D. R., and Spencer, W. D. 2008. "Forks in the road: Choices in procedures for designing wildland linkages." *Conservation Biology,* 22: 836–851.

Bennett, A. F. 2003. *Linkages in the Landscape: The Role of Corridors and Connectivity in Wildlife Conservation,* 2nd ed. Gland, Switzerland: IUCN.

Bormann, B. T., Haynes, R. W., and Martin, J. R. 2007. "Adaptive management of forest ecosystems: Did some rubber hit the road?" *BioScience,* 57: 186–191.

Botkin, D. B., Saxe, H., Araújo, M. G., Betts, R., Bradshaw, R. H. W., Cedhage, T., Chesson, P., et al. 2007. "Forecasting the effects of global warming on biodiversity." *BioScience,* 57: 227–236.

Brown, R., and Harris, G. 2005. "Comanagement of wildlife corridors: The case for citizen participation in the Algonquin to Adirondack proposal." *Journal of Environmental Management,* 74: 97–106.

CBD. 2002. "Global Strategy for Plant Conservation." *Convention on Biological Diversity*, Montréal, Québec, Canada. Accessed February 26, 2009. Available at http://www.cbd.int/gspc/

Chan, K. M. A., Shaw, M. R., Cameron, D. R., Underwood, E. C., and Daily, G. C. 2006. "Conservation planning for ecosystem services." *PLoS Biology* 4: 2138–52.

Clark, J. S., Lewis, M., McLachlan, J. S., and HilleRisLamber, J. 2003. "Estimating population spread: What can we forecast and how well?" *Ecology,* 84: 1979–1988.

Conway, W. G. 1986. "The practical difficulties and financial implications of endangered species breeding programmes." *International Zoo Yearbook,* 24/25: 210–219.

Davis, M. B., and Shaw, R. G. 2001. "Range shifts and adaptive responses to Quaternary climate change." *Science,* 292: 673–679.

Dawson, D., and Shogren, J. F. 2001. "An update on priorities and expenditures under the Endangered Species Act." *Land Economics,* 77: 527–532.

Dawson, T. P., Jackson, S. T., House, J. I., Prentice, I. C., and Mace, G. M. 2011. "Beyond prediction: Biodiversity conservation in a changing climate." *Science* 332: 53–58.

Fischman, R. L. 2003. *The National Wildlife Refuges: Coordinating a Conservation System Through Law*. Washington, D.C.: Island Press.

Griffith, B., Scott, J. M., Carpenter, J. W., and Reed, C. 1989. "Translocations as a species conservation tool: Status and strategy." *Science,* 245: 477–480.

Guerrant, E. O., Jr., and Fiedler, P. L. 2004. "Accounting for sample decline during ex situ storage and reintroduction." In *Ex Situ Plant Conservation,* edited by E. O. Guerrant, Jr., K. Havens, and M. Maunder, 365–386. Washington, D.C.: Island Press.

Guerrant, E. O., Havens, K., and Maunder, M. 2004. *Ex Situ Plant Conservation.* Washington, D.C.: Island Press.

Gunderson, L., and Light, S. S. 2006. "Adaptive management and adaptive governance in the Everglades ecosystem." *Policy Science,* 39: 323–334.

Hampe, A., and Petit, R. J. 2005. "Conserving biodiversity under climate change: The rear edge matters." *Ecology Letters,* 8: 461–467.

Hedrick, P. W. 2001. "Conservation genetics: Where are we now?" *Trends in Ecology and Evolution,* 16: 629–636.

Hewitt, G. M. 1996. "Some genetic consequences of ice ages, and their role in divergence and speciation." *Biological Journal of the Linnean Society*, 58: 247–276.

Hoegh-Guldberg, O., Hughes, L., McIntyre, S., Lindenmayer, D. B., Parmesan, C., Possingham, H. P., and Thomas, C. D. 2008. "Assisted colonization and rapid climate change." *Science,* 321: 345–346.

Hole, D. G., Willis, S. G., Pain, D. J., Fishpool, L. D., Butchart, S. H., Collingham, Y. C., Rahbek, C., and Huntley, B. 2009. "Projected impacts of climate change on a continent-wide protected area network." *Ecology Letters* 12: 420–431.

Holling, C. S., ed. 1978. *Adaptive Environmental Assessment and Management.* New York: John Wiley.

Holling, C. S., and Meffe, G. K. 1996. "Command and control and the pathology of natural resource management." *Conservation Biology*, 10: 328–337.

Hunter, M. L. 2007. "Climate change and moving species: Furthering the debate on assisted colonization." *Conservation Biology,* 21: 1356–1358.

IPCC. 2007. *Climate Change 2007: The Physical Science Basis*. Cambridge, UK: Cambridge University Press.

Jacobson, S. K., Morris, J. M., Sanders, J. S., Wiley, E. N., Brooks, M., Bennetts, R. E., Percival, H. F., and Marynowski, S. 2006. "Understanding barriers to implementation of an adaptive land management program." *Conservation Biology,* 20: 1516–1527.

Kohn, M. H., Murphy, W. J., Ostrander, E. A., and Wayne, R. K. 2006. "Genomics and conservation genetics." *Trends in Ecology and Evolution,* 21: 629–637.

Lindenmayer, D. B., and G. E. Likens. 2009. "Adaptive monitoring: A new paradigm for long-term research and monitoring." *Trends in Ecology and Evolution,* 24: 482–486.

Lindenmayer, D. B., Steffen, W., Burbidge, A. A., Hughes, L., Kitching, R. L., Musgrave, W., Smith, M. S., and Werner, P. A. 2010. "Conservation strategies in response to rapid climate change: Australia as a case study." *Biological Conservation* 143: 1587–1593.

Mandelker, D. R. 2006–2008. "NEPA Law and Litigation." *Environmental Law Series*. Deerfield, IL: Clark Boardman Callaghan.

Maunder, M., Guerrant Jr., E. O., Havens, K., and Dixon, K. W. 2004. "Realizing the full potential of ex situ contributions to global plant conservation." In *Ex Situ Plant Conservation,* edited by E. O. Guerrant, Jr., K. Havens, and M. Maunder, 305–324, 389–418. Washington, D.C.: Island Press.

McLachlan, J. S., Hellmann, J. J., and Schwartz, M. W. 2007. "A framework for debate of assisted migration in an era of climate change." *Conservation Biology,* 21: 297–302.

McLaughlin, J. F., Hellmann, J. J., Boggs, C. L., and Ehrlich, P. R. 2002. "Climate change hastens population extinction." *Proceedings of the National Academy of Sciences, USA,* 99: 6070–6074.

Menges, E. S., Guerrant, E. O., and Hamze, S. 2004. "Effects of seed collection on the extinction risk of perennial plants." In *Ex Situ Plant Conservation,* edited by E. O. Guerrant, Jr., K. Havens, and M. Maunder, 305–324. Washington, D.C.: Island Press.

Millar, C. I., Stephenson, N. L., and Stephens, S. L. 2007. "Climate change and forests of the future: Managing in the face of uncertainty." *Ecological Applications,* 17: 2145–2151.

Moorcroft, P. R., Lewis, M. A., and Pacala, S. W. 2006. "Potential role of natural enemies during tree range expansions following climate change." *Journal of Theoretical Biology,* 241: 601–616.

Mueller, J. M., and Hellmann, J. J. 2008. "An assessment of invasion risk from assisted migration." *Conservation Biology,* 22: 562–567.

Naidoo, R., Balmford, A., Constanza, R., Fisher, B., Green, R. E., Lehner, B., Malcolm, T. R., and Ricketts, T. H. 2008. "Global mapping of ecosystem services and conservation priorities." *Proceedings of the National Academy of Sciences of the United States of America* 105: 9495–9500.

Nathan, R. 2006. "Long-distance dispersal of plants." *Science,* 313: 786–788.

Office of Technology Assessment. 1993. *Harmful Non-indigenous Species in the United States*. Washington, D.C.: Office of Technology Assessment.

Parks, C. R., Wendel, J. F., Sewell, M. M., and Qiu, Y.-L. 1994. "The significance of allozyme variation and introgression in the *Liriodendron tulipifera* complex (Magnoliaceae)." *American Journal of Botany,* 81: 878–889.

Parmesan, C. 2006. "Ecological and evolutionary responses to recent climate change." *Annual Review of Ecology, Evolution, and Systematics,* 37: 637–669.

Pelini, S. L., Dzurisin, J. D. K., Prior, K. M., Williams, C. M., Marsico, T. D., Sinclair, B. J., and Hellmann, J. J. 2009. "Translocation experiments with butterflies reveal limits to enhancement of poleward populations under climate change." *Proceedings of the National Academy of Sciences, USA,* 106: 11160–11165.

Phillips, S. J., Williams, P., Midgley, G., and Archer, A. 2008. "Optimizing dispersal corridors for the Cape Proteaceae using network flow." *Ecological Applications,* 18: 1200–1211.

Pound, J. A., Bustamente, M. R., Coloma, L. A., et al. 2006. "Widespread amphibian extinctions from epidemic disease driven by global warming." *Nature*, 439: 161–167.

Ranganathan, J., Daniels, R. J. R., Chandran, M. D. S., Ehrlich, P. R., and Daily, G. C. 2008. "Sustaining biodiversity in ancient tropical countryside." *Proceedings of the National Academy of Sciences USA*. 105: 17852–54.

Ribaudo, M. O., Hoag, D. L., Smith, M. E., and Heimlich, R. 2001. "Environmental indices and the politics of the Conservation Reserve Program." *Ecological Indicators,* 1: 11–20.

Richardson, D., Hellmann, J. J., McLachlan, J., Sax, D. F., Schwartz, M. W., et al. 2009. "Multidimensional evaluation of managed relocation." *Proceedings of the National Academy of Sciences, USA*, 106: 9721–9724.

Root, T. L., and Schneider, S. H. 2002. "Climate change: Overview and implications for wildlife." In *Wildlife Responses to Climate Change: North American Case Studies,* edited by S. H. Schneider and T. L. Root, 1–56. Washington, D.C.: Island Press.

Seddon, P. J., Armstrong, D. P., and Maloney, R. F. 2007. "Developing the science of reintroduction biology." *Conservation Biology,* 21: 303–312.

Sewell, M. M., Parks, C. R., and Chase, M. W. 1996. "Intraspecific chloroplast DNA variation and biogeography of North American *Liriodendron* L. (Magnoliaceae)." *Evolution,* 50: 1147–1154.

Snyder, N. F. R., Derrickson, S. R., Beissinger, S. R., Wiley, J. W., Smith, T. B., Toone, W. B., and Miller, B. 1996. "Limitations of captive breeding in endangered species recovery." *Conservation Biology,* 10: 338–348.

Soltis, D. E., Morris, A. B., McLachlan, J. S., Manos, P. S., and Solts, P. S. 2006. "Comparative phylogeography of unglaciated eastern North America." *Molecular Ecology,* 15: 4261–4293.

Tewksbury, J. J., Levey, D. J., Haddad, N. M., Sargent, S., Orrock, J. L., Danielson, B. J., Brinkerhoff, J., Samschen, E. I., and Townsend, P. 2002. "Corridors affect plants, animals, and their interactions in fragmented landscapes." *Proceedings of the National Academy of Sciences, USA,* 99: 12923–12926.

Theobald, D. M., and Hobbs, N. T. 2002. "A framework for evaluating land use planning alternatives: Protecting biodiversity on private land." *Conservation Ecology,* 6 (1): article 5 [online] http://www.consecol.org/vol6/iss1/art5

van der Brugge, R., and van Raak, R. 2007. "Facing the adaptive management challenge: Insights from transition management." *Ecology and Society* 12 (2): article 33 [online] URL: http://www.ecologyandsociety.org/vol12/iss2/art33/

Walters, C. J. 2006. "Is adaptive management helping to solve fisheries problems?" *Ambio,* 36: 304–307.

Walther, G.-R., Post, E., Convey, P., et al. 2002. "Ecological responses to recent climate change." *Nature*, 416: 389–395.

Webster, M. S., Marra, P. P., Haig, S. M., Bensch, S., and Holmes, R. T. 2002. "Links between worlds: Unraveling migratory connectivity." *Trends in Ecology and Evolution,* 17: 76–83.

Williams, J. W., Jackson, S. T., and Kutzbach, J. E. 2007. "Projected distributions of novel and disappearing climates by 2100AD." *Proceedings of the National Academy of Sciences, USA,* 104: 5738–5742.

Williams, P., Hannah, L., Andelman, S., Midgley, G., Araujo, M., Hughes, G., Manne, L. Martinez-Meyer, E., and Pearson, R. 2005. "Planning for climate change: Identifying minimum-dispersal corridors for the Cape Proteaceae." *Conservation Biology,* 19: 1063–1074.

Zakharov, E. V., and Hellmann, J. J. 2008. "Genetic differentiation across a latitudinal gradient in two co-occurring butterfly species: Revealing population differences in a context of climate change." *Molecular Ecology,* 17: 189–220.

# Chapter 20

## *Saving a Million Species*

LEE HANNAH

Are a million species at risk of extinction from climate change? This book has explored the analyses required to answer that question. From estimates of the number of species on Earth, to their imperilment by climate change, we have looked across taxa, across methods, and across geographies to estimate risk.

The threat from climate change is clear. Massive coral bleaching and terrestrial extinctions are already linked to warming. The number of species whose ranges are known to be shifting due to climate change are mounting each year. In the oceans, acidification poses a threat that is still being assessed, but that clearly will affect millions of species for centuries to come.

Teasing out the species at risk of extinction and quantifying their number is more daunting. We have seen that the initial estimates were based on methods that are still relevant today, despite much intervening debate. Thomas et al. (2004) used climate models and biological models similar to those used in climate change biology today. The species assessed represented multiple taxa and multiple regions of the world.

Refinements of the initial 2004 estimates have been slow to arrive. This is probably because global analyses require major investments in data gathering and modeling, and are unlikely to substantially change the conclusions of the Thomas et al. analysis. The single global, systematic follow-on estimate (Malcolm et al., 2006) confirmed the general findings of Thomas et al. The Malcolm et al. results were based on species endemic to global biodiversity hotspots and properly used the endemic area relationship to estimate extinction risk, thus addressing some of the most important criticisms of the Thomas et al. methods. We have seen that among several possible sources of bias in the Thomas et al. methods, most would tend to produce underestimates of risk. The MaxEnt theory of ecology now offers possible improvements on these methods, making it possible to assess extinction risk from modeling results with a stronger theoretical foundation. These results (table 5-1) suggest that for species with modest population sizes (less than 10,000) and a minimum viable population size of 1,000, extinction risk calculated from MaxEnt theory and the species-area relationship approach used by Thomas et al. would be of a similar order of magnitude.

Current extinctions offer less insight into possible magnitude of risk, but expand the pool of species affected into the marine realm. The hundreds of terrestrial extinctions linked to climate change involve amphibians killed by a fungal pathogen. The contributory role of climate change in these extinctions is a subject of current controversy. The first marine extinction linked to climate change was driven by coral bleaching, reminding us that millions of marine species are at risk from climate change; those associated with coral reefs are at most immediate risk. The first coral extinction was proved to be local, rather than global, when a range extension of the species was discovered, highlighting the difficulties of documenting and estimating extinctions in the face of limited taxonomic information. Other heavily bleached corals may become functionally extinct, rather than biologically extinct, hanging on in tiny patches but losing their critical mass for reef-building.

The past shows major episodes of extinction linked to climate change. Climate change is implicated as an ultimate or proximal cause in all of the five major extinction events that occurred from 65 million to 500 million years ago. More recent events such as the Paleocene-Eocene Thermal Maximum have seen large numbers of extinctions due to rapid climate perturbation. Another wave of extinctions occurred as Earth descended into the ice ages, but subsequent glacial-interglacial transitions saw few extinctions until humans arrived on the scene. The Pleistocene extinctions at the end of the last glacial period showed the power of climate change and human impact working together.

Predicting future extinctions without models is difficult. We know that large numbers of tropical, freshwater, and marine species are vulnerable to the impacts of climate change. Knowing which of these species might go extinct is limited first by our lack of knowledge of the species themselves and then by our emerging understanding of climate change impacts. For most of the species on Earth, estimation of extinction risk can be approximated only by extrapolation from the risk faced by better known (modeled) species. We know that most species are insects, but we don't know exactly how many insect species there are, much less their climatic tolerances. We know that species numbers are greatest in the tropics and that tropical mountains with large numbers of species arrayed across steep climatic gradients are particularly at risk. The Andes, Himalayas, and mountains of New Guinea stand out. In the marine realm, species are at risk from both climate change and the direct effects of increasing carbon dioxide on ocean chemistry. Corals are immediately at risk and of greatest concern, but millions of other species face this double threat.

Subsequent modeling and other lines of evidence tend to confirm the general magnitude of the terrestrial extinction risk estimated by Thomas et al., and suggest that millions of additional marine and freshwater species need to be included in the assessment of global extinction risk due to climate change. Specifying the extinction risk of these species, and the distribution of risk across taxa and regions, are major research priorities. It is clear that millions of species are at elevated risk of extinction due to climate change and losses to genetic biodiversity may be even higher (Bálint et al 2011). What proportion of those species might actually be driven to extinction by climate change alone is a much more difficult, and largely academic, question.

The real question is not the risk of extinction due to climate change alone, but the risk due to climate change in association with all of the other things people are doing to the planet. To focus on the biggest parts of the problem, climate change and land use change/habitat loss will interact in ways that exacerbate probabilities of extinction. Species whose ranges are shifting will interact with declining habitat, making persistence difficult in the face of dynamics.

To prevent extinctions, we must therefore act on *both* habitat loss and climate change, and do so quickly. Most of the world's most biologically unique areas have already lost more than 70 percent of their high quality habitat. Range shifts in these biodiversity hotspots will face tremendous difficulties. Climate change will elevate already high extinction risk in these areas.

When climate change and habitat loss are considered together, the number of species at risk of extinction is clearly in the millions, even at low-end global species counts, if present trends continue. Millions of marine organisms are affected by coral bleaching, acidification, and overfishing. Millions of terrestrial organisms are affected by forest clearing, habitat loss, and range shifts.

Waiting for precise numbers of species at risk misses the point. We understand the threats, we understand many of the actions needed to greatly reduce these threats. What then, does a world look like in which risk of extinction is declining? Climate change may hold the answer on both fronts. Regulatory approaches to halting habitat loss have had limited success, especially in the major frontiers of loss in the tropics. An incentive-based complement to regulation seems needed to stabilize the frontiers. Financing from climate change carbon markets can provide just this incentive. Those markets will become fully active when global action on climate change mitigation is implemented at scale. Reduced Emissions from Deforestation and Degradation (REDD) under the United Nations Framework Convention on Climate Change (Grainger et al., 2009) will therefore herald both reduced habitat loss and a broader global effort to reduce greenhouse gas emissions.

The other side of climate change is dealing with the biological changes that will take place due to greenhouse gases already in the atmosphere. We are already committed to substantial change in climate and substantial biological reorganization as a result. Improved conservation strategies are needed that contribute to further reduction in habitat loss at the same time that they enable conservation of species whose ranges are shifting. This is a large, but not intractable, task. Protected area additions and connectivity have been shown to be effective in "climate-proofing" existing conservation networks (Hannah et al., 2007; Phillips et al., 2008; Williams et al., 2005). An international framework to support planning and cross-border implementation is needed to ensure cost-effective deployment of these tools (Hannah, 2009). Suites of complementary, more intensive management actions may be needed (Hoegh-Guldberg et al., 2008; McLachlan et al., 2007). Earlier reductions in greenhouse gases will both reduce the biological damage to be dealt with and limit the need for intensive management options.

Climate change is the environmental issue of this century. Through REDD, solutions to climate change can help address habitat loss, the environmental issue of the last century that plagues us still. Innovative conservation strategies can reduce the risk of extinction due to climate change and habitat loss individually and as interacting forces. Exact quantification of the extinction risk of climate change may elude us for some time. But we know that the risk is large, and magnified by habitat loss. The important action is therefore not the quantification of the risk, but its reduction—through vigorous climate change policy action, including REDD, and through adaptation actions that make our conservation systems robust to climate change and habitat loss.

## REFERENCES


Bálint, M., S. Domisch, C. H. M. Engelhardt, P. Haase, S. Lehrian, J. Sauer, K. Theissinger, S. U. Pauls and C. Nowak. 2011. "Cryptic biodiversity loss linked to global climate change." *Nature Climate Change* 1: 313–318.

Hannah, L. 2009. "A global conservation system for climate-change adaptation." *Conservation Biology* 24: 70–77.

Hannah, L., G. Midgley, S. Andelman, M. Araujo, G. Hughes, E. Martinez-Meyer, R. Pearson, and P. Williams. 2007. "Protected area needs in a changing climate." *Frontiers in Ecology and the Environment* 5: 131–138.

Hoegh-Guldberg, O., L. Hughes, S. McIntyre, D. B. Lindenmayer, C. Parmesan, H. P. Possingham, and C. D. Thomas. 2008. "Assisted colonization and rapid climate change." *Science* 321: 345–346.

Grainger, A., D. H. Boucher, P. C. Frumhoff, W. F. Laurance, T. Lovejoy, J. McNeely, M. Niekisch, et al. 2009. "Biodiversity and REDD at Copenhagen." *Current Biology* 19 (21): R974–R976.

Malcolm, J. R., C. R. Liu, R. P. Neilson, L. Hansen, and L. Hannah. 2006. "Global warming and extinctions of endemic species from biodiversity hotspots." *Conservation Biology* 20: 538–548.

McLachlan, J. S., J. J. Hellmann, and M. W. Schwartz. 2007. "A framework for debate of assisted migration in an era of climate change." *Conservation Biology* 21: 297–302.

Phillips, S. J., P. Williams, G. Midgley, and A. Archer. 2008. "Optimizing dispersal corridors for the Cape proteaceae using network flow." *Ecological Applications* 18: 1200–1211.

Thomas, C. D., A. Cameron, R. E. Green, M. Bakkenes, L. Beaumont, A. Grainger, Y. Collingham, et al. 2004. "Extinction risk from climate change." *Nature* 427: 145–148.

Williams, P., L. Hannah, S. Andelman, G. Midgley, M. Araujo, G. Hughes, L. Manne, E. Martinez-Meyer, and R. Pearson. 2005. "Planning for climate change: Identifying minimum-dispersal corridors for the Cape proteaceae." *Conservation Biology* 19: 1063–1074.

## Anthony D. Barnosky

Anthony Barnosky is a biologist, paleoecologist, and author of scientific and popular publications. On the faculty of the University of California at Berkeley Department of Integrative Biology, Museum of Paleontology, and Museum of Vertebrate Zoology, he studies how climate changes impact Earth's ecosystems and the evolution and extinction of species. Anthony's recent book, *Heatstroke: Nature in an Age of Global Warming,* explores what global warming means for nature itself, for the wild places we love, and for our future.

## Jedediah Brodie

Jedediah Brodie is broadly interested in ecology and evolution, but much of his work addresses two main themes: (i) the evolution and demographic importance of species interactions, and (ii) the cascading impacts of anthropogenic stressors such as climate change and hunting on terrestrial vertebrate populations and communities. In general he works at the population level, combining large-scale natural experiments and controlled field manipulations with quantitative analysis. Jedediah has explored six continents and done fieldwork in a variety of ecosystems, including the rain forests of Southeast Asia, the Arctic tundra, and the deserts of southern Africa.

## Barry W. Brook

Barry W. Brook, a leading environmental scientist and modeler, is professor and director of climate science at the University of Adelaide's Environment Institute. He has published three books and more than

170 refereed scientific papers and regularly writes popular articles for the media. Barry has received a number of distinguished awards for his research excellence and public outreach, including the Australian Academy of Science Fenner Medal and the 2010 Community Science Educator of the Year. His research interests are climate change impacts (past, present, and future), simulation modeling, energy systems analysis (with a focus on nuclear power), and synergistic human impacts on the biosphere. Barry runs a popular climate science and energy options blog at http://bravenewclimate.com, which has seen more than 2.7 million hits.

## Mark B. Bush

Mark Bush (Ph.D., University of Hull, UK, 1986) is professor and chair of the Conservation Biology and Ecology Program at the Florida Institute of Technology. He has more than 30 years' experience of working on the biogeography and paleoecology of tropical systems. His research focuses on fossil pollen and charcoal analysis of Neotropical settings and environmental reconstructions of past climates, fire histories, and vegetation communities. Mark also investigates pre-Columbian influences on the environment and responses to past climate change. He has published two books and more than 100 papers on tropical ecology and climate change.

## Alison Cameron

Alison Cameron is a lecturer in climate change adaptation and mitigation at Queen's University Belfast. She has a B.Sc. in tropical environmental science from the University of Aberdeen, an M.Sc. in conservation biology from the University of Cape Town, and a Ph.D. in biology from the University of Leeds, and has held postdoctoral research positions at Princeton University, University of California-Berkeley, and the Max Planck Institute for Ornithology. Her work focuses on the biological consequences of climate change, particularly on threatened species and protected areas. Much of Alison's research has been conducted in Madagascar, a highly threatened global biodiversity hotspot.

## William C. Clyde

William Clyde is an earth historian interested in how climatic and tectonic changes affected mammalian evolution during the geologic past. In particular, he uses a combination of paleontological, geophysical, and geochemical methods to develop integrated stratigraphic records of continental environmental change. He received his A.B. in geology from Princeton University in 1990 and his M.S. and Ph.D. from University of Michigan in 1993 and 1997, respectively. He is now associate professor of geology and chair of the Earth Sciences Department at University of New Hampshire.

## Robert R. Dunn

Rob Dunn is a writer and biogeographer at North Carolina State University. In his science he is interested in the biogeography of societies (be they ants, wolves, or humans) and the species that influence and are associated with them. In his writing, he is interested in the broad stories of science and scientists and the limits to our knowledge and discovery. Rob's most recent book, to be released by HarperCollins in July 2011, is *The Wild Life of Our Bodies.* It tells the stories of human interactions with other species and how they have changed in light of several hundred thousand years of local and global change.

## Matthew C. Fitzpatrick

Matt Fitzpatrick is interested in modeling species distributions and patterns of biodiversity in the context of climate change and in quantifying niche dynamics across space and through time, with an emphasis on plants, ants, and invasive species. His current research focuses on developing dynamic stochastic models to predict range expansion of species across heterogeneous landscapes and the use of new statistical approaches to model species-environment relationships, including hierarchical Bayesian models, site occupancy models, and generalized dissimilarity models. Matt is an assistant professor at the University of Maryland Center for Environmental Science, Appalachian Laboratory.

## Peter W. Glynn

Peter W. Glynn is a professor in the Division of Marine Biology and Fisheries, Rosenstiel School of Marine and Atmospheric Science, University of Miami, Florida. Peter graduated from Stanford University in 1963 with a Ph.D. degree in the biological sciences. Following graduation, he joined the Institute of Marine Biology, University of Puerto Rico (Mayagüez, 1960–1967), then moved to the Smithsonian Tropical Research Institute (Panamá, 1967–1983). In 1983 Peter joined the faculty of the Rosenstiel marine school. Peter's research interests have centered on coral reef ecology, specializing in biotic interactions (predation, mutualisms), coral community structure, species diversity, reef accretion and distribution, biogeography, and disturbances to reefs, especially coral reef bleaching and El Niño disturbances. In 1992 he received the Charles Darwin Medal from the International Society of Reef Studies in recognition of a record of sustained, highly significant contributions to coral reef studies.

## Benjamin S. Halpern

Ben Halpern is a research biologist at the University of California-Santa Barbara, the project lead for a research initiative to evaluate and better inform efforts to do marine ecosystem-based management, and a lead scientist for the Ocean Health Index project. He received his Ph.D. in marine ecology from UC-Santa Barbara and then held a joint postdoctoral fellowship at the National Center for Ecological Analysis and Synthesis, and the Smith Fellowship. Ben has led and participated in several key synthetic research projects, in particular addressing marine protected area effectiveness and cumulative impact assessment and mapping. He has also conducted field expeditions in tropical and temperate systems in the Caribbean, Red Sea, Mediterranean, Solomon Islands, Indonesia, and various parts of the South Pacific, California, and Chile.

## Lee Hannah

Lee Hannah is senior researcher in climate change biology at Conservation International. He heads their efforts to develop conservation responses to climate change. His research has appeared in journals including *Nature, Conservation Biology,* and *BioScience*. Lee edited the award-winning book *Climate Change and Biodiversity* (Yale: 2005) with Thomas E. Lovejoy. He authored the first undergraduate textbook on the biological impacts of climate change, *Climate Change Biology* (Elsevier: 2011). Lee and Guy Midgley of the National Botanical Institute at Kirstenbosch (Cape Town) have a long-standing research collaboration to model biotic change and develop adaptation responses in the Succulent Karoo and Cape hotspots of South Africa. Lee has led research collaborations to improve conservation planning tools for climate change adaptation, including the development of conservation planning software and very high resolution climatologies suitable for species distribution modeling. These efforts have included founding a National Center for Ecological Analysis and Synthesis working group on fine-scale biological modeling. His broader research interests include the role of climate change in conservation planning and methods of corridor design. Lee has previously written on the global extent of wilderness and the role of communities in protected area management. He holds an undergraduate degree in biology from the University of California-Berkeley and a doctorate in environmental science from University of California-Los Angeles.

## John Harte

John Harte is a professor of ecosystem sciences at the University of California-Berkeley. With a B.A. from Harvard and a Ph.D. from the University of Wisconsin, he was a National Science Foundation postdoctoral fellow at CERN, Geneva, and an assistant professor of physics at Yale. His research interests include climate-ecosystem interactions, theoretical ecology, and environmental policy. John is the recipient of a Pew Scholars Prize in Conservation and the Environment, a Guggenheim Fellowship, the 2001 Leo Szilard prize from the American Physical Society, the 2004 UC Berkeley Graduate Mentorship Award, and a Miller Professorship, and is a corecipient of the 2006 George Polk Award in journalism. He is an elected Fellow of the California Academy of Sciences and the American Physical Society. John has also served on six National Academy of Sciences committees and has authored more than 190 scientific publications, including eight books.

## Jessica J. Hellmann

Jessica Hellmann is associate professor of biological sciences at the University of Notre Dame (Ph.D., Stanford University). She studies the effects of global change on populations, species, and species interactions using mathematical models, large-scale ecological experiments, and genomic biology. Recently, her group has focused on explaining how evolutionary divergence of populations helps or hinders a species' responses to climatic change. Jessica also leads an interdisciplinary effort to develop management strategies that may help species persist through climate change.

## Ove Hoegh-Guldberg

Ove Hoegh-Guldberg is professor of marine studies and director of the Global Change Institute at the University of Queensland in Brisbane, Australia. In addition to his role as director, Ove heads a research laboratory with more than thirty researchers and students. The team is pursuing a greater understanding of how global warming and ocean acidification are affecting and will affect coral reefs (www.coralreefecosystems.org). He was recognized in 1999 with the Eureka Prize for scientific research and is currently the Queensland Smart State Premier's Fellow. Ove has published works that include more than 180 refereed publications and book chapters. He is one of the most cited authors within the peer-reviewed literature on climate change and its impacts on natural ecosystems.

## Lesley Hughes

Lesley Hughes is an ecologist in the Department of Biological Sciences at Macquarie University, Sydney. Her research focuses on the potential impacts of climate change on species and ecosystems and the implications of climate change for conservation policy. Lesley is a lead author on the Intergovernmental Panel on Climate Change's Fourth and Fifth Assessment Reports and is currently one of Australia's climate change commissioners. She is a founding member of Climate Scientists Australia and has also served as an advisor to World Wildlife Fund, the New South Wales and Australian governments, Earth

Watch, International Union for Conservation of Nature, and the Convention on Biological Diversity.

## Carrie V. Kappel

Carrie Kappel is a project scientist at the National Center for Ecological Analysis and Synthesis in Santa Barbara, California. A marine conservation biologist and community ecologist, she received her Ph.D. from Stanford University. Major themes of her work include quantifying the ways humans affect marine species, habitats, and ecosystems; understanding the spatial distribution of ecological and human components of ecosystems to inform conservation and management; and developing ways to integrate biophysical and socioeconomic data to support environmental decision making in coastal ecosystems. Her research has been aimed at informing marine protected area design, ecosystem-based management, and marine spatial planning.

## Justin Kitzes

Justin Kitzes is a Ph.D. candidate in environmental science, policy, and management at the University of California-Berkeley. His research focuses on the development of theoretical and simulation models to predict the distribution of biodiversity in fragmented landscapes, as well as the application of these models to conservation planning.

## Rebecca LeCain

Rebecca LeCain graduated with her M.S. in geology from the University of New Hampshire in 2010 after receiving her B.S. in geology from University of Missouri in 2006. She completed her thesis on the magnetostratigraphy of the Hell Creek and Fort Union Formations in the fossiliferous exposures of northeastern Montana.

## Thomas E. Lovejoy

Thomas Lovejoy became University Professor in environmental science and policy at George Mason University in January 1980. He also holds the Biodiversity Chair at the Heinz Center for Science, Economics, and the Environment based in Washington, D.C., where he was president from 2002 to 2008. Prior to that, he was the World Bank's chief biodiversity advisor and lead specialist for environment for Latin America and the Caribbean, as well as senior advisor to the president of the United Nations Foundation. He chairs the science and technical advisory panel for the Global Environment Facility. Thomas also conceived of the now ubiquitous "debt-for-nature" swap programs, and created the largest experiment in landscape ecology on forest fragmentation in the Amazon (now in its thirty-third year). He also founded the series *Nature,* the popular long-term series on public television. In 2001, Thomas was awarded the prestigious Tyler Prize for Environmental Achievement. In 2009 he was the winner of BBVA Foundation Frontiers of Knowledge Award in the ecology and conservation biology category. In 2009 he was appointed conservation fellow by *National Geographic*. Thomas holds B.S. and Ph.D. (biology) degrees from Yale University.

## Yadvinder Malhi

Yadvinder Malhi is professor of ecosystem science at the University of Oxford, UK. His research interests focus on the functioning and ecology of tropical forests, and the interactions between tropical ecosystems and global atmospheric change. He has been conducting long-term field research across Amazonian and Andean forests for over a decade, and more recently has extended this field research to the African and Asian tropics. In addition to field research, he applies ecosystem modeling and satellite remote sensing to better understand and manage the future of tropical forests in the twenty-first century.

## Jonathan Mawdsley

Jonathan Mawdsley is program director at The Heinz Center, a non-profit research institution in Washington, D.C. He directs the center's wildlife and ecosystems conservation program, and also oversees climate adaptation activities for the center. Before joining the center, Jonathan worked for the National Fish and Wildlife Foundation. In addition to his work at the center, he maintains an active research program at the National Museum of Natural History (Smithsonian Institution) that focuses on insect and pollinator conservation in the United States and southern Africa. He received his B.A. from Harvard University (biology) and his Ph.D. from Cornell University (entomology).

## Peter J. Mayhew

Peter Mayhew is an evolutionary ecologist with a special interest in insects, focusing on life history evolution, conservation biology, and macroevolution. He has studied at the universities of Oxford, London, and Leiden, and is currently senior lecturer in ecology in the Department of Biology at the University of York, UK. Peter has written more than fifty academic papers and is the author of *Discovering Evolutionary Ecology* (Oxford, 2006).

## Jason S. McLachlan

Jason McLachlan is assistant professor of biological sciences at the University of Notre Dame (Ph.D., Duke University). To study geographic and evolutionary responses of plants to environmental change, he reconstructs past genetic structure and historic landscapes and performs common garden experiments. Jason also collaborates with legal scholars to understand the relative role of ecological and social barriers to species' movements under modern climate change. He recently co-led an interdisciplinary assessment group on managed relocation.

## Sarah K. McMenamin

Sarah McMenamin is a National Institutes of Health–National Research Service Award postdoctoral fellow at the University of Washington. McMenamin's Ph.D. work at Stanford University focused on *Ambystoma* salamanders in Yellowstone National Park. Her dissertation research examined amphibian population genetics, ecological morphology, and population-level responses to climate change. McMenamin's postdoctoral work examines developmental endocrinology in teleosts.

## Vicky J. Meretsky

Vicky Meretsky is associate professor of environmental science at the School of Public and Environmental Affairs at Indiana University (Ph.D., University of Arizona). She studies conservation biology at single-species and landscape scales, and the impacts of law and policy on conservation of biodiversity. Vicky and her students are presently examining existing policy constraints on using managed relocation of plant species in forward-looking vegetation restoration, and mining historical wildlife data for lessons concerning managed relocation of wildlife species.

## Guy Midgley

Guy Midgley leads the Climate Change and BioAdaptation Division at the South African National Biodiversity Institute. His academic background is in ecology and ecophysiology, with a Ph.D. awarded in 1996. He has published extensively on climate change, biodiversity, and policy topics. Guy co-led the Intergovernmental Panel on Climate Change (IPCC) in 2007, and has led several national and international assessment reports. He co-leads a chapter on "Adaptation, opportunities, constraints and limits" for the IPCC fifth assessment report, due in 2014. He also serves on the South Africa National Climate Change Committee, the South African Scientific Committee on Global Change, and the international scientific steering committee for the Global Carbon Project. Guy is an expert advisor for South African negotiators of the United Nations Framework Convention on Climate Change and Convention on Biological Diversity. He coauthored the illustrated popular book *A Climate for Life,* published in 2008.

## Nicole A. S. Mosblech

Nicole Mosblech is a doctoral candidate in the Department of Biological Sciences at the Florida Institute of Technology. Her dissertation research investigates the effect of climate change and human occupation on Andean plant communities through the analysis of paleoecological proxies, including pollen and charcoal, in comparison with isotopic speleothem records.

## Julian D. Olden

Julian Olden is an assistant professor in the School of Aquatic and Fishery Sciences at the University of Washington. His research focuses on the ecology and management of invasive species, environmental flows, biogeography of freshwater fishes, and the development of conservation strategies in natural and built environments.

## N. LeRoy Poff

LeRoy Poff is a riverine ecologist who seeks to quantify the role of natural and human-modified environmental variation in shaping the structure and function of aquatic and riparian ecosystems at local to landscape scales. His primary research lies at the interface of ecology, hydrology, and geomorphology. He is an international leader in the scientific effort to develop a strong hydroecological foundation to inform sustainable management of freshwater ecosystems in the face of growing human demands and climate change. Leroy is a professor in the Department of Biology and director of the graduate degree program in ecology at Colorado State University.

## Eric Post

Eric Post is a professor of biology at Penn State University. He earned his Ph.D. at the University of Alaska, Fairbanks, spent four years as a postdoctoral researcher at the University of Oslo, Norway, and has conducted research on the ecology of animals and plants in the Arctic since 1990. Eric's research focuses on the consequences of climate change for wildlife conservation and population and community ecology.

## David L. Strayer

David Strayer is a senior scientist at the Cary Institute of Ecosystem Studies in Millbrook, New York. His research is focused on the ecology of biological invasions, conservation of freshwater ecosystems, the ecology and management of the Hudson River, and the ecology of pearly mussels (Unionidae). His most recent book is *Freshwater Mussel Ecology: A Multifactor Approach to Distribution and Abundance*.

## Chris D. Thomas

Chris Thomas is an ecologist and conservation biologist. In the 1990s, most of his research was on the effects of habitat fragmentation on the survival of species. More recently, he has mainly worked on the impacts of climate change on species' distributions. Chris has published more than 170 refereed journal articles, about thirty book chapters, and many popular articles, and he has been a coeditor of nine scientific journals. Chris is based in the Department of Biology at the University of York, where he is professor of conservation biology.

# INDEX

Note: page numbers followed by 'f' or 't' refer to figures or tables, respectively.

# Island Press | Board of Directors